Building Cost Techniques: New Directions

Building Cost Techniques: New Directions

Edited by

P. S. Brandon
Head of Department of Surveying
Portsmouth Polytechnic

LONDON NEWYORK

E. & F. N. SPON

First published 1982
by E. & F. N. Spon Ltd
11 New Fetter Lane
London EC4P 4EE

Published in the USA
by E. & F. N. Spon
733 Third Avenue, New York NY 10017

Printed in Great Britain at the University Press,
Cambridge

ISBN 0 419 12940 5

British Library cataloguing in publication data

Building Cost Research Conference (*1982:
Portsmouth Polytechnic*)
Building cost techniques.
1. Building—Estimates—Congresses
I. Title   II. Brandon, P. S.
692'.5   TH435

ISBN 0-419-12940-5

TRANSACTIONS

OF

BUILDING   COST   RESEARCH   CONFERENCE

Sponsored by

The Association of Heads of Departments of Surveying in Polytechnics

held at

PORTSMOUTH   POLYTECHNIC

(Department of Surveying)

23RD – 25TH SEPTEMBER 1982

CONTENTS

SECTION 1

EDITORIAL

BUILDING COST RESEARCH CONFERENCE
PORTSMOUTH POLYTECHNIC
23rd – 25th September 1982

Preface

These transactions are the proceedings of a conference on Building
Cost Research to be held at Portsmouth Polytechnic in September 1982
under the sponsorship of the Association of Heads of Department of
Surveying in Polytechnics. The conference is believed to be the
first of its kind within the subject discipline in the United
Kingdom.

Recent years have seen a rapid increase in research activity
related to building cost and this has been generated by an increasing
demand from clients to know whether traditional patterns of design,
specification and procedure are providing the economic solutions to
their problem. The need for improved advice will be emphasised as
time progresses and the non-renewable resources of the world become
more scarce. It is likely that this scarcity of raw materials will
have an impact on the life expectancy of buildings and future costs
will play a larger part in influencing design decisions. There is
some evidence that this change of view is already underway. At the
same time the traditional deterministic models for estimating have
been refined to the point where they appear to offer very little
more in the understanding of cost occurrence. Fortunately the
development of information technology, and in particular the
increased power of computers for comparative cost, opens a wide
range of new possibilities which have scarcely been investigated.

This conference has therefore been called at an opportune time to
review how those who are investigating building cost are responding
to these challenges. A quick glance at the titles of the papers
will reveal the shift in emphasis which has occurred in even the
last decade. It is hoped that through the formal papers and discussions a consensus may be established on priorities, trends and
weighting to be given to those matters which may yield the most
benefit within the limited resources available.

The organisers would like to thank all those who have supplied
keynote addresses and papers, the administrative staff of the
Department of Surveying at Portsmouth Polytechnic, and the
publishers, for making a gathering such as this possible. We trust
that this will be the start of several similar gatherings in the
years that lie ahead in order that formal and informal discussion
can be used to accelerate the advancement of the subject matter.

Peter Brandon
John McWilliam
Alan Spedding

BUILDING COST RESEARCH - NEED FOR A PARADIGM SHIFT?

PETER S. BRANDON, Portsmouth Polytechnic

An Editorial Viewpoint on Cost Research

In 1977 a publication appeared called the Encyclopaedia of Ignorance;
it contained fifty essays from prominent writers outlining the lack
of knowledge in the natural sciences. Scientists, who are sometimes
in the habit of dropping the word 'theory' from their ideas, were
confronted yet again with the enormous number of unknowns in a
subject area in which measurement is considered a very precise art.
As the editors remarked in their foreword (1) "Compared to the pond
of knowledge, our ignorance remains atlantic. Indeed the horizon of
the unknown recedes as we approach it".

If the natural sciences with a formal discipline of research
extending over several centuries have problems, then the social
sciences with a comparatively short history have even more. The
psychological factors that influence the behaviour of human beings
have hardly begun to be investigated. (Evidence of the problem can
be seen in the variety of results produced by the econometric models
which at the moment are predicting the upturn, downturn, stabilis-
ation of the U.K.'s current economic position!) Yet it is this
behaviour, whether it be in the production, procedures or procurement
of buildings that has a major impact on its cost. With this high
degree of 'ignorance' coupled with an observed high degree of
variability in measured performance there is obviously a fundamental
need to establish a more substantial body of theory, and better
models, upon which to base our practice.

Problems in Initiating Building Cost Research

The building professions in general, have shown only a token willingness to invest in research and thereby extend their knowledge and
discover an improved methodology. There are a number of reasons for
this lack of enthusiasm including:-
(a)   the practices and firms in the industry are widespread, generally
small in size, and are competing in a commercial market against each
other. The incentive to gain a common pool of knowledge *from which
all would benefit* is therefore diminished, and the gathering of funds
for substantial research is difficult to achieve.

(b)   firms investing in research tend to have higher expectations than the researcher can fulfill. As with other complex areas of study it is not possible for one research project to provide a panacea for all the problems faced in say, building cost forecasting. In the natural sciences, very narrow specialisms are the order of the day which when placed together advance the general body of knowledge in the subject discipline. Practitioners should recognise the need for many small bites at the cherry before it is eaten;
(c)   there is a natural tendency for conservatism and resistance to change, which does not provide a fertile ground for the implement- ation of research. This subject will be developed later in this paper;
(d)   the subject matter of specialisms such as building cost falls between a number of well established disciplines e.g. management, design and economics. This has been particularly apparent to those applying for government research grants although credit must be given to the Science and Engineering Research Council for recognising this problem in recent years. The specially promoted programme in Construction Management, recently initiated, is a very welcome encouragement to those involved in management and cost research.

In addition the nature of the production process and its diversity does not make building cost an easy subject to study. This may well discourage many would be researchers from immersing themselves in what is sometimes called a 'messy' problem.

Progress

It appears that the progress that has been achieved has arisen largely from the refinement of current models. This refinement has taken place over a long period of time through the light of experience. Occasionally a research project will speed up the process, although if it proceeds too fast a gulf seems to arise between the new model and current practice which is too great for industry to bridge. There is a wide body of opinion, even found among the research funding bodies, that we should be 'working at the leading edge of current practice'. This is understandable as it simplifies the objectives of a research exercise and more importantly avoids the gulf arising between practice and research.

The problem is, of course, what if practice is using the wrong model?

Is there not a danger that we might be pouring our limited resources down a well which can never yield a satisfactory spring of new ideas that will allow us to substantially progress? The methodology of experience has served us well for a number of years but it is an unwieldy inflexible tool which cannot respond quickly to a challenge. Coupled with the inertia towards change it could represent a stumbling block to advancement.

Bias in Research

Experience contains within it an enormous bias towards current
practice. Now this is not always undesirable. It has been recognised for some time that as stated by Yerrell (2) "the commitment
and bias of psychologists, and more generally scientists, are
necessary for the scientific process". However this can only apply
where there is a considerable body of parallel research which can
investigate and test the ideas and bias of a single research group.
(The emphasis here must be on research rather than practice i.e. the
discovery and generation of new knowledge rather than the implementation of knowledge within well defined rules.)

Where research is sparsely sprinkled across a very complex subject area then there is a great danger of research being isolated and
dismissed because it does not conform to conventional wisdom, or,
being adopted without being fully tested. Such a problem could
exist within the building industry and examples are available which
suggest that with hindsight both possibilities have occurred.

A study by Mitroff (3) investigated the question of bias with 42
lunar scientists. He reported the scientists view that "the
objectivity of science is a result .... not of each individual
scientists unbiased outlook, but of the scientific communities
examination and debate over the merits of respective biases". In
another paper (4) he goes on to say that "to remove commitment and
even bias from scientific inquiry may be to remove one of the
strongest sustaining forces for both the discovery of scientific
ideas and their subsequent testing".

Because of the lack of research into building cost, any new
model or methodology is tested mainly by practitioners rather than
those undertaking parallel research. There is therefore a strong
bias towards the status quo and the current stock of models. It is
not in the nature of practice to constantly challenge the basis of
the techniques which they apply.

Scientific and Technological Research

A qualification should be introduced at this stage. There is of
course a subtle difference between the objectives of scientific and
technological research. The objective of gaining knowledge in
technology is as summarized by Cross et al (5) to "know how, improve
performance and establish products of skill and quality. Science on
the other hand is directed towards error-free explanation and
scientific 'truth'". In this sense technology may more readily be
answerable to practice than to the research community. The difficulty lies in persuading practice to apply the rigorous tests for a
long enough period to satisfactorily substantiate or reject the
research findings. There may be considerable resistance to even the
consideration of a new model which threatens to overturn the status
quo. As Yerrell (2) points out ".... within an accepted paradigm
the merits of respective biases are readily debated; in contrast to
the discussion of biases inherent in the paradigm itself. Between
paradigm debate is an uncommon activity for the scientific
community". (The definition of paradigm in this context would no

doubt be similar to that of Kuhn (6) i.e. ".... universally recognised scientific achievements that for a time provide model problems
and solutions to a community of practitioners".) If debate of this
kind is uncommon for the scientific community it is seldom found
between research and practice. Yet this dialogue must exist if rapid
progress is to be achieved. As a late starter, building cost
research has a long way to go before it has a satisfactory collection
of models and theory, upon which to build.

Paradigm Shift

However there are pressures which could well bring about a change in
the current techniques used by practitioners and produce what Kuhn
called a "paradigm shift". In his well known historical study (6) he
presented a picture of science (and it may well apply to technology)
as a process involving two very distinct phases. The first is
"normal science", a phase in which a community of scientists who
subscribe to a shared 'paradigm' engage in routine puzzle-solving by
seeking to apply what are agreed-upon theories within the paradigm to
problems observed in the physical world. The second phase called
"revolutions" is a very active period in which the paradigm is found
to be inadequate in resolving the problems; the discipline is thrown
into crisis; and the old paradigm is overthrown and replaced by a new
one which then forms the basis of the next phase of "normal science".
A parallel exists in technology where a change in the tools, machine
or technique forces change upon the current paradigm being used. The
difference between science and technology in this respect is that
science tends to change from within the subject discipline, e.g. a
new discovery, idea etc. which alters the scientists view of the
world albeit with the assistance of technology; whereas in technology
itself the motivation to change is often imposed from outside the
immediate discipline. The provision of electricity, the internal
combustion engine and the silicone chip are examples of innovations
which forced the pace of change in technology over a very wide range
of activities. If the study of building cost can be considered as a
technology rather than a science it is appropriate to look for factors outside as well as inside the discipline which may provide a
catalyst for change.
     This brings us back to the central question of "are we using the
right models" and if not, is there any pressure to change.

The Use of Models

The attractiveness of models lies in their attempt to provide simple
and straightforward answers to our very difficult and complex
problems. They tend to be rules of thumb which unfortunately sometimes become of more importance than the solution itself. To quote
just one example, David Russell (7) writing in the Chartered
Quantity Surveyor with regard to Engineering services said
"Contractors normally arrive at a lump sum total estimate and then
allocate this to the individual items in the Bill of Quantities". If
this is the case then the current model contained within a Bill of

Quantities for the pricing of Engineering services is obviously
suspect. The situation is probably even more complex than this
because the estimators own method of pricing may have little resemblance to the manner in which costs are incurred on site. In addition
the person who prepared the cost plan of the services probably used a
different model, yet again, based upon a completely different set of
criteria. There is a touch of unreality about these models and what
the participants in this forecasting process appear to be doing is
trying to match one simplified model, with another model that is
slightly more complex. Modelling the reality, i.e. the way costs are
incurred on site, does not enter into the process until operational
costs are considered, usually at the post contract stage.

The Nature of Models

The simple nature of models, which in one sense is their attractiveness, is also their weakness. Powell and Russell (8) in a design
context state that ".... to the naive, the persuasive simplicity of
categorisations used by most models produces in itself a sort of
model blindness – an unquestioning and unthinking acceptance of the
model because it happens to support the exploitation of a particular
idea, concept or system. .... in their minds, the categorisations
used in these models can become almost 'absolute', tight boundaries
not to be transgressed .... For this group of model users, the model
itself does not open up new vistas but encourages particular acts of
distinction and boundary definition, which can perpetuate a current
stance, or worse prevent thought concerning other meta-level design
considerations".

They define naive as any person unused to thinking in detail
about systematic and systematic considerations of design and
designing. In building cost research it could be that the slavish
referencing of our research data to insensitive deterministic models
such as the bills of quantities has resulted in a kind of model
blindness which has slowed advancement. It may be that with elemental
costs, abbreviated quantities, unit prices and other bills of
quantities derivatives, we have reached the highest degree of
improvement we can expect from these models. Beeston (9) in his new
book Statistical Methods for Building Price Data says that "there are
two distinct ways of representing costs: the realistic and the
'black box'. Realistic methods are derived from attempts to represent
costs in the ways in which they arise. 'Black box' methods do not
attempt to represent the ways in which costs arise. Their only
justification is that they work". He goes on to discuss their
strengths and weaknesses but suggests that with the development of
computer technology, simulation i.e. the 'realistic' method, is likely
to provide a better basis for development.

Black Box v Realistic Models

The problem with black box methods is that so many of the assumptions
upon which the data is based, are missing. The degree to which the
information derived from the model can be analysed is extremely
limited. They do not contribute to an understanding of *why* certain
costs arise and because they rely so heavily on descriptive information to explain the numbers, they are subject to the vagaries of
language. Descriptive information can seldom be precise and therefore the variability we see in estimates may be largely the result of
the inadequate 'word model' found within the black box technique.

On the other hand the 'realistic methods' place much greater
reliance on numerical information of site performance. Not only does
simulation offer the chance for greater analysis and experimentation
because we can now retrace our steps and investigate the point where
the cost is incurred but it can in the words of Powell and Russell(7)
"open new vistas". Human performance is a variable and the interaction of human performance on site has a major impact on cost. With
simulation we can investigate the results of this interaction
repetitively and provide probability ranges. Unlike the unit cost,
element, bill of quantities, operation cost procedure there is not a
changing model structure as information is refined. Communication is
thereby enhanced because the components of the model have a similar
basis from sketch design to construction. The problem of matching
models should no longer be a major problem although prior to sketch
design the black box method would still need to be used.

What is true for the initial cost of building is also true for the
understanding of future costs. Established techniques such as cost-
in-use, even with sensitivity analysis are essentially deterministic
in nature. The interaction of human beings with the building, has a
major impact on life cycle costs. Simulation may allow a better
understanding of this relationship. (Already the technique is being
used to help understand the effect of human behaviour in energy usage
in buildings.) However the problem of prediction over a long
building life, the unreal assumptions of cost-in-use and the lack of
post prediction control are still with us! While many scenarios can
be postulated it is doubtful whether iterative processes can provide
a satisfactory conclusion that can be held to.

Simulation, the Way Forward

It would be foolish to suggest that we are anywhere near providing
the simulation models we require for the various degrees of design
refinement. However if we can identify a path which will provide a
better opportunity for understanding cost then perhaps research can
be directed towards this goal. Some of the research papers presented
at this conference have already begun to take this particular line.

In terms of traditional quantity surveying practice such a move
would certainly be a 'paradigm shift' (although it would be less of a
change to a contracts manager). Fortunately any change will occur
over decades rather than years and the trauma of a rapid revolution
is unlikely to occur. If, as was stated earlier, the motivation for

a change in the models in technology arises from outside as well as
inside the discipline, then where is the pressure for change to come
from? Undoubtedly the major key to change is the computer. Our
models that we use today have been tailored to suit the limitations
of the human brain and limbs. Both of these limitations have been
reduced considerably by machine power, and enhancement of these
human faculties is likely to continue for many years to come through
the development of computer hardware. The possibility of constructing large data bases; the untiring ability to calculate repetitively
various options; the rapid output of results; will tend to force us
to look outside the present models, which take their present form in
order to cope with mans inadequacies. The difficult problem will,
not necessarily be in providing the software to undertake the
calculations, but in the collection and inputting of data to feed
the new models. It may well result in a highly skilled technical
team working parallel to the professionals or a centralised
information service which all can access.

Client Pressure

Simulation of course will not only be within the province of the
cost adviser. Already architects and engineers are using the
technique for studying human behaviour in buildings, and relating
this to such matters as energy usage, acoustic behaviour and
pedestrian movement. As the use becomes more widespread then clients
themselves will become familiar with the technique. Some portfolio
managers already use simulation in their risk analysis and will
begin to demand an equivalent level of investigation from their
building cost advisers. It may even be that those responsible for
public accountability will also require a more detailed explanation
for why certain figures have been presented.

Despite these outside influences, it is likely that cost advisers
will themselves wish to improve their service to their client in
order to remain competitive. There appears to be a move in Quantity
Surveying circles towards project management and this inevitably
means a closer involvement with resources. Simulation of those
resources will be an important aspect of future development that may
well leave the old models behind.

The argument presented above may sound like the presentation of a
panacea for all our ills. It is not intended to be so. The problems
of handling data in the magnitude required to simulate cost
occurrence is enormous. Simulation does however offer opportunities,
not available in pre-computer days, to change our model to assist
our understanding of how costs are incurred in the building process
and in occupation. It must be worth pursuing this line of thought
although initial expectations should be low in the early years.

Summary

As all researchers into building cost know there is a considerable
lack of knowledge in the subject area. Each research project seems
to generate the need for a score or more of new projects. At the
present time building cost research is not well funded and this
reduces the amount of research activity. This in turn means that
the work that is undertaken is sometimes not subject to the same
degree of scrutiny by other researchers as found in the natural
sciences. Testing is often left to practitioners who have a natural
bias toward the status quo. There is a danger that this may result
in model-blindness which may prevent potentially more fruitful models
being developed. One such technique that has not yet been fully
exploited is simulation. By harnessing computer power this method of
modelling site and building performance may well hold the key to a
better understanding of building cost.

Finally

Whatever research is undertaken in the future there is a need for a
much greater formal and informal dialogue between researchers. It is
hoped that this conference has played its part in this respect and
that future conferences will follow. The new refereed journal to be
published by Spon is also a most welcome addition to the dissemination and discussion of building cost research at an international
level.

References

1. Duncan, R. & Weston-Smith, M. (1977) The Encyclopaedia of
      Ignorance, (Pergamon Press), 9

2. Yerrell, P. (1982) 'Bias in Research – A Methodological
      Interface', Portsmouth Polytechnic News, April, 16–18

3. Mitroff, I.I. (1974) 'Studying the Lunar Rock Scientist',
      Saturday Review World, 2nd November, 64–65

4. Mitroff, I.I. (1973) 'The Disinterested Scientist: fact or
      fiction?', Social Education, December, 761–765

5. Cross, N., Naughton, J. & Walker, D. (1981) 'Design Method and
      Scientific Method', Design Studies, Vol.2, No.4, October, 200

6. Kuhn, T. (1970) The Structure of Scientific Revolutions,
      (2nd edition, University of Chicago Press)

7. Russell, D. (1982) 'Where Computers can help', Chartered
      Quantity Surveyor, May, 304

8. Powell, J.A., & Russell, B. (1982) <u>Model Blindness and its</u>
    <u>Implications for some Aspects of Architectural Design Research</u>
    <u>and Education</u>, (Proceedings of the Conference of Problems of
    Levels and Boundaries, Edited by Professor G. de Zeeuw and
    A. Pedretti. University of Amsterdam, Netherlands)

9. Beeston, D. (1982) <u>Statistical Methods for Building Price Data</u>,
    (Spon , ch.8 to be published 1983)

SECTION 11

KEYNOTE ADDRESSES

COST PLANNING AND COMPUTERS

JOHN BENNETT, University of Reading.

Computers and information technology will cause major changes to the way we work. These effects are widely discussed in general terms. It is now possible to see more of the detail.

I will start with two quotations. The first from James Martin's "Telematic Society" a book which was made mandatory reading for senior civil servants by the incoming Thatcher Goverment.

> "Data-base technology imposes a measure of precision on the way data are defined and referred to, and data-base administrators have often been surprised by how different departments or managers call the same data by different names or different data by the same name. When communication takes place between parties using a common data base, there is less chance of imprecision".

Secondly an excellent review of the state of the art in computer graphics in the Economist of 17th April, 1982, included this:

> "Beoing, which first made large-scale use of CAD in the design of the 747 airliner in the 1960's, was able to get its 757 and 767 aircraft flying on time and under budget by souping up the existing CAD computer programmes used by designers so they could automatically swap data with programmes used by the firm's manufacturing planners. Like other engineering-based firms, Boeing is now working on a system that will store all the information about a design in a single integrated database. That data will replace the engineering drawing as the final authority on the shape of each design. Salesmen, production engineers, designers and after-sales service teams will then all draw on a shared pool of information about each project".

Systems with the capability of providing the same facility, of replacing drawings as the primary information source for building and other construction projects, are available commercially.

Such design systems which also automatically instruct computer controlled manufacturing plant to produce the required components are in use by firms manufacturing standardised buildings. We must add that optic-fibre cables and communication satellites will make it possible and economical in the near future to transmit such information to wherever it is needed.

The task now is to decide how to use this capability. I had the good fortune in 1978 to be awarded a research contract by the DOE Property Services Agency to assess the feasibility of providing quantity surveyors with the construction cost data they require for effective cost planning. The final report on that work was published by PSA in June 1981 under the title "Cost Planning and Computers". It is supported by three annual reports also published by PSA which contain detailed descriptions of the theories and evidence which led to specific conclusions.

I would like to pay tribute to the P.S.A. for having the foresight to commission the study of cost planning in the light of computer developments. That they provided sufficient funds to enable myself and colleagues to work on this one subject for three years has been an essential factor in developing a sufficient understanding to allow a best approach to cost planning for the future to be recommended.

There are three points which need to be emphasised. The first is that the work is based on a very careful study of the best of current practice.

Our starting point was to carry out interviews and case studies in 24 practices who for various reasons we and our advisory panel of experienced practitioners believed were amongst the leaders in Cost Planning practice. From the results we built a model of current practice and we measured its performance.

As a slight aside I would like to mention that I see this as a very important role of QS reseach work; to look at the best of current practice, the leading edge, and then to understand and explain it. In doing so I believe we will make good practice more widely available and by building up our theoretical explanations we may also speed the development of practice.

However at this point I want to look at just one aspect of this first stage of our work. (see Table 1). This is our measurement of QS's ability to predict or estimate the lowest tender. Estimating is fundamental to Cost Planning and the data in Table 1 gives us a measure of QS's performance.It is generally considered that a mean accuracy of plus or minus 12% is not good enough to provide the basis for good cost planning.

The second point I want to make is that we used computer models to simulate practice. This was essential to enable us to understand current practice. Our case studies and data from live projects tell us what happens in practice but models are necessary to help us understand why estimating accuracy and other features of QS's work are as they are.

We cannot rely directly on live data because construction projects are all so individual, so conditioned by the client, the architect, the suppliers, the weather, the market, total demand and much else, that live data is too messy to support direct conclusions. What we can do with live data is to use it to validate our models. So we measured estimating accuracy and collected tender lists. We then produced and developed a model based on QS's methods and on studies of contractor's tendering in competition, which matched and could predict the nature of the live data.

Having achieved a valid model we ran it many times so that we could see what the effect of poor cost data is, what the effect of contractors estimating methods is, what the effect of the way contractors price bills is and several other things besides. All this is in our Third Annual Report which was published by DOE in 1981.

In simple terms we have shown that the accuracy, some may say the inaccuracy of QS estimating is as much a function of the structure of the selective tendering system based on priced bills of quantities, as it is of poor cost data. In other words if we wish to improve we need different methods of estimating as well as better cost data.

Like any good model ours is not a slice of life encapsulated on computer disk. It is a very careful selection of the important characteristics and relationships in QS estimating, tendering and contractors pricing. We have got the structure right and so we can explore how cost data behaves in various situations (see Table 2). The data in Table 2 is absolutely dependent on the use of a model and could be produced in no other way. The figures have important implications for QS practice.

If you take the view that present methods of estimating are not very good then Table 2 suggests that better methods can be made available. In my view this is therefore a matter of professional competence and I do not see that Q.S. have any real alternative but to move to methods which on the basis of very strong evidence are better than those currently used.

The third introductory point I want to make is that in Chapter 4 of our Final Report we, for the first time, explore in some detail the kind of differences the new computer and information technology is likely to make to QS practice. The report was written 18 months ago and as I indicated in my opening remarks technology continues to move faster than the profession can respond. There is however some progress to report.

I would like to continue by summarising our main conclusions before discussing the progress and the changes which I now think are needed to these conclusions.

Our Final Report sets down, as clearly as possible, given the new computer and information technology, our view of the most promising available approach to cost planning.

We envisage QS being involved in projects from the outset. They will provide advice directly to clients and designers in a manner which matches the client's needs for financial advice on initial and life cycle costs taking full account of the costs and timing of finance, of taxation implications, grants and allowances. All this will be produced in a manner which matches the clients own objectives. At the same time designers will be given cost, construction and organisation advice, budgets and estimates in a form which matches their needs. We see QS as being responsible for organisation decisions leaving the designers free to design.

In my view the style and form of our cost advice and the documents we use to plan and control costs are very important. We need to be aware of the skills of marketing men when we are selling the idea of cost planning to clients and designers. Marketing means discovering what clients want, matching our service to their needs and continually reinforcing the fact that we are working on their behalf to make costs predictable and controlled and to give them good value for money in their buildings.

It also means being very competent. I believe an important part of the use of computer and information technology is that it helps to create an image of competence. It is vital that this is not just an image and that there is a solid reality to back it up. That is all part of good modern marketing methods.

The approach we are suggesting in order to achieve that kind of competence is based on the use of work stations linked to a central computer databank. There are two reasons for this.

Firstly we believe that this will be the normal style of working for all business activity in the future. QS cannot afford to be left out or to work in some way different from the rest of the business community.

Secondly we have suggested computer use because the approach identified by our model as producing the best results could not be used manually. It is totally dependant on the computer processing of large amounts of data. This will give QS a better understanding of construction costs and allow us to match new projects to the information in the database much more accurately. The result will be more accurate estimates which will form the basis for more effective Cost Planning.

So we have assumed QS working in their own office, building up cost analyses of their own projects in a standard pattern which identifies the 100 most cost significant items on each project. This is not the same as a standard form of cost analysis which is a purely manual device and given computers is not necessary and indeed prevents us having really good cost data.

Each QS practice will build up their own database of their own projects and at the same time these cost analyses will be transmitted over the British Telecom Network to the central database. Here each new analysis will be matched and compared with all the others. The centre will produce sets of unit rates in a form which makes it easy to choose the right one for any particular item on a new project. These unit rates will be transmitted back to individual QS as they need them.

Within the framework of roles and responsibilities which we have envisaged, the QS with his work station will rely as at present on his own data when this is sufficiently robust. His cost planning system will tell him when he has enough personal data to produce good estimates or good cost advice. When this fails he will look at his viewdata screen to provide an up-to-date summary of good cost data. His system will tell him when this is inadequate for his current project. He will then be able to search the central data base and carry out, perhaps with the advice and help of experts at the centre, studies of the full set of data to establish some new cost fact.

At each stage the data will be handled and calculations made by computer. Statistical tests and a wide range of estimating methods will help the QS to get a feel for the right level of costs and of the decisions which really matter in the design. The methods will take account of the influences of time as well as quantity. They will relate directly to the management of projects.So QS will have the systems and the data needed to manage projects.

To allow all this to happen we need cost analyses produced by computer. They will sub-divide costs into about 100 work sections of roughly equal value. (see Figure 3). We have proposed a hierarchical data structure which will allow individual projects to be analysed to the required depth in a consistent yet flexible manner. Part of this data structure is given in Figure 3. In producing it we kept in mind the 200 work sections of the new SMM7 and the classification table approach which it will use. It is of course important that these data structures are linked. Using the proposed data structure will allow each cost analysis to contain only information which is cost significant. At present our standard form of cost analysis produces very uneven data with individual unit rates varying in significance by 1 to 400 or more. We need instead a standard approach in which the pattern of data is dictated by the individual project.

We also need better descriptions of the cost significant characteristics of projects. I see computers helping to provide this. All information handling is going to become much easier in the near future. So if we have an insitu concrete frame with a certain grid pattern of a certain height to carry defined loads, we should be able to look at cost analyses of projects with similar structural frames. Similar search routines are needed for all work sections.The basis of search must be based on the actual pattern of costs in practice not on any standard form.

An important element of the better descriptions of projects envisaged, is the programme or as they say in America, the schedule. Time has a large and direct influence on many costs. This is particularly true of Preliminaries costs. It is very important and that QS should develop good ways of considering method and time related costs. I do not see how we can advise design teams if they do not understand how buildings are built, well enough to estimate the costs of Preliminaries in reasonable detail.(see Table 3).

Our present data is so variable as to provide almost no basis at all for prediction. This is mainly because many of the costs are time related and QS have not developed methods of dealing with time. Table 3 shows the variability in Preliminaries costs which totally invalidates the normal QS approach of making a simple percentage addition to allow for these costs.

The other important reason for needing effective ways of handling time is that clients are increasingly concerned to get their buildings quickly. If QS have no means of considering or costing time, it is not surprising that clients should form the view that QS cannot help in achieving fast construction and in fact merely cause delay with over complicated bills of quantities.

That may sound like a rather opportunist reason for promoting my ideas on cost data. It is not. Clients want cost and time to be managed. Cost and time are interlinked and are the essential raw material of project management. Todays inflation rates and interest rates merely add urgency to these facts. I see time planning and control as an essential feature of QS cost planning in the future.

So that there is no doubt about the conclusions of our PSA research, let me make it clear that the system which our report describes does not yet exist. Our model of current practice and the advice we have received on computers and information technology point in the same direction. What we have done is to map out the steps necessary to develop the kind of system which the profession needs. I would like to illustrate this point and I will take as an example that hoary old question of whether QS should use operational estimating or quantity related estimating.

We are saying that it is not clear whether and in what circumstances a quantity related or an operational method of estimating is most appropriate. Beyond that we are saying that we need a large data base in order to answer such questions and in order to improve our understanding of costs. The range of projects and the variety within them is such that our present manual methods or any simple computer version of them just will not do.

What we have done is to define the general nature of the system which QS need as rigorously as is currently possible in terms of its level of detail and its organisation and arrangement. The key parameters are that we should aim at an accuracy of about plus or minus 5%, we need to use about 100 evenly sized items per project and we can accept a coefficient of variation of 36% in individual unit rates.

These figures are internally consistent and we believe reflect the variability inherent in construction pricing.

So our general approach is to monitor data from large numbers of projects in terms of its variability. By doing this we can identify catagories of data which vary beyond the limits we have identified. Such data is unpredictable within acceptable limits. That means we cannot estimate its cost. So we need to study the category concerned, perhaps undertake some new research and maybe change the classification of the data or the method of estimating. In this way the best method of estimating the cost of each kind of building work will be identified. This will change over time as construction methods or the relative costs of resources change. So our cost data and our estimating methods will grow and develop in response to real changes in design or construction costs. It is vital that this should happen and that we should not lock ourselves into an arbitrary standard, fixed structure of data.

This approach will lead to great simplicity in use. As far as the individual QS at his personal work station is concerned all of this is going on in the background. The individual user will be presented with the best current data and methods in a easily useable form. He will also have access to the whole system so that he can determine how and when he uses what data. In other words the QS remains in control and responsible for his own professional judgements.

I believe that if we adopt something like this approach we have a real opportunity for QS to take a firm hold of the building industry's information methods and procedures. In doing so we will be serving our clients better. The danger is that each practice sets up its own data base and gets marginal benefits from doing so but misses the big improvements which we can achieve together. In this field we already have the excellent example of BCIS in a manual form. I believe the profession now needs computerised BCIS at the heart of its everyday working methods more or less in the form which I have described. I see this as a very important responsibility for BCIS at this moment in QS history.

Since that conclusion was formed BCIS have indeed taken the decision to produce a computer database and a computer estimating system capable of using the database. I believe this to be an important decision in the history of QS. What now needs to be done?

Well first and most obviously, it is doubtful whether BCIS will be able to stop at producing an estimating system. They will need a cost analysis program to produce data in the right form. Cost analysis will almost certainly need to be directly linked to bills of quantities production. This inevitably leads on to a wide range of QS work.

It will in other words be difficult to avoid producing a complete QS suite of programs. In addition given the facility of transferring data to a central point, analysing it, processing it and distributing the results plus the generally available library search facilities and information intelligence systems which already exist, and which can therefore be monitored centrally and related to the data base, it is probable that BCIS will be able to provide a comprehensive technical management system for QS.

For this to happen in an orderly and efficient way I believe we need a clear statement of how the new technology should be used. This will raise political issues. There is a danger generally that political lobbying based on the vested interests of existing monopolies will lead to ill advised regulation which will rob us of part of the benefits which technology could bring. There will be examples of this within the building industry. This is especially likely when individual professions or organisations confront the need to change their methods and procedures.

Bearing this in mind a well informed debate within the profession is vital. The report "Cost Planning and Computers" and the supporting annual reports plus the computer model on which much of the conclusion is based, provide a start. That work is in my view already out of date. Two important new factors are the recent CAD/CAM developments and a recognition that SMM7 can and should have a much more direct influence on the cost data structure than is recognised in the report. In particular, the SMM7 proposals for preliminaries should be central to future cost planning.

Taken together these additional factors promise to provide better and more readily available descriptions of projects linked to a data structure which closely reflects construction costs. Given that, QS will be well placed to provide cost management, not just cost planning, and time management as well.

The exact mechanism by which project databases can be drawn into a central database needs to be thought out carefully and flexibly. It may be that all that is needed centrally is an index in sufficient detail to enable design teams to identify projects similar to their own. It will also need to be sufficiently detailed to enable costs to be monitored and generalised principles and cost levels to be established.

Given a competent central index then detailed cost estimating can in the future be based on a direct use of priced bills of quantities for similar projects transmitted electronically. Should contrived issues of commercial secrecy prevent this, the small private professional practice will be severely disadvantaged.

Since bills in the near future will provide for quantity related, time related and fixed costs to be dealt with separately and in a way which reflects the implications of construction method, the professional QS will be able to provide cost planning based on the construction implications of designs.

This is of course yet another issue where narrow short-term professional self-interest may prevent clients being provided with the service to which they are entitled.

I am confident that the market place will resolve these issues and that clients will insist, as Boeing did with their 757 and 767 airliners, on getting construction on time and under budget.

Meanwhile the BCIS proposals are a sensible next step. They will lead in the near future to a great burst of cost research. This is because the computer database will allow many cost relationships to be studied and tested with an ease and economy never previously possible. This is an enterprise in which many should join and this Conference could with advantage resolve to exploit the new BCIS facility to the full, by planning a major and co-ordinated programme of cost research.

TABLE 2 QUANTITY SURVEYORS ESTIMATING ACCURACY IN THEORY

ESTIMATING METHOD	MEAN DEVIATION OF ESTIMATES FROM TENDERS (%)	COEFFICIENT OF VARIATION OF ERRORS (%)
1. Cost per square metre taken from one previous project	18	22·5
2. Cost per square metre derived by averaging rates from a number of previous projects	15·5	19
3. Elemental estimating based on rates taken from one previous project	10	13
4. Elemental estimating based on rates derived by averaging the rates taken from a number of previous projects	9	11
5. Elemental estimating based on statistical analysis of all relevant data in the data-base	6	7·5
6. Resource use and costs based on contractors estimating methods	5·5	6·5

TABLE 3 Variability in Preliminaries Costs

Project value range £000's	Number in range	Preliminaries % range*	Mean	Coefficient of Variation
0 – 100	47	12·6 – 64·4	33·0	38·5
100 – 1000	114	12·6 – 54·6	25·4	32.3
1000 – 5000	53	12·6 – 58·8	18·3	40·4
5000 – 10000	2	15·4 – 19·6	16·5	15·2

*Preliminaries % calculated on remainder of cost

FIGURE 3   AN EXAMPLE OF THE PROPOSED ELEMENTAL BREAKDOWN

1. Substructure

2. Superstructure
   - 2.1 Frame
   - 2,2 Upper Floors
   - 2.3 Roof
   - 2.4 Stairs
   - 2.5 External Walls
     - 2.5.1 Loadbearing Brickwork
     - 2.5.2 Non-Loadbearing Brickwork
     - 2.5.3 Blockwork
     - 2.5.4 Stone
     - 2.5.5 Precast Concrete Panels
     - 2.5.6 Asbestos Cement Cladding
     - 2.5.7 Metal Sheet Cladding
     - 2.5.8 Plastic Sheet Cladding
     - 2.5.9 Curtain Walling
     - 2.5.10 Insitu Concrete
       - 2.5.10.1 Formwork
       - 2.5.10.2 Concrete
       - 2.5.10.3 Reinforcement
     - 2.5.11 Infill Panels
   - 2.6 Windows and External Doors
   - 2.7 Internal Walls and Partitions
   - 2.8 Internal Doors

3. Internal Fittings

4. Fittings & Furnishings

5. Services

6. External Works

7. Preliminaries

PRODUCTIVITY: WHOSE RESPONSIBILITY?

DONALD BISHOP, University College London.

The focus of this Conference is cost forecasting, with particular
emphasis on cost modelling and on the factors that ought to enter
into the construction of models. As far as can be inferred from the
titles, cost (to the industry's clients) is accepted as a proxy both
for the cost of production, "... that which produces wealth" and the
value of the wealth so produced to individuals and society -
possibly the most important aspect of this subject area. In practice
tenders acceptable to the industry's clients are the recognised
measure of both, and hence the importance given to cost forecasting
and to cost control throughout the successive phases of projects. ·

This paper is addressed to the strategic question of whether the
conventional procedures for cost forecasting and cost control
necessarily promote high productivity* in the industry. To the
extent factors of production are efficiently deployed an industry makes
greater or smaller claims on society's resources for the same output.
An efficient industry will draw fewer resources from other activities
and vice versa. Therefore, despite the current state of the economy
high productivity continues to be important; inefficiency is not
costless. This examination does not aver that the established
procedures fail to serve clients' interests, project by project.
However methods producing a 'best buy' for individual projects may
not lead necessarily to a 'best buy' for clients in general, that is
to high productivity.

Cost forecasting and cost control should contribute to an industry's
productivity by providing relevant information to would be clients
and feedback to design and construction. The procedures do not stand
alone. They reflect the industry's processes; other command
arrangements, other criteria for investment, other way of mobilising
design and construction resources would entail different roles for
cost forecasting and control. The titles of the papers to be
presented indicate most authors have assumed the conventional arrangements for the industry; i.e. essentially bespoke design, separation
of the several stages of design from responsibilities for construction,
 *Economic Progress Report HMSO.
 *Productivity is a measure of the quantity of output of goods
  and services that can be produced for a given input of
  factors of production (land, labour, capital, energy, entrepreneurial skills, for instance). A major long-run aim
  of policy is to increase the stand of living of the community,
  and raising productivity is the main way of achieving this.

and the employment of main contractors supported by many sub-
contractors. This fragmented system makes cost estimating and control
more complex than it might be in other circumstances. It also
diffuses responsibility for productivity which at first sight clearly
rests with operatives and their immediate supervisors, and with
contracts managers. However the character of the work to be executed
is determined elsewhere by designers and by component manufacturers
and all involved work in the context of an industry called into being
by clients' demand for construction. The activities of these groups
both influence and are influenced by each other to an extent that the
question "Productivity, whose responsibility?"can be answered only by
examining the process as a whole. Individual site operations form
the starting point of this paper which successively considers
site managements' and design teams' responsibilities|for productivity
before examining the relationship between cost forecasting and
productivity.

Productivity of single operations

In every industry, productivity comparisons are fraught with difficulties and estimates of inputs and outputs are uncertain. At first
sight these problems should evaporate for tasks that are apparently
as independent of other trades as is bricklaying; teams are clearly
defined (some combination of bricklayers and labourers) and the
output is countable – bricks laid. Surely, therefore, the
productivity achieved is a matter only for the team's dexterity, skill,
and will to work? This is not the case because bricklaying
requires a 'workplace' created by others eg a foundation, or concrete
slab, or the previous lift.

Therefore many necessary conditions have to be met before a gang can
make progress: eg

- the workplace must be ready;

- access provided (and re-provided as each lift is completed);

- materials available (which may require service from a crane
                        or forklift);

- instructions issued and reference setting-out completed.

If any of these (or other relevant) conditions are not satisfied
either work will not be able to start at that workplace or it may
progress slowly. It will be seen that bricklayers' work is dependent
on

- the arrangements made by management;
  (for materials, scaffolding, instructions, setting out);

- the fulfilment of these arrangements, often by organisations
  not under management's direct control;

- the work of preceding trades (which create the workplace);

- (sometimes) competition for scarce resources i.e. cranes or
  fork-lift.

It is too much to expect that all these conditions (and similar
conditions for other operations) will be met each time a gang of
bricklayers needs to start the next task. Nearly always, unless a
project is very small or otherwise constrained, a safety net is
provided by ensuring that each gang has the choice of several workplaces as they progress from one task to the next. Therefore, in the
ordinary run of events, one workplace is likely (no more than likely)
to be ready for their attention. This over provision of workplaces
is one of the characteristic features of building operations.

Suppose, however, all conditions are met so that a craftsman of craft
team can move without delay from one workplace to the next. Despite
this there will usually be a setting up cost whilst materials and
tools are moved, new instructions are absorbed, and a start made.
Even when these are discounted, many studies have confirmed the great
variability of output of individuals tackling apparently similar
operations.

There is less understanding than might be expected of the factors
influencing output at the workplace. This might be explained at
least in part by the varied nature of the tasks, the constantly
shifting workplace, and by the many possible sequences of activities
and their interaction. Studies based on test pieces are seldom
convincing because the inevitable bitteness of building operations
and the interactions of one task with another are lost; studies of
whole tasks require extended andintensive activity sampling – an
expensive undertaking unless part of management processes.

From past studies the following are important: –

- techniques and skill demanded by a task (obviously);
- quality (an essentially elusive property);
- whether a task (of whatever nature) leads to the development
  of a rhythm so that the individual movements flow;
- repetitiveness within phases of a project (this leads to
  higher output as operatives and management 'learn' how best
  to tackle their tasks);
- whether the time required to complete a stage of work fits the
  intervals determined by workbreaks;
- methods and (for some tasks) physical stamina.
- 'hygiene' (in a Hertzog sense) of the site.
- operatives' motivation.

It is self-evident that these factors act in concert so that no one is
likely to be dominant except in unusual circumstances. Moreover one
team often needs the help of another, therefore the progress may
depend as much on the co-operativeness of the workforce as on each
team working at its optimum output, with little regard for the progress

of others. (For example bricklayers were observed to spend as much
as 30% of total time in helping forward other operations and
working on tasks that could not have featured in any bonus target).
Also the output of any gang may be largely determined by the nature
of the workplace created by the preceding gang. 'Hygiene' and
motivation factors, whilst important, are not well understood.

Managements' responsibilities; the productivity of sites.

Management's general responsibilities for productivity are to ensure
that a skilled labour force is both trained and maintained, to
obtain conditions where high motivation and willing co-operation
naturally flourish, to select working methods that will generate
repetitive sequences of tasks (to the extent permitted by the design),
to provide resources and to make workplaces ready. All lay at the
heart of managements' tasks, for the purpose of this paper we will
concentrate on only one feature, the provision of workplaces, which
has a decisive influence on productivity, because a high proportion
of labour input is not directly concerned with activities at
workplaces.

Forbes and others have shown that the labour input on site may be
analysed into three main components;

- directly concerned with making a building grow;
- concerned with preparing for work to make a building
  grow; and
- non-productive time

Extended studies of site operations have shown the extent of the
variation of the three components, and the degree to which the
incidence of non-productive time is influenced by factors outwith
the control of operatives and, often, of site management. For the
time being it may be helpful to note that inputs directly incurred
with making a building (project) grow have been recorded as ranging
from roughly 10% (1) to over 80% (2) and that for the ordinary
run of housing Forbes estimated the three components to be roughly
equal. It is therefore unhelpful to focus on operations at the
workplace when these almost always account for less than half of
labour input and sometimes account for only a small proportion of the
labour input on any site.

The nature of the problem is perhaps illustrated best by an
idealised line of balance diagram (fig 1, 2) over –

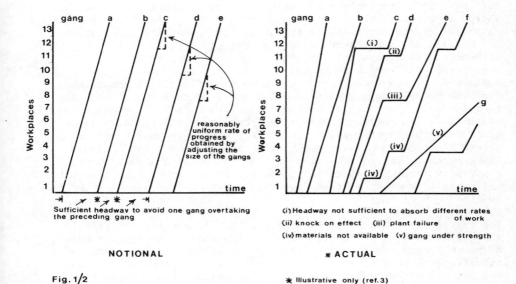

NOTIONAL   * ACTUAL

Fig. 1/2   * Illustrative only (ref. 3)

Implicitly diagrams such as these assume that tasks do not change
significantly as gangs progress through a project, that progress is
sufficiently unvarying to ensure that successive tasks do not collide
providing each is given sufficient headway, and that sufficiently
similar rates of progress may be obtained by adjusting gang sizes.
In practice none of these assumptions is realised. Even in
repetitive construction each workplace (and therefore the task at it)
is unique; tasks change (and therefore the elapsed time) as work
progresses; some tasks – often those done by sub-contractors – are
completed quickly so that gangs may either depart for other sites
if no suitable work remains to be done or claim for non-productive
time; many factors intervene to delay progress of some gangs but
not others. Therefore any attempt to achieve a very rapid rate of
construction is likely to result in high non-productive times as
gangs queue for workplaces. Matters are much more difficult when a
contractor tackles non-repetitive construction. There the tasks to
be completed change in nature and extent from one stage to the next.
Not only do the size of gangs and the headway between successive
gangs change frequently, but the anticipation of requirements for
materials, access, and information is much more difficult.

There are two extreme possibilities distinguished below by the terms
"tight" and "slack" control – the latter in a technical not a
pejorative sense. Contractors may attempt either to impose tight
control over all the variables or to reduce the problem by creating

many more workplaces than there are gangs to man them. The first is
difficult fully to implement – unless a project has a 'most favoured
nation status'. Tight control runs against the grain of the industry.
Many factors may intervene to frustrate it; essential information
may be missing, queries not answered, key equipment may fail,
allowances for dimensional tolerances may be insufficient.
Importantly the temporary nature of site organisations and the
employment of many sub-contractors make it difficult to ensure every
resource is available when needed and stays on site for as long as
is required. There are of course many other factors, including the
weather. It is not surprising, therefore, that the industry usually
adopts the tactic of building relatively slowly so that there are
always many more available workplaces than there are gangs to man
them, perhaps five to ten times as many. That is the workplaces
apparently ready for attention queue for gangs rather than vice versa,
and gangs often seek a "ready" workplace rather than are directed to
one.

The tactic is robust. It accommodates the variability implicit in
the structure of the industry, reduces the task of management which
often no longer has to ensure that that workplace is made ready at a
specific time, and ensures that delays that arise from all the many
factors are not as damaging as they might be because equipment and
gangs can be quickly redeployed on other productive tasks or diverted
to some other sites requiring their attention. In this way waiting
time is reduced and productivity increased. Moreover the management
effort that would be required to impose tight control is not necessary.
Of course the tactic is not costless; establishment charges are
borne by contractors, and the cost of the capital unproductivity
invested in work-in-progress by clients:

It is not self-evident whether this solution – building relatively
slowly – to the problem of the organisation of work on sites is cost-
effective. It is always possible to ensure particularly urgent
projects are built quickly. In effect such projects have "favoured
nation status" when resources are allocated by design teams, or
contractors, or sub-contractors, or by materials suppliers. When the
industry is busy other neighbouring projects were probably delayed,
but probably not to any great extent unless the favoured project was
very large eg a major power station. If, however, an attempt was
made to build quickly the number of potential workplaces for the
local labour force would be sharply reduced. Therefore all projects
would have to attempt to impose tight control to ensure a close match
between construction resources and the available workplace: to be
effective this would require at the very least

- a simplified approach to design as with industrial buildings
  so that there was a limited and well-understood vocabulary
  of technical solutions;

- a smaller number of sub-contractors on any project;

- higher stockholding costs in the building materials and
  components industries to reduce the risk of stock out
  and
- higher management costs

The tactic of building relatively slowly cannot solve the whole
problem of non-productive time which is influenced by other human
factors eg motivation in its many aspects and by the complexity of
design. The latter exerts a pervasive influence. Repetition within
and between operations fosters high productivity from several
sources; 'improvement' at the work face, a gradual resolution of
the problems of working methods and provision of materials and plant,
and a much reduced requirement for specific instructions. In the
heyday of industrialised building detailed studies were made of one
method based on timber panels and a light steel frame. On one site
where no single block of low rise dwellings repeated, non-productive
time was approximately twice that of another which was composed of
40 identical blocks – largely because of the difficulty of ensuring
that the right panel was at the right place at the right time.

Finally reduced non-productive time is only one of the ways
contractors may achieve high productivity. Making buildings grow
may be made less costly by better working methods, mechanisation, and
by improved operative skills. All require investment and to some
extent specialisation. These have been made possible by the growth
of specialist sub-contracting, plant hire, and by the ever increasing
range of component systems for eg. roofing, claddings and flooring.
Inevitably there are snags. Whilst the range of skills demanded of
many operatives may be narrower than in the past, the work to be
done still requires the understanding of the process, dexterity, a
knowledge of materials, and the care that were the hallmark of craft
work. It is far from certain that the industry now provides an
adequate training base. Another problem is that the potential
advantages of these components are probably realised best when
manufacturers, designers and contractors have repeated experience of
working together so that a proper range of application can be defined,
operational problems resolved, and the cost of special solutions
known. This leads to a consideration of the design teams responsibility
for productivity.

Design teams' responsibilities

Should high productivity be an overt objective of design? The
foregoing discussion has made plain some of the ways this might be
sought – and none is particularly sophisticated; eg by

- considering each project as a whole from the standpoint of
  site processes;

- defining tasks that can be tackled by a gang without
  interaction with other gangs;

- evolving designs calling for relatively few distinct
  specialisms (and, hence, few interruptions to the sequence of

construction;

- ensuring, given other considerations, repetition;
- using proven and compatible components;
- providing adequate tolerances; and
- defining tasks within the competence of the operatives likely to be engaged on the work.

Suppose high productivity were made a prime objective, in current circumstances would the effort be rewarded? Probably not, for two reasons: first the tender may not attract lower bids and secondly the difficulties of designing for productivity.

Although the intention to produce a design capable of yielding high productivity might be made plain in the invitation to tender, the potential advantages to the successful contractor are not readily displayed in the bills of quantities. Moreover, even if they were, the contractors concerned would have to be sufficiently convinced that they would be able to realise the advantages to bid lower prices. Whilst contractors may be keen to price themselves out of obviously difficult jobs, it is probable the benefits stemming from a design capable of being built with high productivity would be submerged by other considerations eg the local market, lack of confidence that the advantages would be realised and so on.

The second reason is more decisive. Experience in many industries has shown the extreme difficulty of designing for production. Even when there are repetitive processes at fixed workplaces, only continuous product development can achieve high productivity. By following a repeating cycle of design, production, monitoring, evaluation and re-design the product, production methods, work sequencing, and quality control can be gradually improved. Studies have shown that this improvement continues for many years even when processes are largely automated. The case for continuous product development in the construction industry is even stronger because, without repeated experience, it is difficult to determine the consequences of design decisions on site operations.

The possibilities for continuous product development exist in some sectors of the industry, for example specialist firms tackling eg foundations or tunnelling, some parts of the component industry, firms marketing off-the-peg industrial buildings and, especially, private housebuilding. A recent study of public and private house-building illustrated the consequences of these two very different systems for productivity. Public sector housing has to satisfy many objectives, the balance of which is determined for each project by a complex and fluid interaction of elected members with several departments of the local authority and often with the public. Each project is essentially unique; frequently many house types have to be included in a scheme which may be designed to read as a whole with public rather than private space at the focus. Consequently

Consequently, although the technical specifications for the two
markets are very similar public housing layouts are often complex
and plan forms vary more than might be expected from the brief.

In contrast the objectives of private sector housing are relatively
simple, to build acceptable houses that will sell in a particular
market at a predetermined cost. The outcome is often achieved by
houses (or flats) drawn from a portfolio of designs which are
transposed to a site with the minimum of modification. Typically
these evolve slowly in the light of a firm's accumulated experience
of marketing, production, and making good defects. In this way
unpopular, difficult, or fault-prone features are eliminated and
buildable designs emerge which use materials and methods familiar to
a firm's managers and foremen and the subcontractors they employ.
This essentially implicit process, made possible by the short chains
of command, is responsive to changing consumer demands on the one
hand and to current site experience on the other; the latter being
constantly refreshed by subcontractors' reactions to the work required
of them.

Without labouring this point too much, private housebuilding has two
of the important features of continuous product design; repeated
experience with a continuously evolving product and timely feedback of
information on which to base decisions about the next generation of
the product. In the main these features stem from the nature of the
operations rather than any sophisticated attack on the problem.

Designers in the mainstream of the industry are differently placed.
With the exception of some public sector programmes of the recent
past, few have the opportunity to hone finely a limited range of
techniques applied to a single building type. Nor do they often have
the benefit of continuous feedback from projects similar to that on
which they are engaged.

Cost forecasting and productivity.
From the standpoint of productivity feedback currently takes the form
of cost advice. In the long run, taken over many projects there is
no reason to suppose that the level of pricing of works sections in
bills of quantities does not reflect the cost of carrying out the
work. It is far less certain that the implications of specific
designs are similarly reflected. Consider the problems. The
productivity achieved will depend on the character of the individual
tasks defined by a design, by the relative complexity of the work,
by the way the tasks comprising a works section mesh with the
remainder of the work, and by the quality that will be demanded. This
paper has identified some of the principal factors. They include: -

Principal factors affecting productivity

factors	input directed to making a building grow	preparing to make a building grow	non-productive time
nature of tasks			
- skills demanded	design (mainly) and choice of method	design and method	
- quality demanded	design: available operative skills	-	
- repetition within tasks	design and to some extent operatives skills	design and method	some tasks and badly balanced gangs create within gang non-productive time
- repetition between tasks	design and choice of method	design sometimes exacerbated by management decisions	complex designs will generate high non-productive time as will inadequate management, planning and instructions
- meshing of work/ rest intervals	management decisions	management decisions	management decisions
operatives skills	industry's training base; within firm training; operative selection		
human relations motivation and 'hygiene' factors:	management responsibility with management/ union/operative/ interactions.		
working methods	management responsibility within the possibilities defined by a design and by the market eg for specialisation or ownership of specialised plant.		

Cost forecasting is rooted in methods of measurement that describe work-in-place in greater or lesser detail. This principle is well established – for example the masons' accounts exhibited in the Jewel Tower, Palace of Westminster. Prima facie this basis for measurement is well justified; clients buy work-in-place. It cannot however represent the character of site operations or the overall and detailed features of a design that make for high productivity.

This has consequences both for individual projects and for improved productivity in the long term. Contractors' estimators may fail to react to the implications of a particular design by high or low prices as the case may be: i.e. designs that are conducive to high productivity may be over priced and vice-versa. From the standpoint of an individual project this may lead to windfall profits or losses, to a project that runs with few problems, or to one that is dogged by disputes and claims. From the standpoint of the improvement of productivity in the longer term the consequences are more serious – at best feedback is weak. Therefore any drive to achieve higher productivity by continuous product development is blunted because on the one hand the four groups concerned, product manufacturers, the design team, contractors and sub-contractors do not have a continuing relationship and on the other because there is no operationally based cost feedback in common. The proposals currently before the industry to improve project information and to revise SMMG will not materially improve the situation.

What might be done to make cost feedback more attuned to productivity considerations? Little, one suspects, given the arrangements encapsulated in the JCT form of contracts, except perhaps to ensure tender documents are not misleading and to reduce the effort involved in pre-and post-tender stages. Clearly these arrangements are not sacroscant. Already they are challenged on several fronts. The industry is moving perceptibly to component based construction with manufacturers marketing entire systems complete with eg fastenings, joints, flashings, auxiliary components. Project management and fee contracting are well established.

Taken together these developments could lead to new alignments of the industry bridging design and construction. Suppose, for example, a project management organisation became expert in a building type or types. Those concerned would then be able to monitor a succession of projects and thus obtain feedback to inform future design teams and the selection of ranges of compatible components. In these circumstances the information provided to tenderers for the main sub-contracts could be attuned to the operations of that project and set in the context of the overall programme. Clearly some of the freedom currently enjoyed by design teams and by main contractors would be reduced but this may be inevitable if the longer term benefits of continuous product development are to be reaped. Relevant feedback

will be essential. Currently the systems discussed at this
Conference serve either design teams or contractors. They will not
be adequate to inform these wider and continuing functions of
project management. The techniques to be used might be informal,
such as those employed in private sector housing where the main
input is the prices quoted by the sub-contractors serving this
market. For other more complex and less repetitive building types
more than one type of feedback is likely to be needed, for example,
sub-contractors bids supplemented by activity sampling data to
provide the operational context of sequences, delays and non-
productive time. This could create a new and operationally based
role for those concerned with cost advice and one that would make a
direct contribution to improving the industry's productivity. It
is perhaps significant that these possibilities have arisen when the
two streams of the profession are to coalesce.

References (1)   Engineering construction performance
                (The Mortimer Report) NEDO HMSO 1976.

          (2)   Forbes W S et al BR studies of productivity
                eg Golden Jubilee Open Days Papers 1971
                BRE DOE.

          (3)   Roderick I. F. Examination of the use of
                critical part methods in building.
                Building Rsearch Station Current Paper
                12/77 DOE.

DESIGN, ECONOMICS AND QUALITY

PROFESSOR GEOFFREY BROADBENT, Portsmouth Polytechnic

Ten years ago (1973) I wrote a book called *Design in Architecture*.
It was concerned with what people need of their buildings, and the
processes by which designers might try to satisfy those needs. It
included design methods, computer-aided design and many other things;
some of the techniques which I (and others) described have now been
absorbed so thoroughly into design practice that they are hardly
worthy of further comment. But I also tried to predict things to do
with the future of our professions, and some of those predictions
are now coming true.

I tried to explore the limits of what could be achieved by
systematic design methods: getting the appropriate 'fit' - it may
be close or it may be loose - between rooms and the activities
within them; putting the rooms in appropriate relationships with
each other; designing walls for good thermal and acoustic insulation, not to mention thermal capacity; making windows appropriate
in size, shape and location for the particular climate (not so big
as to permit excessive solar overheating, not so small as to keep
out the daylight) and so on. Then I tried to show all this in
action with a worked example (Chapters 19 & 20) by which I concluded
that; try as one might to analyse the problems of designing a
building, one's analysis can only give fragments of a solution. It
can hardly ever determine what the building should be like as a
three-dimensional construction.

I realised that, whatever their intentions, architects would
bring their preconceptions to bear at the stage of deciding what
kind of building they were designing - usually within the going
style; Arts and Crafts, Modern, prefabricated concrete, steel and
glass, High Tech, Vernacular, Neo-Classical or whatever. The
history of architecture is a history of changing styles, and if
there were perfect solutions to the problems of its design, they
would have been found, sometime in the past 10,000 years.

At first sight this matter of changing styles seems very superficial, but you only have to read the Philosophers of science, such
as Thomas Kuhn (1960) to realise that fashion permeates everything
that humans do; not just the arts, dress design and car styling,
but also engineering and the hardest line sciences.

So how can we cope with that? How do we get something sensible, enduring - and economic - from what starts as mere fashion or whim?

That's easy, as I said in my book. You submit your initial conceptions, or even your preconceptions to a series of practical checks. Does your concept work in such a way? If it does, then pass on to the next check. If your checks are practical enough, and you apply them rigorously enough, then only a good solution can get through.

It worked, and in the last ten years or so, this whole approach has gained great intellectual - and practical - credibility from the pioneering work of people like Anderson and Landau (both 1966), Hillier, Musgrove and O'Sullivan (1973). They, and I, have drawn on the work of Karl Popper, another philosopher of science, who has insisted, on a number of occasions, that the building of scientific theories is a matter of conception and check, of hypothesis and test; of what Popper calls *Conjectures and Refutations* (1963). But if the hardest line scientists have to work by Conjectures and Refutations, how can designers hope to be more scientific than that?

There are those who wish it wasn't so. They want design to be more motivated than that. They want design to be more creative than science, but far from inhibiting creativity, Conjectures and Refutations actually free it. Your Conjectures can be as wild and free as you like; all you have to do is to submit them to proper refutation procedures.

So despite some determined efforts, no one, so far, has shown me a design which was not done in this Popperian way.

I mentioned various ways of conjecturing designs in my book, and I submitted them to testing procedures, but the best framework I know for such testing was first described by Hillier, Musgrove and O'Sullivan (1973) who asked, and answered, the basic question: "What, despite anyone's intentions, does the building actually do?"

I have developed and extended their answers on a number of occasions (1979). Here is a recent version:

Despite anyone's intentions, any building:

1. Will enclose internal spaces which, by their size, shape and arrangement will permit a range of activities. It may in practice encourage certain activities whilst inhibiting others.
2. Will act as a climatic filter which, by the substance of its walls, floors, roofs, partitions and so on may insulate the internal activities from each other in visual and acoustic terms. It may or may not provide a pleasurable internal climate in visual, acoustic and thermal terms, whatever the weather outside, from the hottest of summer days to the coldest of winter nights.
3. Will act as a cultural symbol, expressing the economic, social, political, religious, aesthetic or other status of whoever built it and whoever they built it for. This will help determine its aesthetic qualities.
4. Will have economic implications, in terms of land values, of its own and adjacent sites, the capital costs of design consultancies, of the materials, equipment and labour for its

building, of running costs for heating, lighting and even cooling
systems; the costs of cleaning, maintenance, repairs and
rehabilitation.
5. Will have an impact on the environment into which it is built
which will certainly be visual, climatic (shading, wind vortices)
social (generating traffic) and so on.

As surveyors, presumably you are interested in No. 4: Economic
Implications and these are very important, but I want to make it clear
that things can go very badly wrong if you concentrate on economics
to the exclusion of everything else. They go particularly wrong if
you try to pretend that No. 3 is unimportant, that is the Symbolism
of the building. But why am I suggesting that you would?

We all know the carricatures that architects and surveyors draw
of each other. I say draw, but architects are supposed to do the
drawing: surveyors to measure what they have drawn, and put prices
onto it!

And so the stereotypes have grown: of the architect as a creative
designer, wilful and arrogant but possibly feckless, and certainly
irresponsible when the spending of other people's money interferes
with the visions he wants to build. The stereotype surveyor, by
contrast, is a sober citizen: honest, reliable, upright and trust-
worthy. He sses it as his job to inject reality, to curb the
architect's wilder flights of fantasy; to bring him down to earth,
to cut him down to size.

Like all stereotypes they convey certain truths. They even determine how sixth formers choose one profession or the other. Once they
have decided that they aspire to one stereotype or the other they
may even be disappointed - and disorientated - if their professional
education blurs the distinctions too much and fails to confirm them
in that role.

But it seems to me that such roles, by now, may have served their
historic purposes. They might have been appropriate in the 19th
Century, they are changing fast in the 20th and they will have little
place in the 21st. I believe this for a number of reasons which
have to do with the professions themselves: architecture, surveying
and economics, the ways they are changing anyway, and changes which,
like it or not, the computer is forcing onto them.

I shall deal with each of these in the course of my paper, but
with less emphasis on those things which I believe others will be
covering more fully. So let's start with one of those, the computer:

Implications of the Computer

The computer has been around for forty years now and whilst
architects are supposed to be the creative innovators, surveyors,
in fact, embraced it earlier, and with greater enthusiasm, than
architects did.

This is partly to do with the nature of our different kinds of
work, but also because of those roles we insist on playing. *Design
in Architecture* compared human intelligence with artificial intelligence in a chapter on computer-aided design. And that led to some

thoughts in my Introduction:

> *More and more, over the past two hundred years, the various*
> *functions of building design have been separated out. As*
> *the theory of structures developed, so the engineer became*
> *a separate practitioner, whose work, on the whole, could be*
> *quantified. He believed himself, for this reason, to be*
> *doing a tougher job than the architect and so, as they*
> *emerged, did the quantity surveyor, the heating, ventilat-*
> *ing and electrical engineers. The architect's task gradually*
> *shrank; he was left with the 'soft-edged' aspects of*
> *building design.*

I then went on to discuss attempts to quantify those 'soft edged'
aspects: human response to colour, to forms, proportions, textures
and so on, which led to one of my central points:

> *Curiously enough, the device which makes such quantifi-*
> *cation possible, the computer, seems likely also to*
> *rehabilitate the architect or, at least, the architec-*
> *tural mode of thinking ...*

I took the example of a bridge, pointing out that the engineer may
start by deciding to do a suspension bridge whilst the architect
would look for strategies: how to cross a river - by tunnel, by
ferry, by bridge on piers - rather than merely presupposing a
suspension bridge and then designing the cables. As I said:

> *Designing cables, not to mention columns and beams, can now*
> *be programmed and thus made the subject of routine compu-*
> *tation, as can the equivalents for cost, the design of*
> *heating, ventilating and lighting systems. Those who,*
> *in the search for 'tougher' more certain jobs opted for*
> *the quantitative aspects of design are increasingly*
> *likely to find themselves redundant. But the architect's*
> *'soft edged' skills - those which require personal*
> *judgments in the study of human needs - are becoming*
> *valuable again.*

This was all summed up, beautifully, in a cartoon review of the book
by Louis Hellman (1973) *(Fig 1)*.

Those views may have been prophetic at the time, but now they
are the literal truth. For the fact is, given a big enough computer
system, you can sketch out your plans with a light pen, get the lines
straightened up by the computer, get draft print-outs of your plans,
elevations and sections, you can engage in dialogue with it about
the structure: load bearing, concrete framed, steel-framed and so
on; size the columns and the beams, get them drawn with reinforcement and whatever schedules you need.

You can have your building's environmental performance checked,
in thermal, acoustic, day - and artificial - lighting terms. You can

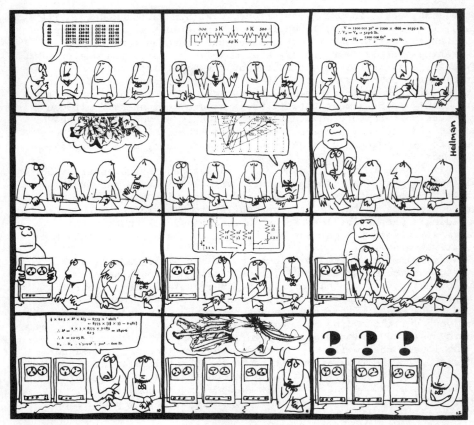

Fig 1

have it checked against the Building Regulations, simulate how a
fire might spread and so on. You can call up your standard details
and get them drawn, as many times as you like to make perfect production drawings of astounding accuracy. You can call up your
standard specification clauses and, of course, you can have the
whole thing quantified and costed. You can have your perspectives
drawn.

All this presupposes that your office is successful enough to buy
computers on this scale, starting at around £100,000. I have to
confess that whilst I have seen all those things in action, they
were happening in a number of places: in architects' offices,
certainly, but also in engineers' offices of various kinds, and
surveyors' offices, not to mention academic research units.

Most architects still would see them on that scale as a distant
threat - or a promise.Yet some of them are your stock-in-trade.

When I look at the topics of a Conference like this: Probabilistic
Cost Data, Cost Data Banks, Cost Data Bases, Algorithms in Price

Modelling, Costs in Use Calculations, Cost Simulation Models, Life
Cycle Costing Analysis, I realise that these, not to mention papers
on such topics as User Interfaces suggest that the computer indeed
is taking over the very things that make your profession what it is.
I won't be as brutal as the Editor of the Royal Institute of British
Architects' Journal who asks (1982): "Will architects be battling
it out with quantity surveyors, whose jobs, in their present form,
are likely to be almost totally eradicated by computers?" If you
work out your computer systems as efficiently as you seem to be
doing, and any fool architect can operate them, then what is there
left for you - and him to do - apart from dreaming up Hellman's
flowers?

Well, there is more. In my book I described the computer as our
saviour from routine drudgery. But, of course, it is far from
that: on the contrary you need a particular kind of patience to
write programs in the first place whilst feeding in the data on a
building: each point located by three-dimensional coordinates,
makes taking-off by contrast seem like a bold exploration into the
creative unknown!

We may, before long, be able to point our video cameras at a
rough, three-dimensional model, thus feeding in coordinates
analogically, rather than digitally. But at the moment, your
computer by no means reduces the amount of routine, repetitive
work to be done. It merely changes the kind. You reduce the need
for your dull, plodding engineers (I'd better not say anything about
surveyors) and replace them by even duller programmers instead!

Obviously that will help determine who you, and we, should be
educating in the future, but it seems to me inevitable that those
whose boring lives have been based on routine, manual calculation
will, literally, be out of a job, except for the comparatively
small number required to tell the programmers what to programme.
Of course, we shall be concerned with costs, but with even more
emphasis on the exercise of judgment as to the actual basis of
costing. It should be clear by now that, in my view, such judgments
will have to be creative.

We have evidence around us in most of what has been built in this
country over the past thirty-five years of what happens when you
take a routine "economy equals minimum cost" approach to building.
It all looks very cheap and very nasty. It doesn't work very well,
and, these days, it is incurring the most horrendous maintenance
costs. We even have examples such as the high-rise flats built
with prefabricated concrete panels, in which architects and surveyors
between them, in their anxious desire to get buildings that looked
cheap, actually made buildings that were more expensive, even in
capital cost, than traditional building would have been. When one
adds the cost of repair and rehabilitation, not to mention the
demolition of the most unacceptable examples (one recent estimate
suggests that they will cost £300 million nationwide), then one
realises that that indeed was false economy.

So what do I mean by creative economy?

Creative Economics

One of the most influential of the world's economists, J. L.
Galbraith, has some pertinent things to say (1974) about the complex
relationships between art and economics. Economists, he says, tend
to treat science and technology seriously, whilst thinking of
painting, sculpture, music, theatre and design as merely frivolous.
The economist might well be interested in the supply of artists'
materials, but he will by no means understand the artists' products.
Indeed, he may well find the artist - or more particularly the
designer - an uncooperative fellow, whos very creativity makes him
difficult to fit into an organisation concerned with the economics
of making and marketing.

Within such an organisation, anyone who shows independence of
thought is condemned as 'something of an artist' - a brilliant
eccentric who refuses to be accommodated within any organisational
structure. That is why, according to Galbraith, innovation is the
province of small firms in which creative enthusiasm is not
swamped by organisational hierarchies. They innovate and sell a
good quality product, expensively, to the discriminating. Then, if
they see mass markets in what they have done, the large firms copy
their successes.

Galbraith suggests that perception of the 'good' is a function of
period and class. The commonest ways of displaying your affluence
in the past, as a private individual, as a ruler, as a government,
as a church or whatever was by "architecture and its embellishment".
The private household displayed its wealth in the form of the house
itself, with the paintings, the sculpture, and the furniture by
which it was embellished. Rulers built palaces on an even grander
scale, governments built Parliament houses, religious bodies built
temples, churches, cathedrals, mosques and so on.

At the moment, such things don't buy much prestige. That function
has passed to scientific and technological achievements: to space
rockets, lunar modules, Skylab or even Concorde. Most individuals
buy cars, boats and even aeroplanes instead of paintings or other
works of art. Of course they still buy and furnish houses, indeed
Galbraith suggests that at higher income levels, architecture,
interior design, furniture and landscape - not to mention food and
drink - are still enjoyed partly for their own sake, as conspicuous
consumption and partly, still, for the esteem they attract. Such
things as the pace-setters enjoy at the moment, he suggests, will
permeate to the rest of society, thus "artistic accomplishments
will ... be increasingly central to cultural development."

Surprisingly enough, Galbraith omits any mention of art as an
investment. Yet shrewd investors - such as the administrations of
the various Union Pension Funds know that few things - not even
gold - appreciate at quite the rate of art, or at least of such art
as is approved by the critics.

In 1980 for instance, Picasso's 'Saltimbanque' was sold at auction
for $3 m.   Yet, in 1923 when it was painted, the materials
could not have cost more than a dollar or two, nor for that matter,
could Picasso's time.

Some kinds of architecture obviously have a similar, if less exaggerated, investment value. A well-built Georgian house, for instance, is worth very much more now than it was 100, 20 or even five years ago and that is explained in part by aesthetic quality. Other things, obviously, come into the value of a building: location, the number, size and arrangement of the rooms, condition of the structure and so on. All the things I mentioned in my five points at the start.

But, other things being equal, aesthetic quality will make the difference between high prices and astronomical ones. We know that, so why do we disregard it in our discussion of building economy?

Aesthetic judgment, of course, is the paradigm of all judgment. Indeed exchanges exist for currency, insurance, produce and shares because brokers, of all people, insist of making aesthetic-type judgments about national economies, public companies, wheat or copper, accidents and Acts of God. There are no hard facts at these levels, merely informed guesses: economics becomes intangible as art.

I hope it is clear by now that I am concerned with real economics, which puts the value of things above the price rather than that spurious kind which equates cheapness with "economy". I don't even have to speculate as to how the economic value of quality in architecture might be demonstrated. For the fact it is has been demonstrated, by John Portman, architect, of Atlanta, Georgia. Portman has said on more than one occasion (1976, 1980, 1982) that as an architect at first he felt constrained. Realtors (estate agents) cost accounts, and above all, clients each naturally had views as to what kinds of building were wanted, and how much each building should cost.

Portman felt this unduly restricting, so having read the fine print of the American Institute of Architects' Code of Conduct, he decided to become his own developer, or, as Barnett put it (1976): "To think of real estate architecturally, and architecture entrepreneurially."

Portman took his first tentative steps in this direction over a small medical office building - we should call it a Health Centre. The design was good enough to win a *Progressive Architecture* award, and Portman put the letting of the individual surgeries in the hands of a real estate firm. They failed to find enough tenants and Portman lost $7,500. He resolved therefore that in future he would control both the architectural quality and the real estate aspects of his designs. He went into partnership with one of his old teachers - an authority on building specifications - on the grounds that, if his approach was to work at all, he should start with the practicalities.

Portman's problem then became to choose a building type, and a location, which had higher chances of success in real estate terms. Having lived and worked in Atlanta since the early 1940s, Portman had 'absorbed' the city by looking around, thinking about it, feeling its structure, as it were, through the soles of his feet. As a schoolboy he had worked as parking attendant, and the garage -

having been requisitioned during the war - had been freed in the
late 1950s. It was at the edge of the city centre and Portman,
without even consulting the owners, began to conceive new uses. It
occurred to him that Atlanta - a marketing centre for the whole
of the deep South - needed a furniture mart where wholesalers could
rent space to display their wares.

Given his intimate knowledge of the building, Portman saw that
the garage could be converted easily into a Mart and he proposed
this to the trustees. As garagers, they were not interested, but
they offered him a partial lease. He leased 40,000 square feet,
and took some furniture people into partnership. They persuaded
friends in the industry to exhibit in the mart, which was so
successful that within a year, they had taken over all 240,000
square feet of the garage.

That initial success persuaded Portman Atlanta could absorb a
mart of 2 million square feet provided that it was accessible, on
foot, from the main city centre hotels. Portman found a suitable
site, at Peachtree and Harris and persuaded the owner to sell for
what at the time was the highest price ever paid for such land in
America.

Having succeeded at a smaller scale, Portman and his colleagues
were able to raise a massive mortgage, to get help in buying the
site and to find additional equity.

Building started on the first million square feet of the two
million square foot Mart in 1959. It was an immediate success
which meant that Portman and his colleagues could plan for expansion.

Fig 2
Merchandise Mart

The Mart attracted thousands of visitors to Atlanta. There were by
no means enough hotel rooms of the right quality; more office space
was needed as were shops, restaurants and so on. Portman planned
a comprehensive development: the Peachtree Center, combining all
these and more, for sites in quite different ownerships. Portman
felt that if he offered the right package of architectural quality
and real estate potential, then the sites, and the necessary finance,

would come his way, which they did!

The Peachtree Center now amounts to some nine blocks of
real estate, comprising the (doubled) Merchandise Mart, five office
towers connected by bridges, the Hyatt Regency Hotel, a shopping
gallery, a dinner-theatre, the Peachtree Plaza Hotel (a 700 foot
cylindrical tower), parking garages and the (new) Apparel Mart.

Whilst the Merchandise Mart is unexceptional - if somewhat
bland - architecturally the Regency Hotel presented Portman with
his first opportunity to use architectural quality as an essential
component in his economic equation.

Originally Portman had designed a fairly "normal" Hotel with a
slab containing several storeys of corridors with bedrooms down
either side, supported over a "podium" containing the lobby, public
rooms and so on. But then, as he said (1976):

> *I didn't want the hotel to be just another set of bedrooms
> ... a cramped thing with ... a dull and dreary lobby ...
> elevators over in the corner ... a hotel room with a bed,
> a chair and a hole in the outside wall.*

Instead of that:

> *I wanted to explode the hotel: to open it up, to create
> grandeur of space, almost a resort, in the center of the
> city.*

And so he did, with the famous 22-storey atrium, a vast, rectangular
open space with glass elevators gliding up the face of a concrete
wall to a roof top restaurant and open balcony access to the bedrooms.

Atlanta Hyatt

Fig 3
Exterior

Fig 4
Interior

Portman showed his designs to Conrad Hilton, whose considered
comment was: "That concrete monster will never fly". The Sheraton,
Loews and Western Hotel chains were equally pessimistic for they
knew intimately, the economics of running hotels. They really did

not see how Portman's could be viable. For one thing it was strange, which, they felt, might deter potential customers and, for another, they could not see any possibility of financial return on that vast, unuseable atrium space.

Portman argued that he had saved on finishes – no marble, no terrazzo, just concrete – and eventually a West Coast Motel chain – Hyatt House – decided to gamble on the Regency as their first venture into city centre hotels.

Of course, it was an instant success and within three months demand for rooms was such (94.6 per cent occupancy) that Portman had to design an extension. The available site was too long and too narrow to permit another atrium, so Portman designed a cylindrical glass tower, extremely economical on plan, with its central service core and radial bedrooms.

I have laboured Portman's first successes because that is the best way I know of introducing his particular and unusual methods. Further and more spectacular examples of the atrium hotel are in Chicago (1971), San Francisco (1974) and, under construction, on Times Square, New York. I could describe further variations on the glass cylinder hotel: the Atlanta, the Peachtree Plaza (1976), the Los Angeles Bon Aventure (1977) and the Detroit Renaissance Centre. I could go on to describe the further comprehensive developments of which some of these are part, such as the Embarcardero Center in San Francisco (1976) and the Renaissance Center in Detroit (1977). But whilst each of these would throw further light on Portman's entrepreneurial activities and the methods, including ten years of political, economic and legal fighting (through 7 courts and 16 judges) which delayed his Times Square Hotel, in the long run all this would merely demonstrate some fascinating and varied repetitions of Portman's basic strategy.

So let us now examine those basic strategies, bearing in mind the sheer persistence by which they are motivated. Portman's first success, as we have seen consisted of assessing – from his knowledge of Atlanta – that a particular building type was needed which no one else had identified – a Merchandise Mart, deciding what kind of location would be most viable, at the edge of the city centre; identifying an existing building which would be suitable in terms of internal spaces for his purposes and then exercising his considerable powers of persuasion.

Naturally, with experience, Portman's methods have become more formalised. He describes seven aspects of the development process which architects must master if they are to become successful developers. These paraphrased from Barnett (1976) are:

1. Study of the growth pattern of the city and its present structural organisation.

2. Study of the real estate market, also the effects of design and cost on the marketability of buildings.

3. Prepare studies, quantified where possible, of the economic, social and political feasibility of the building.

4. Project the development costs, which will be a substantial percentage of the total for in addition to the costs of building they will include the costs of land, design fees and consultancies, legal and financial costs, furniture, fittings and equipment, the pre-opening salaries of those who will work the building and so on. (see Appendix 1).

5. Prepare a financial pro-forma, that is a projection of probable income from the building and expenditure on it over time (Appendix 2).

6. Study the financial market, including possible sources of immediate loans for constructing the building and longer-term mortgages for paying back those initial construction loans.

7. Plan the renting and general operating of the completed building.

Portman, naturally enough, has much more to say on each of these and a number of points emerge. These are: firstly, that whilst he quotes, and obviously believes the old developers' saying that the three most important factors in real estate are "location, location and location" he also spends a considerable amount of time literally walking the streets of the city, as the only way of understanding personally its urban structure and function. Portman notes that city centre sites are likely to be expensive anyway so it simply is not worth putting anything <u>but</u> good and expensive new buildings on them. Where it seems appropriate to develop a new location, then the development itself must be large enough, and diverse enough in itself to attract business without any support from its surroundings. Once it has proved its viability of course, then the surrounding sites will also become desirable.

In one case, the Hyatt Regency Hotel in San Francisco, Portman found himself obliged to build on what he considered a most inappropriate site. The Embarcadero Center of shops, offices and so on, strings away from the city centre and his hotel had to go on the most remote part of the site. His answer, literally, was to build the most spectacular atrium hotel in the world. Naturally this was expensive, but he has been rewarded by an occupancy rate of 96 per cent ever since the Regency was opened.

Roche and Rock (both 1981) have pointed out how British developers tend to be much more cautious over such things. As Roche says, their main criteria for investment are: (1) location, (2) location (3) location, (4) location and (5) location - which is somewhat more insistent than Portman. Thus they will invest in what they see as prime sites, whilst ignoring those they see as marginal and therefore risky. This leads - according to Roche - to a polarisation in the quality of environment.

That is partly because the developers themselves are dependent on institutions - such as the pensions funds. As he says:

There is now a failsafe view of investment possibilities, with the institutions always looking for a large, established, traditional company, conventional sites, conventional

Fig 5

San Francisco Hyatt
Interior

*architecture, single use buildings and a limited range
of building types, such as offices and factory estates.*

So,

*The lively entrepreneur is now virtually stripped of his
vitality, acting as an agent for the blinkered investment
world.*

This links, of course, to Portman's second major point, concerning
the relationship of architectural quality to marketability. Most
developers, in preparing an initial market analysis, assume an
"average" building. They simply have not realised that marketability can be enhanced by good design. Of course, it was
difficult at the start for Portman to quantify the enhancement
value which might be attributed to quality. But he has now built
up an extensive enough track record to quantify with considerable
accuracy the marketing advantages of good design.

     There are, of course, times when he has to cut down. And that,
Portman insists, is an architect's job, for only an architect can
decide where to strike the balance between cost and quality. It's
a matter of deciding on each detail, whether to cut down on wall
finishes (he's quite happy with concrete in certain public spaces)
or bath taps (his bathrooms feel very luxurious).

     As Portman suggests, market analysis can never be an exact
science: it is a matter of informed judgment. One can quantify
many things: the number of families living within a particular
trading area, the statistics on age, sex, occupation, income
levels, etc., not to mention travelling distances: a five minute
walk, a ten minute car journey and so on. But once those things
accessible to quantification have been quantified, one must exercise
one's judgment, basing it, as far as possible, on precedents and
experience.

     As Portman says, the value of any building is related to the
long term income available from it, rather than to its basic

construction costs. Thus a cheap structure, which returns a long
term high income may be worth many times the cost of its building,
whilst an expensive one, or even a building of great historic
value, may be worth almost nothing. If it has no forseeable use,
then according to Portman, the only sensible thing in economic
terms, will be to pull it down.

Of course, there is more to it than that. "Worth" in this
sense depends on many things. In *Design in Architecture* for instance,
I reported on work by Cowan (1964) who showed that some 70 per cent
of <u>all</u> human activity: eating, sleeping, meeting and so on can take
place, comfortably and conveniently in rooms of around 15 square
metres. If your historic building contains many such rooms, then
your chances of finding new uses for it are greatly enhanced. If,
on the other hand, it contains one large space - especially one
divided by pillars - the number of potential uses is very greatly
reduced. That is why it has become so difficult to find uses for
redundant churches. Even so the Church Commissioners report a dozen
different <u>kinds</u> of use for redundant Anglican churches, ranging
from arts centre to light industrial (quoted Ridge, 1982).

Conversely, if a building is truly historic, in the sense that
historic events took place there, then that, too, has its value
as an attractant. That is true, for instance, of the Alamo in San
Antonio. As a <u>space</u> it is not very useable, but because of its
place in American history, it has been turned into a museum which,
in itself, assures viability of the shops, restaurants, hotels and
other commercial structures on nearby sites.

Increasingly, architectural character is being seen to have
effects. One recent example is Charles Moore's Piazza d'Italia
in New Orleans (1979). It was built as an urban open space with
fountains, in the form of a map of Italy surrounded by classical
colonnades. Its purpose was to trigger economic renewal in a
particularly squalid part of the city and it is having precisely
that effect.

Fig 6

<u>Piazza d'Italia</u>

As far as development costs are concerned, Portman finds it quite
unrealistic to look at recent buildings of a similar type and to
derive from them average construction costs. As he says (quoted
Barnett, 1976):

*If the site selection, market studies, and construction cost
estimates are all based on averages, it is not surprising
that the architect ... ends up by having to design an
average building.*

He sees it as important, therefore, to take the actual design,
worked out in sufficient detail for accurate estimates to be made.
He then insists on a "maximum guaranteed construction price" from
his contractor and it is this, above all, or so he believes, that
has allowed him to build his unconventional designs.

Building costs in any case, amount to only some three-fifths
of the total development costs - see Table   - which in themselves represent only a small proportion of the total life-cycle
costs. Given the shifting relationship between construction costs
and staffing costs, for instance, it is now cheaper to stay in a
well-planned Portman Hotel than its routine equivalent in London,
planned in the days when staffing was cheap!

Unlike most developers - and architects - who 'leave' the building
at the end of the contract, Portman continues to care about the
management of his buildings. Indeed, he was so disgusted at what
he felt to be bad management of the Atlanta Hyatt that for the
Los Angeles Bonaventure, the Atlanta Peachtree Plaza and other
hotels, he shifted his Embarcadero to Western International.

There is no space here to describe the 44-part management
structure of the Portman firms. Suffice it to say that this has
two main divisions, comprising John Portman Associates, the design
firm and Portman Properties, developers. The design firm includes
architectural, structural and construction management consultants,
interior design, model-making and other sections, whilst Portman
Properties has sections devoted to Project Development, Feasibility
Studies, Financing, Land Acquisition, Legal, Project Administration,
Project Management, Public Relations and so on, together with
management offices for the Embarcadero Center, the various Marts, the
Hotels and the various restaurants, not to mention a bulk-purchasing
office which buys furniture and fittings at favourable discounts.

Successful as he evidently is, Portman presents only one model
of how things might be going. This, in my view, has particular
attractions in that Portman demonstrates those relationships
between marketability and design quality which have been forgotten
for too long by those who have equated "economy" with "cheapness".
He has proved his point, beyond reasonable doubt, but of course,
there are other possibilities. The Codes of Professional Conduct in
Britain now allow my profession to become not just architects and
developers in Portman's sense, but also architects and builders,
architects and manufacturers, or any combination of these. Few of
us will do them with Portman's flair, but the question which springs

to mind at a Conference such as this is: "Where does that leave
you as Quantity Surveyors?"

My answer, of course, is "threatened" unless you too are
prepared to change your ways. You literally will not survive into
the future if you see your job as cutting the architects down to
size. For if the architect is his own developer and his own
contractor, he will decide on the relationship between capital costs
and life-cycle costs, between quality and marketability.

Suppose, for a start, you took Portman's point, that there is more
money to be made, over a longer period, by building, with imagina-
tion, the right building in the right place, rather than an average
building on the cheapest location you can find.

So, with the computer telling you from one side to dream flowers,
on the grounds that it can do the number-crunching better than you
can, with the architect aspiring to quality again, on the grounds
that quality makes economic sense, you will have to move with the
times, if you are not to go under.

It's a matter of generosity. We have tried the other approach
in this country: of building the maximum amount of building,
housing, schools, hospitals, offices and so on at minimum cost,
nearly bankrupting ourselves in the process. We have also destroyed
our cities, turning them into bleak and hostile urban wastelands.
We have lumbered ourselves with the horrendous maintenace – and
demolition – bills. It doesn't make even economic sense to continue
with that mean-spirited, penny-pinching way.

Society demands better architecture and that will cost a great
deal of money. If you are prepared to help us spend that money on
quality, then I think you have a brilliant future. But if you
aren't, then we shall dream the flowers and get out computers to do
the number-crunching.

ILLUSTRATIVE BUDGET FOR A 1,500-ROOM HOTEL WITH 100,000 SQUARE FEET OF REATIL SPACE

PROJECT COST PRO FORMA For a construction period of thirty-one months.

1.	Land	$ 7,000,000
2.	Base Building	70,000,000
3.	Architect's fee	4,200,000
4.	Property tax during construction	2,000,000
5.	Material testing	500,000
6.	Project administration	1,500,000
7.	Financial, legal and closing	4,000,000
8.	Technical consulting	400,000
9.	Miscellaneous	1,700,000
10.	Furniture, fixtures and equipment	9,500,000
11.	Retail space finish	1,500,000
12.	Preopening and expendables	5,000,000
13.	Contingency	6,000,000
14.	Interim interest	8,000,000
	Total	$121,300,000

Source: Portman, J. & Barnett, J., *The Architect as Developer*, New York, McGraw Hill.

ILLUSTRATIVE OPERATING PRO FORMA FOR A 1,500-ROOM CONVENTION HOTEL

	Stabilized Year	
Attained transient rate	$53.50	
Occupancy	82%	
Gross operating revenue	Ratio (%)	($000)
Rooms	48.2	24,019
Food and beverage	46.8	23,321
Telephone	1.9	947
Laundry and valet	1.2	598
Health club and pool	.2	100
Hotel retail: net	.6	299
Parking	1.1	548
Total gross revenues	100.0	49,832
Cost of sales		
Rooms	11.1	5,531
Food and beverage	34.1	16,993
Telephone	2.7	1,345
Laundry and valet	.8	399
Health club and pool	.2	100
Parking	.3	149
Total costs	49.2	24,517
Other operating expenses		
General and administrative	5.0	2,492
Advertising and promotion	3.0	1,495
Maintenance and utilities	7.0	3,488
Total other	15.0	7,475
Total operating expenses	64.2	31,992
Gross operating profit	35.8	17,840
Fixed expense		
Insurance	.3	149
Real estate tax	10.8	5,382
Total fixed expenses	11.1	5,531
Net house profit	24.7	12,309
Other revenues		
Shopping gallery	5.0	2,492
Theatre	.2	100
Net before fees and reserves	29.9	14,901

Years 1 and 2 also predicted

Source: adapted from: Portman, J. & Barnett, J., *The Architect as Developer*, New York, McGraw Hill.

REFERENCES

Anderson, C., (1965) The Context for Decision-Making 4 : Traditional that isn't 'Trad Dad' in *Architectural Association Journal*, May 1965.

Barnett, J. (1976) Architecture for Buildings People Use Every Day & Architecture as Investment, in Portman, J. & Barnett, J. *The Architect as Developer*, New York, McGraw Hill.

Broadbent, G., (1979) Recent Developments in Design Method Studies, in *Open House*, Vol. 4., No. 3., 1979, Stichting Architecten Research Group, Eindhoven, Holland.

Broadbent, G., (1973) *Design in Architecture*, John Wiley & Sons Ltd., Chichester.

Bridge, K., (1982) Redundant Churches - Destruction or Re-Use?, Un-published Dissertation, Oxford Polytechnic.

Cowan, P., (1964), Studies in the Growth, Change and Ageing of Buildings, *Transactions of the Bartlett Society 3*, The Bartlett School of Architecture, London.

Dicks, T., (1982) quoted by Wallace, M., More Council Blocks are Doomed : Bill may be 3,000 Million, in *Sunday Times*, 13 April 1982.

Galbraith, J. K., (1974) *Economics and Purpose*, London, Andre Deutsch.

Hellman, L., (1963) Untitled Cartoon in *Architects' Journal*, 4 July 1973.

Hillier, W. R. G., Musgrove, J., & O'Sullivan, P. (1972) Knowledge and Design, in *EDRA 3, The Proceedings of the Environmental Design Research Association Conference No. 3.*, (Ed., Mitchell, W.) 1972.

Kuhn, T., (1960) *The Structure of Scientific Revolutions* (1962 Edn. consulted), Chicago, University of Chicago Press.

Landau, T., (1965), The Context for Decision-Making 5 : Towards a Structure for Architectural Ideas, in *Architectural Association Journal*, June 1965.

Popper, K. (1959), *The Logic of Scientific Discovery*, Hutchinson, London.

Popper, K., (1963) *Conjectures and Refutations*, Routledge, London.

Portman, J. (1976) Architecture as a Social Art, in Portman, J., & Barnett, J., *The Architect as Developer*, New York, McGraw Hill.

Portman, J., (1982) The Architect as Developer, in *Transactions*, No.1,

1982, RIBA Publications Limited, London.

Roche, F. L., (1981) The Changing Face of Patronage, Paper to RIBA Conference, London, October 1981.

Rock, D., (1979) *The Grassroots Developers - A Handbook for Development Trusts*, RIBA Conference Fund.

The Use of Simulation as a Research Tool.

B. Fine, B.Sc., A.R.C.S., F.I.M.A.
Messrs Fine, Curtis and Gross
Consultants in Planning, Computing and Management Science.

There are three parts to this paper. The first part considers uncertainty and indicates the necessity to consider it. The second
part describes some aspects of simulation processes. The third
part is a very brief indication of what we should expect from those
processes.

Uncertainty

The Greeks knew about chaos. They invented the word. Their mytho-
logy religion with its warring gods causing chaos allowed them to
discuss the subject. Chaos was an earthly sign that the gods were
in conflict on Mount Olympus.

Earlier than this the Jews knew about uncertainty and chance. The
drawing of lots was an acceptable form of decision making, just as
it was to be later in Greece. Their literature contains many
references to the subject, for example the writer of Ecclesiastes
knew -

> "the race is not to the swift, nor the battle to the
> strong, neither yet bread to the wise, nor yet riches
> to men of understanding
> nor yet favour to men of skill : but time and chance
> happeneth to them all. For man also knoweth not his
> time- as the fishes caught in the snare, so are the
> sons of man snared in an evil time, when it falleth
> suddenly upon them."

The rise of Christendom and then of Islam altered the view held
generally in society. A monotheistic religion does not have gods
in conflict, and an all powerful God is not easily accomodated
alongside uncertainty. The closest approach to uncertainty was via
ignorance. God knew - man did not. Our uncertainty is an
expression of our ignorance and is not a representation of God's
uncertainty. For most of the past 2000 years religion has been a
major element in Western Philosophy; scientists are as influenced by

religious philosophy as is the general population.

Our language of science contains some historical clues about its
philosophical origins. Most of the lay public and a large part of
the scientific community regard science as the progressive
exposure of Laws that have always existed.

Newton's Laws, Maxwell's Laws, Ohm's Law, Boyle's Law, Charles' Law
etc.

The "Law" was originally viewed as "God's Law", but later as non-
believers contributed more to science these origins were forgotten
and they became mere "Laws".

These concepts are deterministic. A law gives an answer, always
the same answer. Of course our early attempts at law formulation
may be faulty - but ultimately we can discover the truth.

Descartes had no doubts about determinism. God had wound up the
clock and we played out our predetermined acts as it unwound.

Religion now plays almost no part in these views, for Marx-
Leninism has the same laws, the same answers, and perhaps a more
rigidly deterministic approach.

Marx-Leninism assumes that immutable laws control the behaviour of
society and drive it inexorably towards socialism.

As a contrast to these approaches, Rabbi Menachem Mendel of Kotsk
1787-1859 an atheist on occasion, produced a challenging aphorism.

> *If I am what I am because you are what you are*
> *and if you are what you are because I am what I am*
> *then I am not me and you are not you.*

Uncertainty does now exist in our science - but it became introduced
in an interesting manner. Perhaps the first acceptance of it was
in the Gas Laws. Here the random motion of the molecules of a gas
provide the deeper cause of the gas laws that we observe. This
sort of uncertainty was accepted only with difficulty by physicists.
One view was that the uncertainty calculations performed by us were
merely artificial methods of performing even more lengthy deterministic calculations. We could have done these sums given time,
but uncertainty was a shortcut. An alternative view was that the
uncertainty which produced a definite answer was a demonstration
from God of his "Mysterious Way".

Quantum Mechanics provided the first real emotional challenge.
Heisenberg's Uncertainty Law has in its title the links between the
old determinism and the new uncertainty.

Those who thought deeply about the structure of our surroundings had cause to worry. Einstein could not accept the challenge that this posed. He put his view succinctly.

"God does not play dice with the universe".

Science was and still is often accepted as dealing with the materials that exist in the universe, and it is certainly from the science of physics, engineering or that of chemistry that our most powerful intellectual models are drawn.

The Region of Divergence

Attempts to apply the methods of natural science, and in particular mathematical methods, to the behaviour of people, often cause people to object, and sometimes with good reason. Many such efforts are purely deterministic calculations using simple formulae and providing no space for our human freedoms.

When I first started to talk about uncertainty in planning and in construction processes generally in the early 1960's it was clear that I was seriously offending many of the audience. Some indicated this in no uncertain terms and told me that I was talking offensive rubbish. I think that by a fortunate accident I had landed on a sensitive intersection of the routes of science. These routes diverged at the location marked by the Bill of Quantities and those models of construction processes that underlie it.

The BoQ does not talk about people or mathematical models of the behaviour of people, it just multiplies quantities by rates to give a price.

IT IS OF COURSE ACCEPTABLE!

I spoke about the uncertainties in the behaviour of people, and about problems in forecasting their behaviour.

TO MENTION PEOPLE IN FORECASTING IS CLEARLY UNACCEPTABLE!

It has been my usual behaviour in papers on this topic to argue the case for the acceptance of uncertainty. This time I choose to ask questions instead. I am stating the principle quite simply, and I expect you to accept it.

WE DO NOT KNOW WHAT TOMORROW WILL BRING.

I can do this now because in the last 15 years almost every University Department dealing with Construction Management has been exposed to the concepts of uncertainty.

The trigger for this change in approach was provided by the recognition that there are now others in the field attempting and sometimes succeeding in selling the concepts to clients. The audience response is also muted, uncertainty is now only daring whereas formerly it was pornographic.

The Questions

I)   For the behavioural scientists

Why has the change occurred now? What conditions were required to change the academic view of the problem? Can we learn from this, what it is that we need to do in order to affect the non-academic bulk of the industry that we serve?

2)   For the professional advisors to industry

How do we start to talk to the client about uncertainty in cost and duration? What sort of guidance should we give to our clients? How can we rescue ourselves from an environment in which our clients view us with distrust?

3)   For the academics

How do we apportion risk? What data is needed to deal with uncertainty? What calculation tools can be provided to participants in the process? How do we start to use quality control techniques in our advisory processes?

A Comment

Changing from flat earth to spherical earth views created new navigational opportunities. Which of the opportunities resulting from the recognition of the central position of uncertainty in our processes do we wish to grasp first?

The Use of Simulation as a Research Tool

There are many ways of describing the world about us. Each has its own specific areas of utility.

There are moral tales, fables, social and descriptive literature of all sorts having utility at all levels. Certainly these types of description are sufficiently important for governments in many parts of the world to ban and burn the books involved, and often to imprison, torture and murder the authors and distributors.

These imitations of the world as it is (or as it may become) are generally non-numerical. Because of this it is sometimes difficult to translate the analysis made by the authors to other locations or to other conditions.

Mathematical models of the world about us can be reprocessed with new numerical parameters so that relocation, retiming or new conditions may be analysed with ease.

There are a few obvious differences between the two types of description mentioned. Authors of books generally talk about people and write directly for the general public. The appeal is primarily to the heart and secondarily to the intellect. Mathematical models are generally about things and are devised for small groups of expert users. The appeal is intellectual and unemotional.

This often raises doubts in my mind about research into construction management. Can we reasonably expect the construction industries to be run via cerebration rather than by exhortation? Perhaps as a society we should cease to pay mathematical researchers and instead commission competent writers to illuminate the industries for us. What about getting the Quantity Surveyors to commission Geoffrey Archer to write a new version of Not a Penny More, Not a Penny Less.

Mathematical models can be regarded as the intellectual embodiment of our technological progress. (I do not here regard mathematical models as being restricted to the narrow algebraic tools of school mathematics. These models encompass such descriptive tools as those used in chemistry, where initially numbers appear to play little part).

The power of mathematics lies in the ability of mathematicians to use one model in several different situations. Often in different technologies. Lessons learned in one field can then be directly transferred to another, often producing astonishing perdictions, or much clearer understanding.

Mathematicians do not talk about reality but only about their analogue of reality. Predictions made for the model fit each of the situations to which the analogue is applied.

For example for a whole class of situations we use a set of equations applying to wave motions. The same equations may be used for stress analysis, for sound vibrations, for electromagnetic waves and in some situations which non-mathematicians would not recognise as having anything to do with wave motions. Indeed we regard the examples given as wave motions BECAUSE the mathematics of wave motions fits.

As another example the inverse square law is widely applied. It has applications in gravitational attraction, electromagnetic attraction, lighting intensity calculations, pressure wave calculations and elsewhere. We do not need to re-analyse the situation each time we see it, instead we become bold and invoke the magic words INVERSE-SQUARE-LAW. We can of course sometimes be wrong, but I have no intention of talking about these errors here.

Most of the mathematical models that we talk about are well founded
in antiquity. They have gained respectability because of age.
Their originators have become illustrious because their models have
survived. This is the warning note. We tend to forget the very
large number of models that have failed to survive and that have
been superceeded by our current set of respectable models.

Occasionally we do find some reference to models that have failed.
This is often because model inventors often work in many fields, in
some they become illustrious, in others they failed. Sometimes the
models impinged on social or moral views held by a large section of
society and controversy arose because of this. Sometimes we have
conflicting models still in use, time will tell who is to become
illustrious. Some examples : the atomic theory of matter survives,
the idea of continuous matter is dead : Phlogiston is dead, long
live oxygen : Creationism may still be kicking, so are the ideas of
natural selection and evolution.

Processing mathematical models may not be simple. For the solution
of some of the problems posed by the classical models used we still
have no adequate analytical methods. It is only the advent of fast
computers that enables us to produce reasonable numerical solutions.

Most of our training has been directed towards models with simple
solutions expressible in familiar algebraic terms. This has led us
to ignore those classes of problem in which only numerical solutions
are possible so far.

A very important reason for the use of mathematical models is the
predictive nature of these. The models often predict things that
were not specifically incorporated into those models. Of course
this view can be challenged. Once we have described the model, then
an observer could say that it is only our own ignorance that led us
not to expect the unexpected result.

It is worth mentioning two examples of unexpected predictions.

The first concerns the mathematical model used in stress analysis
of a homogeneous elastic medium. Using this model we can calculate
(predict) the way in which an object changes shape under applied
loads. For some shapes of body we can show that a condition of
instability exists and that these body shapes can suffer catastro-
phic changes in shape under very small increases in loan. (Euler
buckling and its analogues in other fields still comes as a
surprise to many engineers).

The second concerns the model of a gas as a set of molecules in
rapid random motion. The applications of this model lead to the
ability to calculate such properties as specific heat and viscosity.

The classical models that we all (or nearly all) accept have the
following properties.

I)   The models are simple, they can be explained in very few words.

2)   The models can be examined in as much mathematical depth as the investigator wishes, using whatever mathematical tools he wishes, and that under these circumstances inconsistencies do not appear.

3)   The analysis leads to unexpected predictions.

The processing of a single model involving only one branch of natural science is generally called a calculation.

The processing of linked models involving more than one branch of natural science is generally called simulation, as is the processing of models incorporating a large element of mathematical uncertainty.

The application of mathematics to things is generally accepted.

The application of mathematics to processes where there is no human element involved is also generally accepted.

The application of mathematics to processes in which human intervention occurs is almost always rejected once it has been recognised. This is the field in which we must make progress if our ability to organise ourselves is to improve.

The models that we must use in this region must recognise that man has a free will. The models to be used must therefore incorporate uncertainty. In order to be useful models they must have the three characteristics described earlier. These are the simulations we must use as research tools.

The help that simulation can offer

Having spoken about simulation and about uncertainty, it is worth questioning whether there is any evidence that the ideas spoken about can help us now or whether they could possibly help us in the future. The doubts are not merely academic. An increase in our knowledge may be of no benefit to us.

There are locations in which knowledge exists or has existed for very long periods, and benefit from that knowledge has not been apparent. For example we have known about the planetary motions for a considerable period. It has been possible for us to make forecasts about the appearance of the sky for a very long time. It is only in the last few years that we have been able to use this knowledge to help our satelites.

The successes of simulation are few but they are significant. This is not the place to produce a catalogue of these, but it is worth mentioning a few fields in which simulation produces understanding and also an improvement in performance.

Bidding theory is one such region. Uncertainty exists in our estimates. The applications of bidding theory simulations enable
contractors to improve their contractual position, and enable
clients to estimate more closely the amount of money they may need
to pay for their projects.

Simulation of construction projects subject to uncertainty is the
only way of tackling such problems as exist in the construction of
projects in stormy or tidal regions. In these cases the uncertainty
is caused largely by the weather. We simulate the projects using
either real or simulated weather conditions so as to examine how
sensitive the project is to these.

For the study of repetitive processes we may use simulation techniques to examine the consequences of various strategies that may be
adopted. Uncertainty exists in the performance of the work, and
external interferences may prevent some activities from being
started at the planned time. Simulations here lead to some surpri-
ses and I will mention just a few of these.

Most economics texts talk about the benefits of scale. Simulations
of repetitive processes show that when interferences and uncertainty exist then there may be very large penalties of scale. There is
some evidence that in our industry it is the larger jobs that go
wrong most frequently.

Most work study experts talk about the benefits of specialisation.
Simulations show that for large projects subject to uncertainty and
interference there are penalties of specialisation.

Most models of construction processes assume that the cost of a
project is the sum of the costs of the activities. Simulations of
repetitive processes show that costs are largely generated by the
interferences and uncertainties that exist, and that simple additive
models like the BoQ seriously under-estimate costs. It is the
failure of these simple additive models that produces much of the
drama in our industry.

There is a clear requirement for simulations to cover the modelling
of the processes from conception to destruction of the projects that
we model. The design, procurement, construction, use and dismant-
ling of our building projects are not disconnected entities that can
be considered totally in isolation. There will be of course regions
of mathematical separability, but there are also strong connections
that cannot be ignored.

The concepts that we have to use are not easy concepts to become
familiar with. We need to change our thought patterns to accomodate
to them.

Normally when we use deterministic calculations it is relatively easy to produce experiments which enable us to calculate the parameters that we need. In this statistical field we have to use a whole range of experiments just to determine a single parameter. As one cynic put it to me, of course, you can prove you are right, your whole theory depends upon proving that the result of the forecast is different from the forecast and that happens every time!

The different concepts provide for us differing freedoms that may not have been expected. The traditional approach to construction would expect financial benefit to accrue from productivity increases and would direct research towards increasing speed of production. The simulations using the concepts of uncertainty direct us towards obtaining benefits from reductions in uncertainty. The benefits to be gained by these reductions in uncertainty can be shown to be very much larger in magnitude than gains made possible by productivity increases. So it may be in our interest to carry on studying uncertainty.

Now the construction industry suffers from some of its uncertainty because society sweeps its entropy onto our carpet. I think that it is wiser to have uncertainty in construction industries than in agricultural industries. If we improve our status by sweeping our entropy elsewhere, where should we sweep it to? Can we sweep our uncertainties into harmless and cost free locations?

ENERGY CONSERVATION: SOME ECONOMIC QUESTIONS

PATRICK O'SULLIVAN, Welsh School of Architecture

The purpose of this paper is (by taking examples from one aspect of
Building Economics - namely the Economics of energy conservation) to
investigate the state of current knowledge, to determine how building
economists have aquitted themselves in the past, to suggest what
lessons can be learnt and to to indicate therefore what the future
might be.

On 1st July, 1982, the fifth report from the Select Committee on
Energy: Energy Conservation in buildings was published. It dwelt
on the economics of Energy conservation as follows.

Witnesses agreed that the potential for saving energy in buildings
through higher efficiency is very large, but estimates differed
according to the way in which the potential was defined. The main
differences depended on whether the technical or the cost-effective
potential was being measured. The *technical* potential indicates how
much energy could be saved technically by the widespread application
of current best-practice technology. The *cost-effective* or economic
potential indicates how much could be saved within specified investment criteria (eg pay back period or rate of return). As a result,
the cost-effective potential is dependent on current fuel prices and
expectations of future energy prices, technological developments, the
cost of capital and the rate of new construction. In principle, the
proportion of the technical potential for energy saving which is cost
effective will increase over time in response to rising real energy
prices. However, the Committee has identified considerable confusion
in the measurement of the cost effective potential. The Building
Research Establishment (BRE) followed the Treasury guidelines that
new investment should be rigorously assessed using discounted cash
flow (DCF) techniques and should meet a minimum Required Rate of
Return of 5% in real terms. Whilst these guidelines might well be
appropriate in long lead time, large-scale investment programmes in
the public sector, it is not apparent that they have much relevance
for the individual householder, or, indeed, that they are used at all
by local authorities and other public sector bodies in assessing the
size of their conservation-related capital programmes. The same
appears to be true for commercial organisations. Most of the
witnesses appeared to use cruder pay back, or rate of return criteria,
which did not make allowance for the cost of capital, arguing that,
whilst this may be unsophisticated, these were likely to be the
primary criteria used by consumers. Some witnesses evaluated

conservation against two, five or ten year pay back periods
(equivalent to crude rates of return of 35%, 14% and 7% per year
respectively). Such results are, therefore, only indirectly comparable with the more formal investment appraisals used by the Treasury.

ETSU made an estimate of the *technical* potential in 1979; they
stated that the technical potential for conservation in buildings of
all kinds was 45% (equivalent to 76mtce of primary energy per years).
BRE stated that low energy houses show a reduction in energy
consumption of between 30% and 70% in comparison with similar houses
built to the current building regulations. Low energy hospitals show
savings of over 50%. Witnesses were more reluctant to provide
evidence for the *cost-effective* potential for saving energy in the
building sector. BGC estimated that for non-domestic buildings 10%
could be saved by good housekeeping and a further 10% through minimal,
cost-effective capital expenditure. BRE stated that 30% was realistic
for buildings as a whole, arguing "that this is a figure that is
realistic in the sense that it combines measures which are cost effective (on the basis of current energy prices)". By 1979 the Property
Services Agency of the Department of the Environment (PSA) had already
achieved a 34% saving in the civil estate since 1974. Other witnesses
stated that very large cost-effective savings had been made in certain
non-domestic buildings. Although lower than the technical potential
for the reasons given earlier, these estimates of cost effective
potential represent very large sums of money savings. For example a
20% saving for domestic buildings alone is worth over one billion
pounds per annum.

Witnesses provided us with estimates they had made of the cost-
effectiveness of a range of specific measures, usually on the basis
of extensive field trials. These results are summarised in Table 1.
The ranges given reflect the fact that cost effectiveness is
influenced by: (a) the type of heating system used; (b) the current
fuel used (and hence different relative fuel prices); (c) whether the
measures are being applied throughout the building or only in some
rooms (eg full or partial double glazing); (d) whether insulation is
being applied for the first time or is enhancing an existing
provision, (eg roof insulation); (e) whether the measure is being
undertaken on a one off basis (eg by a private householder) or as
part of a larger programme (eg by local authorities whose bulk
purchasing power can reduce unit costs).

During the present inquiry, witnesses have stressed the economics
of conservation, emphasising the large potential for undertaking cost-
effective measures. It is thought that cost-effective measures, many
with pay back periods of five years or less, could produce savings of
the order of 30% of present delivered energy consumption (c.31mtce)in
both domestic and non-domestic buildings, using existing technology.
If these savings were to be achieved, the potential market for cost-
effective conservation products and services would be very large.
This potential market, in the domestic and non-domestic sectors, for
insulation materials, controls, more efficient heating systems, double
glazing and their installation and associated consultancy services, is
probably of the order of £10bn, although accurate estimates have yet
to be made. Within this total, the Department of Energy has

calculated that £3-4bn could be spent on the loft, cavity wall and hot water tank insulation and draught-proofing of untreated dwellings producing savings of £720m/year; i.e. with a pay-back of no more than 5 years.

Table 1.

Paybacks on Domestic Conservation Measures

Measure	Period (years)	Source
Draught stripping	1-6	ACE,BGC,DOE
Hot water cylinder insulation	0-5	DOE
Loft Insulation	1-5	ACE,BGC,DOE
Cavity Wall Fill	4-10	ACE,BGC
Double Glazing: Heated Living Room only*	6-25	GGF
Double Glazing (Whole House)*	9-85	GGF,BGC
Adequate Control System	2-5	ACE

*Depending on whether DIY or Professional

Payback on Non-Domestic Conservation Measures
(see PSA; Appendices p 49)

Measure	Period (years)
Optimum Start Control	less than 5
Simple Time Control	less than 3
Internal Space Temperature Control	3-10
Roof Insulation	4-10
Cavity Wall Foam Fill	2-10
Internal Insulation	2-10
Draught Stripping	2-5
Automatic Lighting Control	2-14

Abbreviations-
    ACE: Association for the Conservation of Energy; Domestic Energy Conservation and the U.K. Economy
    BGC: British Gas Corporation, Evidence pp.69-98
    DOE: Department of the Environment, Appendices p.2
    GGF: Glass and Glazing Federation, Appendices No.40 and 42

During the present inquiry we have taken much evidence about the imbalance of national and public expenditure on energy conservation investment and new supply investment. This has been an issue of major concern to the Committee in much of its recent work. In our First Report on nuclear power we stated that "... we were dismayed to find that seven years after the first major oil price increase, the Department of Energy has no clear idea of whether investing around £1,300m. in a single nuclear plant (or a smaller but still important amount in a fossil fuel station) is as cost effective as spending a similar sum to promote energy conservation", and we recommended that: "the Department of Energy should assess in future, as it should have done in the past, the economics of public expenditure to promote energy conservation with the same rigour as that required for the economic appraisal of new generating plant." We were, and remain, concerned about the opportunity cost of capital. In our report on the Department of Energy's Estimates for 1981-82 we noted the "imbalance between the scale of resources devoted to energy supply and the much smaller commitments to reducing demand." In its response to our First Report on the nuclear power programme, the Government made the following comments: "The Department of Energy will carry out further research into the relative costs and benefits of investments in energy conservation and supply." The Government added that there is "a large potential for investment in energy conservation, some of which is more cost-effective from a national point of view than investment to increase supply." As a result of our present inquiry, it is our considered opinion that there are now many conservation measures which are so much more cost-effective than most energy supply investment that the caveats expressed by the Department of Energy (over such matters as the effect of energy conservation measures on supply capacity at times of peak demand) seem quibbles. Such reservations do not represent reasons for not stimulating highly cost-effective energy conservation investment.

We have also learnt that quite different criteria seem to be applied to the appraisal of investment in energy supply as compared with conservation. The nationalised fuel supply industries are required to meet a minimum 5% real rate of return on new investment. Conservation investment, whether undertaken by householders, firms or local authorities, is currently based on crude rates of return, commonly expressed in terms of very short pay back periods, usually under five years. These short pay back periods which are used for conservation investment appraisal are very much more stringent than the 5% real rate of return. Furthermore, industrialists and public sector bodies have reported not investing in conservation measures even with these short pay back periods owing to a lack of capital. Thus, in practice, the shortage of capital leads to the adoption by firms and householders of pay back periods often below two years. This is also true in the public sector. Although the Department of the Environment notionally adopts a 5% real rate of return for appraising public sector conservation investment, in practice this is seldom used, especially by local authorities short of capital; a pay back period of under two or three years is common We are thus concerned at the misallocation of resources which results from invest-ment in supply being appraised at the Required Rate of Return, while many conservation projects are required to satisfy much higher rates of return. We consider it a testimony to the irrationality of present energy policy that investment in additional supply capacity by the coal, electricity, gas and oil industries is, in practice, assessed by different criteria than those applied to investment in energy conservation by Local Authorities and other public sector bodies responsible for buildings.

The Committee have carefully considered why, if investment in conservation is so cost-effective, more investment is not taking place. Much has been learnt about this subject since 1974, and the answer appears to be that, whereas there are no substantial technical obstacles to a conservation programme, there are major barriers to the

working of the market for conservation products which severely limit
the efficacy of a policy based, as is that of the present Government,
mainly on economic pricing. We have concluded that there are two
groups of obstacles to the proper working of the market - economic
and organisational (or institutional).

The fact that home owners find it difficult to gauge the effects
of any conservation investments they undertake. Thus it was argued
that it is difficult for owner-occupiers to measure the energy
savings in physical terms, and because of rising energy prices they
do not even see their energy bills becoming smaller.

Consumers are not aware of the potential for saving money because
they do not know what investments can be undertaken and how much
energy can be saved; moreover, in inflationary times, the economic
benefits are not only difficult to measure but also to perceive,
since, in effect, conservation merely reduces the rate of increase of
energy prices.

The discrepancy in knowledge, effort and achievement amongst local
authorities is striking. A great deal of energy could be saved if all
local authorities could match the achievements of the best.

Other priorities for investment, especially when capital is scarce
and where energy costs are a small proportion of overall costs. (The
position in the private sector has been worsened by the recession and
in the public sector by the reduced availability of finance).

Shortage of cash. Eurisol UK Ltd., the association of insulation
materials manufacturers, told the Committee the housing market was
"severely affected at present by lack of personal disposable income
and by high interest rates".

*High rates of mobility and house prices which do not reflect
expenditure on conservation measures.* Owner occupiers move frequently
every 7 to 8 years on average, making them reluctant to undertake some
conservation measures particularly with payback periods of over 5
years, especially if the cost cannot be recovered in a higher sale
price for the house. Unfortunately the value of most conservation
measures, unlike gas central heating or double glazing, is not at
present fully reflected in the sale price of the house. Therefore
decisions about conservation investment are not based on the fact that
the life of the house might be 60-100 years.

The fact that Building Societies have a mortgage interest in a
large proportion of the owner-occupied housing stock but *no Building
Society requires specific conservation measures to be undertaken as a
condition of receiving a mortgage,* in the same way that rewiring, wood
treatment, and the installation of damp courses are now a common
requirement. The Building Societies' Association took the view that
house prices should be kept to a minimum so that as many people as
possible can afford to buy them. However, some societies do encourage
their borrowers to install insulation.

Insufficient finance. Even authorities which have made savings
could make further cost-effective investment, but have been constrained by a shortage of funds. Very limited council budgets were used to
fund those items, e.g. schools, teachers, which councils considered
to be higher priorities than energy conservation. Consequently only
investments with very high rates of return are made. In addition,
accounting procedures and the definition of resources as either

specific to capital or revenue account produce further barriers to conservation; good housekeeping measures can often be funded under maintenance budgets on revenue account, but more substantial investment is constrained by limitations on capital account.

Changes in the method of financing by Local Government. Recent changes in local government finance were thought to be likely to affect conservation measures adversely. At a time of clashing priorities, the discontinuation of earmarking of funds for the Home Insulation Scheme has reduced Local Authorities' allocation to insulation programmes. Restraints in local government spending, including tighter expenditure targets for individual local authorities and reductions in the Rate Support Grant, have had the effect of reducing spending on energy conservation. Furthermore, if local authorities or Government did not make money available for energy conservation in the public rented housing sector, witnesses thought it unlikely that tenants would. Suggested changes in policy which should be adopted by local authorities for both their housing and other buildings included:-

(i) A return to the ear-marking of funds for conservation; and;

(ii) The need for Government to stress to local authorities the achievements of, and financial advantages accruing to, the "best-practice" local authorities.

Insulation manufacturers have had difficulty in selling their product to industry because conservation expenditure has had to compete with other claims for capital already scarce because of the recession. Mr. Huxley of Welsmere Energy Management Ltd., a subsidiary of Debenhams Ltd., put the problem most succinctly. Capital for energy conservation was a low priority because management was ignorant of the cash savings that could be made by conservation. Management, "however, may be correct in assuming that the return from increasing productive capacity or turnover, will bring greater rewards"

Accounting Practices (i.e. (i) the manner in which conservation investment is depreciated; (ii) the distinction between revenue and capital accounts; (iii) the absence of the life-cycle costing of buildings).

The fact that standard accountancy practice may militate against energy conservation. This was said to apply in two ways. In the first place a conservation investment was likely to be depreciated over a period or years just as if it was a productive machine. No account was usually taken of its far longer life span or of the extent to which buildings were likely to appreciate in value as a result of it. Secondly, the value of energy saved by conservation investment is likely to exceed the rate of inflation over the next few years. The value of most products manufactured by productive machinery on the other hand is likely only to keep pace with inflation. This difference was rarely understood by accountants and did not therefore feature in their assessments. Witnesses thought that successful energy conservation in the commercial sector "will not be achieved until the necessary marked pressures are promoted to encourage (it)".

The fact that the lower running costs of energy-efficient buildings are only partly reflected in either higher rents or higher resale prices for such buildings.

Rents were usually determined more by location than the quality of the building.   In those parts of the country where rents were very low, builders and owners could not get higher rents for more energy efficient buildings.

The fact that the *existing leasing structure* presents a potential clash of interest in that the benefits from a conservation investment might not accrue directly to the party who financed it.   Thus sometimes there was no incentive for the landlord to improve the building.   Similarly, occupiers could not increase the energy efficiency of the building they occupied and did not always see it as being in their interests to do so.   Multiple tenancies exacerbated the problem, because any single tenant could veto a shared conservation investment.

Lack of certainty about future fuel prices and relativities.

The paucity of publicity and advice about energy conservation. Witnesses expressed concern over present marketing policies for conservation materials and appliances.   There is little generic advertising;   household improvements such as new bathrooms and kitchens, and consumer durables are much more heavily advertised than energy conservation as figures from the Institute of Advertising indicate.   Witnesses told the Committee of the lack of cheap, reliable local advice.   It was, moreover, sometimes difficult to buy the product at all.   The Consumers' Association stated that "builders merchants, discounters and department stores have a dismal record of giving energy advice".   It was thought that if consumers are to conserve energy, they must be aware both that they can do so without reductions in comfort and that such investment is cost effective. They should, moreover, be able to find out easily how to do so and at what cost.   Witnesses from consumer bodies added that the utilities' showrooms were a "valuable location from which to provide information to the public", and could be more fully used.

A recent survey on obstacles to energy conservation - based on interviews with energy managers across a wide spectrum of organisations concluded:-

The two obstacles (to energy conservation) considered the most important, namely the low profitability of conservation measures and the lack of investment finance, are certainly not independent but the greater importance attributed to the former justifies their separation. It suggest that even if firms had plenty of money to invest, they still would not be able to find profitable enough energy conservation projects in which to invest.   Because they will have more confidence in their assessment of the likely savings one would expect energy managers to be more likely than the financial management to find technological investments worthwhile (unless the latter give greater weight to likely fuel price increases).   In spite of this the energy managers find the main constraint to be the lack of suitable, profitable investments.   This conflicts with many technical reports on energy conservation and requires some explanation.

From discussions with energy managers and financial managers it seems that the almost universal criterion for profitability of a conservation project is simple pay-back and a pay-back period of 2-3

years is demanded for adaptations to existing plant. This is equivalent to a rate of return of the order of 25 per cent or more. Clearly
inflation may influence the required return but perhaps more important
is the recession and the general uncertainty in many businesses as to
just what, if anything, they will be doing in two years' time, let
alone four or more. It is important to bear in mind that energy
conservation measures are generally adaptations to existing plant:
when plant is replaced a longer pay-back period, at least 4-5 years,
is likely.

As well as different ideas about appropriate pay-back periods (or
different rates of time preference) there may be other factors
causing this difference in view of profitability by industry on the
one hand and by the Department of Energy and in various technical
reports on the other. First, the latter's more theoretical
estimates may exaggerate savings in the way that laboratory achievements so often outperform the same device in the field, or the
theoreticians may have underestimated adjustments and adaptation
costs of the installation or have overestimated the certainty of
savings. There may also be a difference in expected gains caused by
different assumptions about the future price of energy: the simple
pay-back period of industrialists sometimes does not allow for an
increase in even the nominal price of energy (even though the high
rate of return they demand is likely to be affected by inflation).

Probably the industrialists' calculations are rather too
pessimistic and the conservationists' estimates rather too optimistic.
Another, fairly general view given to us was that, if there is
competition for scarce investment capital between a project serving
the company's production or marketing and another one aimed at energy
conservation, the former is almost always the winner.

Obstacles to energy conservation

Ranking[a]	Obstacle	Percentage Responding 'unimportant'	'very important'
1	Low profitability of any relevant conservation measures	19	49
2	Lack of investment finance	34	33
3	Management attitudes	37	31
4	Existing plant not due for replacement	42	20
5	Low share of energy in total costs	45	19
	Uncertainty about future energy supplies	50	21
	Employee attitudes towards change	48	17
	Extra supervision required	47	16
	Lack of confidence within your organisation in savings attainable	45	14
	Uncertainty about future energy prices	50	19

	Lack of conservation measures relevant to your organisation	48	12
12	Laws and regulations	60	14
13	Lack of, or cost of obtaining, information	63	12
14	Trade practices and procedures	63	9
15	New skills required	65	9
16	Expecting technology to change and therefore waiting	70	5

[a]Rankings were worked out by assigning a reply of 'very important' twice the weight of 'moderately important'. Experiments with alternative methods confirm the ranking of the first five and last five factors but the order of those in the middle can be changed markedly: therefore, rankings have not been assigned to this middle group.

Thirdly, a paper presented to the recent IAEE conference on "International Energy Markets - The Chaning Structure", concludes:-
Any effective energy policy must be based on a recognition of the elements of demand as well as supply. No easy task when supply policy traditionally involves only a very limited number of large utilities, whereas demand policy involves influencing the behaviour of a large number of uncoordinated individuals and organisations.
Now the published evidence of the Select Committee on Energy suggests that no direct market machanism exists whereby the value of a demand decision can be directly compared with the value of a supply decision. The Government can and does invest directly in supply: whereas investment in demand reduction depends on the action of others at best under the influence of Government. Yet in the absence of such direct market mechanisms, any demand (conservation) policy requires not only the establishment of such values but also the solutions to the problems of getting the demand (conservation) market to respond to these values once established.
In fact the direct policy options to effect policy are limited and are encorporated in the current UK Energy Policy (and indeed the energy policies of many other countries) namely:

maintaining "correct" energy prices

providing information and incentives

legislation to ensure minimum standards

The conventional wisdom being that the Price mechanism works, only it takes rather a long time - can be confused by short term fluctuations and the role of information is unresolved.
However, Demand Policy has associated with it:-

(i) High Political costs (Adjustment causes frustration).

(ii) High Welfare costs (the adjustment costs to the disadvantaged groups).

(iii) The need for improved Estimation (a better data base is seen to be the solution to a number of demand estimations and therefore by implication policy problems).

For these reasons it is argued that demand policy has to be concerned with the actual behaviour (of Artifacts and people) and that the only way to do this is to understand what is happening.
It can be observed that:

(a) Consumers make enormous investments in double glazing. Such that whatever its real benefit, it is not only energy saving.

(b) Architects observe that standards of workmanship (of insulating roofs etc.) leave a lot to be desired. Which demonstrates that theoretical performance is not a good predictor to in use performance.

(c) Consumers open kitchen windows when they put their cookers on- which demonstrates that people behave sensibly but according to their own lights.

From such observations we can learn that:

(a) The real value to people of useful temperate flow space.

(b) Access to and location of services (Buildability) influences installation practice.

(c) Controls must relate to behaviour.

The good news is that such messages are always easily (but not simply) incorporated into good new design - which reinforces current policy stances that the major trends are in the right direction.
The bad news is that new constructions are only a small percentage of the total, 1-4% per annum. Yet in the case of existing buildings similar improvements only happen in a limited number of 'up-market' properties, i.e. the anomalies are large.
This leaves the householders in the bulk of existing buildings unable to repond effectively to price signals alone because:-

(a) The expectation of future energy prices are variable.

(b) Consumers are ill-informed and do not know how to respond.

(c) They cannot afford to respond.

Government intervention to reduce these anomalies is needed and what is more on both the suppliers and the consumers side.

E.g.: Suppliers side:

(i) Improved standards of workmanship e.g. by Building Society inspectors.

(ii) Encourage control development (Anticipate rather than respond).

Consumers Side:

(i) Useful specific information:- tell how to draught-proof and insulate.

(ii) Proper location and type of metering.

(iii) Specific financial incentives.

My questions to this conference are therefore if this situation is indeed a reasonably accurate on then:

(a) What responsibility do the Building Economists accept for this state of affairs

- Is it their fault that confusion exists in the subject areas of accounting practices, real costs, and investment criteria. Have they been adequately prepared to meet these new challenges or not. Do they hold a proper place in the hierarchy of decision making such that they are consulted in good time, or not. Are they indeed interested in this area of endeavour or not.

(b) If to date you agree that success has eluded you, then what actions do you/should you propose to change the situation in the future.

SECTION III

COST DATA BASES

DATA BASES USED BY PROPERTY SERVICES AGENCY QUANTITY SURVEYORS

DONALD H GREEN, Property Services Agency

The opinions expressed in this paper are those of the author and are not to be interpreted as representing those of the Property Services Agency of the Department of the Environment.

1. Quantity Surveyors in the Property Services Agency use computer data bases for several interrelated fields of building price and cost information.

2. PSA Schedule of Rates for Building Works 1980

The Property Services Agency's Schedule of Rates for Building Works is one of seven schedule of rates published by Her Majesty's Stationery Office. The Schedule, containing some 20,000 rates, provides a basis for estimating, tendering and pricing work in all the main building trades. While it is used by the PSA and other public authorities and their contractors mainly on term contracts for maintenance and small works, it can also serve as a basis for speedy tendering for urgent larger capital works.

The rates reflect costs of resources as at the middle of 1980 and the Schedule is aligned, where appropriate, with the Sixth Edition of the Standard Method of Measurement of Building Works and the latest editions of the PSA General Specification and the PSA BQ system. The rates are synthesized by computer and each resource is linked to Price Adjustment Formulae or Department of Industry indices for labour and materials.

3. Building Tender Prices

The level of pricing contained in Bills of Quantities in tenders for new capital work accepted by the PSA are compared with the rates in the PSA Schedule of Building Works. All projects based on Bills of Quantities are used with the exception of those for housing, civil engineering, mechanical and electrical engineering, and alterations and extensions.

For each accepted tender, items to the value of not less than 25% from each work section of the Bill of Quantity are repriced at rates from the PSA Schedule, each work section is weighted by its value in the tender and an index derived for that tender. The mean of all tenders indexed in a quarter are published as the DQSS Building Tender Price Index.

The essential components used for each tender are recorded on a computer and can be recalled using a computer selection program.

4.   Price Adjustment Formulae

An index, compiled monthly, is used to measure the cost to the contractor of constructing a typical PSA building. The index is based on an average PSA building compiled by analysing the bills of quantities for a range of representative contracts for typical PSA buildings into the work categories of the Building Formula Indices. The average PSA building is compounded from buildings: reinforced concrete framed, steel framed, and in loadbearing brickwork.

The index is calculated by computer from the monthly Building and Specialist Engineering Formula Indices and in effect, measures the movement in the cost of building the whole average building each month.

5.   Measured Term Contract Updating Percentages

The measured term contract updating percentages, issued monthly, reflect the general movement in national price levels of labour, plant and materials contained in the Schedule or the movement in cost to the contractor of work described in the Schedule. The percentages are used for updating the net rates in a Schedule of Rates when placing an order for work under a measured term contract. They are added to the Schedule rates before applying the contractor's primary and secondary percentages.

To compile the percentages, an analysis was made from records of orders for work paid under measured term contracts by all PSA areas. This produced, for each Schedule, a comprehensive range of Schedule items which were further analysed into their labour, plant and material content. Each month the percentages are calculated by computer by applying to these labour, plant and material items the relevant Price Adjustment Formulae, Department of Industry or other appropriate indices and comparing the result with those for the base date of the Schedule.

6.   Measured Term Contract Prices

An index, issued quarterly, measures the price to the PSA, assessed at date of order, of work ordered under measured term contracts. The index can be used as an aid to estimating measured term contracts and used in conjunction with the relevant updating percentages, it enables a comparison to be made between the national average and a contractor's tendered percentage for a local contract.

The index is based on the PSA Schedules of Rates, all current measured term contracts held by the PSA on each Schedule, and the measured term contract updating percentages. A mathematical model of order values at net schedule rates has been derived from an analysis of a large sample of accounts paid to measured term contractors. Each quarter the model is adjusted by the current updating percentages and each contract is evaluated using its primary and secondary percentages to produce an index. The individual indices are combined in accordance with the weights of their standardised estimated annual values to form the quarterly index published. Records of all contracts are recorded on a computer which is used for retrieving and calculating all relevant information.

7. Cost Analyses

Cost analyses are prepared on the Standard Form of Cost Analysis and recorded on a computer. They provide historic cost information for use in preparing estimates, cost plans, reference costs, etc.

Appropriate cost analyses can be selected by parameters for building function, type of construction, internal gross floor area and number of storeys using a computer selection program; the program listing brief details of those cost analyses satisfying the selection parameters. Further developments will include for other selection parameters, printing complete cost analyses in SFCA form and BCIS cost analyses.

Computer programs are available for adjusting data obtained from the data base automatically; either one or several cost analyses being used.

8. Interrelation of Information

Many fields used in the above information are common and interrelated. Appendix A shows general relationships.

The data for this collection of interrelated information is stored on computer to serve one or more applications in an optimal fashion. It is stored so that it is independent of those applications programs which use it. The intention is that decision makers can have far better information than they had before the availability of computers and they can have it immediately they need it.

References

1. Schedule of Rates for Building Works 1980 (ISBN 0 11 671066 7), price £50, and other PSA Schedules of Rates can be obtained from HMSO Bookshops and from other booksellers
2. A Tender Based Building Price Index. R Mitchell. Chartered Surveyor Vol 104 No 1. 1 July 1971
3. Price Adjustment Formulae for Building Contracts (Series 2) – Guide to Application and Procedure (ISBN 0 11 670790 9), price £2.25, and Amendment No 1 dated May 1979 (ISBN 0 11 670835 2), price 90p can be obtained from HMSO Bookshops and from other booksellers
4. Price Adjustment Formulae for Building Contracts (Series 2 Revised) – Description of the Indices (ISBN 0 11 670834 4), price £2.95 can be obtained from HMSO Bookshops and from other booksellers

5. Monthly Bulletin Construction Indices (Series 2), price £2.95
    can be obtained from HMSO Bookshops and from other booksellers
6. Technical Memorandum QS "Quarterly Building Price and Cost
    Information, price 10p per page, from PSA Library Sales Office,
    Room C109, Whitgift Centre, Croydon CR9 3LY (telephone
    01-686-8710 Ext. 4525). Also published in Chartered Quantity
    Surveyor every 3 months
7. Updating Percentage Adjustments for Measured Term Contracts,
    price 50p each or £5.00 per annum, from PSA Library Sales
    Office, Room C109, Whitgift Centre, Croydon CR9 3LY (telephone
    01-686-8710 Ext. 4525)

APPENDIX A

GENERAL RELATIONSHIPS BETWEEN DATA BASES

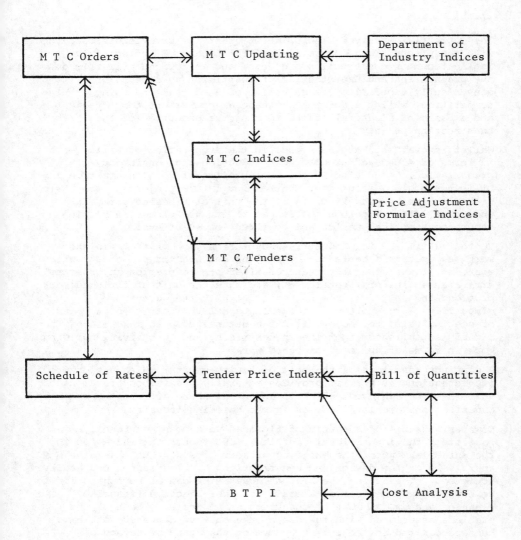

COST DATA BASES AND THEIR GENERATION

DAVID M. JAGGAR,
Liverpool Polytechnic, Department of Surveying.

INTRODUCTION

The currently perceived discipline of quantity surveying is concerned
with the planning and controlling of the financial and economic
implications of construction projects, and is founded primarily on
the preparation and use of bills of quantities as a means of
conveniently expediting the selection of a contractor, using
competition, whilst acquiring a financial commitment for a project
and a basis of financial control during the execution of the
construction project.

Whilst innovations have been made in the preparation of bills, both
in terms of speed and accuracy of production and consistency of
presentation,[1,2] and also in the manipulation of cost information in
bills for use in design cost control for future projects,[3] the basic
mechanism for   carrying cost: the unit of finished work, has
undergone little rigorous testing as to its usefulness in helping
secure more optimal design and construction solutions.

Up to the time of the Industrial Revolution, the unit of finished
work was used as a basis for describing and quantifying construction
work, as it was generally measured after its completion in order to
arrive at a financial settlement, and thus the unit of finished work
provided the only feasible approach.[4]   With the advent of
competitive lump sum tendering during the 19th Century, as a means
of seeking value for money, it was a natural step to make use of the
already established expertise in describing and quantifying building
work in terms of units of finished work.

A distinct advantage of the unit of finished work, which has ensured
its retention, is that it provides an easily understood statement of
what is to be achieved, which is essential where different management
organisations are involved, as in the building industry.

However, the use of the unit of finished work does not readily
facilitate the identification of the constituents that have led to
the establishment of the building project itself, thus objective
appraisals of the design and construction solution cannot be readily
achieved.   Therefore, the bill of quantities cannot be considered to
be entirely appropriate as a basis for the objective financial
planning and controlling of construction projects.   Some of the
problems associated with the use of the unit of finished work have
been recognised in the latest update of the Standard Method of
Measurement for Building Works[5].   This is indicated by the inclusion
of method-related charges for plant items, the costs of which are not
necessarily related to quantity,   but are generally related to time
and non-recurrent aspects.[6] It is suggested, however, that despite
recent improvements in the structure of bills of quantities, they are
not entirely effective, as the prime information source for storing,

retrieving and manipulating cost data as a basis for planning and controlling construction costs. The purpose of this paper is to outline a theoretically sound base for the allocation and control of resources and, thus, costs in the designing and constructing of buildings, and to demonstrate how the theoretical base can be developed into an operational system.

THE ROLE OF THE BILL OF QUANTITIES

Bills of quantities have traditionally been used as a vehicle for selecting a contractor based on fixed price tendering strategies. Although this role is important, it is suggested that the bill of quantities, being essentially the part of the project documentation which reflects the cost implications of the project under construction, should form the basis for the management of resources and costs; a function which is not readily or reliably achieved with current bills of quantities.

The bill of quantities should represent an empirical or causal model of the quality and quantity of the resources required to achieve a particular project stated within the framework which readily allows for the generation of the cost implications of the identified resources.

If this requirement can be achieved, then the bill of quantities can be used objectively as a basis for optimising design and construction solutions in terms of the physical entity of the project and can also be used as the basic tool for the planning and controlling of the task of constructing the project.

Due to the fact that the current bill of quantities does not facilitate the identification of the resources needed, nor their specific costs, and also does not successfully reflect how construction costs are incurred on site, then the bill cannot be used as a reliable data base for either optimisation of a project solution or the implementation of the intended solution.

THEORETICAL CONSIDERATIONS – INTRODUCTION

OPTIMISATION: In order to secure more optimal design/ construction solutions, it is necessary to develop heuristic reiterative modelling techniques to help ascertain the resources required to establish a future construction project.

It is the resources required which generate cost and thus it is essential to ascertain the quantity and quality needed, together with, where appropriate, the time commitment for which they are required.

The achievement of the optimal design/construction solution is that which requires the lease resource commitment to achieve the functional requirements needed in the project to satisfy the performance specification stated by the client.

The identification of the least resource commitment is based on a solution which requires the minimum energy commitment to locate commodity resources, using appropriate amalgams of labour and plant.

Ascertaining energy commitment is based on a knowledge of the mass to be moved, the distance it is to be moved and the time required to move it.   To take account of this, a model facilitating the establishment of information concerning mass, distance and time is necessary in the form of a three dimensional matrix.[7]   Thus, by assigning commodity resources (the mass to be moved) to one dimension and the location to another dimension, then appraisals can be made of the labour and plant (responsible for locating commodity resources) together with the time commitment required, which is then expressed through the third dimension.

It is also necessary that in order to respond to the analytic design process and the synthetic construction process, the project inform- ation documentation must be capable of being expressed at differing levels of detail - this can be achieved by expressing the three dimensional matrix at a level of detail which relates to its position in the design/construction process.

Optimisation in terms of minimum resource commitment is primarily the responsibility of design management systems involved as the commodity resources they select, the geometrical arrangement of the project system and any time impositions have a direct implication on the labour and plant resources required.   It is, therefore, necessary in seeking more objective appraisals that the resource commitment needed to achieve the required functional performance can be object- ively appraised.   Such a process is best satisfied by accessing historic project information sources in order to consider the resource commitment that was needed.   Such information can then be made use of in the formulation of the three dimensional matrix relating to the development of the design and construction solution for the project system under consideration.

The performance of the matrix so established can then be measured by heuristic analysis, and if sub-optimal, a further matrix established by reiteration until the design management groups involved are satisfied that the optimum has been secured.   Such a matrix can then be used by the construction management groups in order that they can appraise the minimum labour and plant resources amalgam which is implicated within the design solution, to locate the designated commodities.

When the matrix has been fully developed in terms of a completed resource model (after appraisal by both design and construction management groups) it then provides the basis for implementation of the construction solution on site and also becomes a historic project information source for use in appraising future projects.

IMPLEMENTATION: In order to execute the task of construction on site it is necessary to have a plan of the intended solution and which can also form the basis of control during the actual construction process.

The matrix, once completed at the end of the analytic design process, represents a resource model in terms of which resources are required, where they are required and when they are required.   Also, by allocating cost to the resource model, it provides cost model (a process which can be achieved, for example, by the process of tendering).

Specific information of relevance to a particular management group at an appropriate point in time can then be readily retrieved from the model as a basis for planning and controlling the construction process.

In order that the above optimisation and implementation roles can be carried out by the various management groups who are involved in the realisation of a project it is necessary that they can readily store, retrieve and manipulate information of concern to them. Thus the project information (cost plans, drawings, specifications, bills, networks, ordering schedules, etc.) of which the bill of quantities forms a sub-set, must form a fully co-ordinated entity in order that the information contained within it can be made use of with the minimum of translation and is also fully cross-referenced within the various documents making up the project information.

It is suggested that the SfB faceted classification System[8] can form an appropriate syntax to organise the project information such that the various parts making up the project information can be readily perceived and comprehended by the management groups involved.

THE PROPOSED SOLUTION

It is possible, using the SfB System, to express a project by means of any of its facets, thus:

                 Facet 1   Functional characteristics
                 Facet 2   Constructional characteristics
                 Facet 3   Resources

and, therefore, the following relationship can be stated

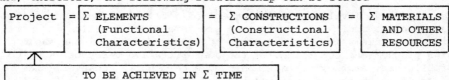

It is, hence, possible to use the SfB System as an essential 'core' to facilitate the description of the project in terms of character- istics (essential for comprehension) using Facets 1 and 2 and to facilitate its description in terms of resources (essential as a means of deriving an optimal project solution) using Facet 3.

So, the three facets can be used to describe the project in terms of either characteristics or resources. In order to facilitate both optimisation and implementation of the project solution, it is necessary to relate the three facets to both location and time dimensions. Such a relationship can be used to meet the require- ments of the three dimensional matrix, as discussed above.

It is at this point where the analytic design process and the syn- thetic construction process commences that the three dimensional matrix needs to be the most detailed, as it is at such an interface that the resources needed can be ascertained deterministically rather than stochastically as is the case where only characteristics are being expressed i.e. at any one point during the analytical design process).

The management group(s) which can provide the minimum plant and labour resources to locate the designated commodities in the project represents the optimal construction solution for that particular project solution.

However, it is important that the management group(s) concerned with the design solution is aware of the implications on plant and labour resources of a particular commodity amalgam in order that the appropriate feasibility band of potential construction solutions is selected.

It is, therefore, necessary that the project information provides an appropriate characteristic description within which resource identity deterministic or stochastic, (In lump sum tendering, commodity resources are deterministic, whereas the remaining resources are stochastic.   Only deterministic resources can be identified and quantified definitively as stochastic resources must be presented in a manner whereby their identity and quantity can be readily derived by the management group(s) responsible for their provision) can be easily ascertained.   To provide such a mechanism it is necessary to look more closely at the synthetic process of construction.   To execute this process, the management group(s) concerned needs to identify the various energy commitment sub-sets needs in the execution of the project.   From such an identification it is then possible to ascertain the appropriate resource amalgam, the time required to execute such sub-sets, and, ultimately, the establish-ment of the overall project programme.

Such a process involves the derivation of activities (defined as the placing of commodity resources in a pre-determined location in a pre-determined time using the resources of labour and plant, the whole being planned and controlled by the administrative resource) and the establishment of their relationship to each other.   However, the main difficulty in facilitating the above, as invariably different management groups are involved, is in allowing the nature and sequence of activities to be derived.   It is also important that the activity can be related to performance appraisal criteria for use during the analytic design process.   This is because an activity does not form a useful basis for comparing one project solution with another as activity content and sequence cannot be universally defined as it is dependent upon a particular project.

It is, therefore, important to identify some methodology whereby the activities and their relationships to each other can be derived from the information generated by those management systems concerned with design.

ACTIVITY DERIVATION

The relationship and content of the three facets of the SfB System can be used to satisfy the analytic designing process and the synthetic constructing process, thus:

Designing (Analytic)   Constructing(Synthetic)

   PROJECT   PROJECT
      ↓   ↑
   1) ELEMENTS   1) ELEMENTS
      ↓   ↑
   2) CONSTRUCTIONS   2) CONSTRUCTIONS
      ↓   ↑
   3) RESOURCES   3) RESOURCES

When designing and constructing are the responsibility of independent management groups and a lump sum tendering strategy is used, then
the above statement can be modified thus:

Designing (Analytic)   Constructing (Synthetic)

   PROJECT   PROJECT
      ↓   ↑
   1) ELEMENTS   1) ELEMENTS
      ↓   ↑
   2) CONSTRUCTIONS   2) CONSTRUCTIONS
      ↓   ↑
   3) COMMODITIES (DETERMINISTIC)   3) RESOURCES (ALL
                                                                DETERMINISTIC)

      LABOUR   )
      PLANT   ) (STOCHASTIC)
      ADMINISTRATION)

As already considered, it is essential that the project information
is capable of transposition from a statement of what is to be
achieved to how it is to be achieved, and it is the activity which
is the key to satisfying this need. It is proposed that the SfB
System can be made use of in order to allow activity identification
and sequence to be achieved in a way which can easily be related to
the three facets. Thus, the interests of the various management
groups can be represented without having to translate the project
information into a different form.

Although an activity is a dynamic concept dependent on the particular
project and the management group concerned, it has been proposed
within the SfB Group (the author's work within the SfB Development
Group has contributed to the concept of Elemental Constructions
being incorporated into the SfB System) that an activity can be
bounded by an Element and a Construction.

The concept of deriving the dynamic activity from within such a
framework is developed further as discussed below to form an integral
component of the three dimensional matrix.

If the characteristics of the project system to be achieved are
stated in terms of Facet 1 and 2 combinations, then it is possible
to derive, as a sub-set of such a combination, the various
activities.

A combination of these two facets can be considered as an abstract planning unit with general application. When specific information is added to describe the project under consideration, then specific activities can be derived.

So, at the planning stage, it is possible to derive the activities using the combinations of Elements and Construction   The completion of the various activities will then lead to the establishment of the various Element and Construction combinations, the total of which represent the completed project.

The process of Activity derivation is shown below:

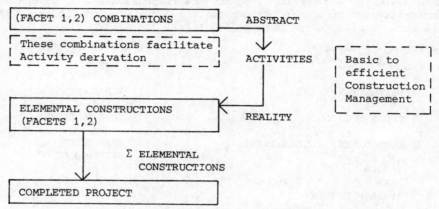

The activities so derived can be used by the management groups concerned as a basis for planning and controlling the construction of the project.   The resources needed to achieve the characteristics identified in the proposed Elemental Constructions must also be stated in order that they can be allocated to the activities so derived.   Where a management group is concerned only with design and not construction, then only deterministic resources (commodities) can be definitely stated within the Elemental Construction combination.   The remaining stochastic resources must be presented in a manner which allows their ready determination by the management group providing them.   However, when all the resources have been identified, then this information will be related to the activities. The project information at this stage will represent the resources required to achieve the characteristics required.

Thus, by adapting the SfB System as described above, the relationship between characteristics and resources can readily be identified and, further, by appropriate combinations of the facets, specific information can be retrieved from the project information to suit the requirements of a particular management group.

Further, it is the resources which are responsible for cost generation and, therefore, resource identification is essential to allow true financial evaluations to be carried out.   Also, as the resources, together with their financial implications, can be easily related to the characteristics of the project, then more objective appraisals can be achieved.

Typical examples of the matrix arrangements expressing the above concepts are described below: (see examples 1, 2 & 3)

Explanation

The dimensions of the matrix are as follows:

Dimension (i) represents a particular facet of the SfB System
Dimension (j) represents location (floor levels in these examples)
Dimension (k) represents the time required to complete the project system.
Each dimension is considered as being made up of an arbitrary 8 units, the summary of each dimension being represented by the digit 9.

Further, the various resources required to achieve the project can also be related to the three dimensional matrix, thus:

(a) (i) = Commodities (part of Facet 3)

      M (1, 1, 1) Cement, below lowest floor level, first unit of time.
      M (3, 4, 5) Bricks, 4th floor, 5th unit of time
      M (9, 9, 9) Total commodities, total locations, total time

(b) (i) = Labour (part of facet 3)

      M (1, 1, 1) 5 men, below lowest floor level, first unit of time
      M (4, 6, 5) 20 men, 6th floor, 5th unit of time
      M (9, 9, 9) Total labour, total locations, total time.

(c) (i) = Plant (part of facet 3)

      M (1, 1, 1) Excavator, below lowest floor level, first unit of time
      M (4, 5, 5) Tower Crane, 5th floor, 5th unit of time
      M (9, 9, 9) Total plant, total locations, total time.

(d) (i) = Administration (part of facet 3)

      M (1, 1, 1) Concreting Administration, below lowest floor level, first unit of time
      M (6, 7, 5) Carpentry Administration, 7th floor, 5th unit of time
      M (9, 9, 9) Total Administration, total locations, total time.

Observation of the summaries of each three dimensional matrix depicted in examples 1, 2 and 3 shows that they represent the same fundamental relationships expressed in the SfB System as shown on p4 of this paper, so:

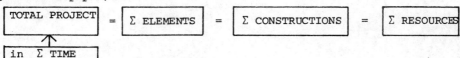

$$\boxed{\text{TOTAL PROJECT}} = \boxed{\Sigma \ \text{ELEMENTS}} = \boxed{\Sigma \ \text{CONSTRUCTIONS}} = \boxed{\Sigma \ \text{RESOURCES}}$$
$$\boxed{\text{in } \Sigma \text{ TIME}}$$

EXAMPLE 1   ELEMENTS (FACET 1)/LOCATION/TIME/MATRIX

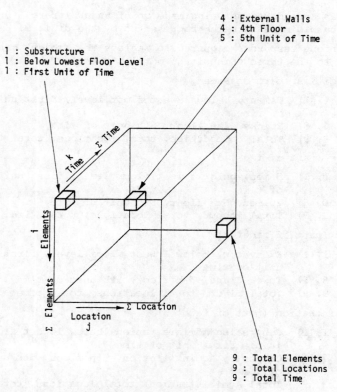

4 : External Walls
4 : 4th Floor
5 : 5th Unit of Time

1 : Substructure
1 : Below Lowest Floor Level
1 : First Unit of Time

9 : Total Elements
9 : Total Locations
9 : Total Time

or, expressed mathematically

M {1, 1, 1} Substructure, below lowest floor level, first unit of time.

M {4, 4, 5} External walls, 4th floor, 5th unit of time

M {9, 9, 9} Total elements, total locations, total time.

EXAMPLE 2   CONSTRUCTIONS (FACET 2)/LOCATION/TIME/MATRIX

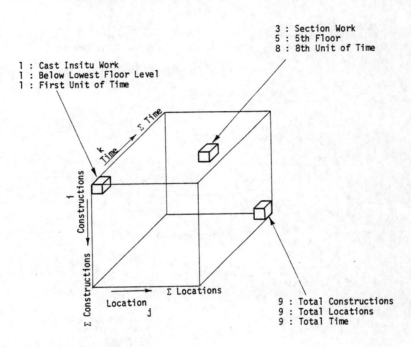

3 : Section Work
5 : 5th Floor
8 : 8th Unit of Time

1 : Cast Insitu Work
1 : Below Lowest Floor Level
1 : First Unit of Time

9 : Total Constructions
9 : Total Locations
9 : Total Time

or, expressed mathematically:

M   {1, 1, 1}   Cast insitu work, below lowest floor level, first unit of time

M   {3, 5, 8}   Section work, 5th floor, 8th unit of time

M   {9, 9, 9}   Total constructions, total locations, total time.

EXAMPLE 3   RESOURCES(FACET 3)/LOCATION/TIME/MATRIX

1 : Resources   6 : Resources
1 : Below Lowest Floor Level   5 : 5th Floor
1 : First Unit of Time   5 : 5th Unit of Time

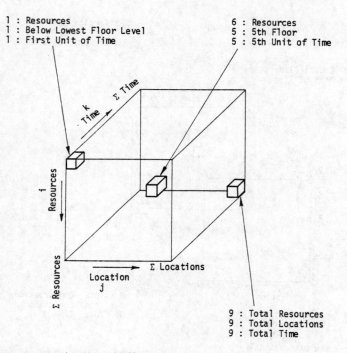

9 : Total Resources
9 : Total Locations
9 : Total Time

or, expressed mathematically

M   {1, 1, 1}   Resources,(cement, 3 men, 1 excavator, 1 foreman) below
                lowest floor level, first unit of time

M   {6, 5, 5}   Resources (bricks, 4 men, 2 hoists, 1 foreman) 5th floor,
                5th unit of time

M   {9, 9, 9}   Total resources (commodities, labour, plant, administration)
                total locations, total time.

Or from the matrices

$$S \text{ (Matrix Summary)} = \sum_{(i)=1}^{(i)=8} \sum_{(j)=1}^{(j)=8} \sum_{(k)=1}^{(k)=8} M(i, j, k) = \text{Total Project}$$

$$\text{Therefore, } \sum_{(n)=1}^{(n)=3} Sn = \text{Total Project}$$

where   (i) = facet 1, 2 or 3
          (j) = location
          (k) = time

It is, therefore, proposed that the project information should be structured upon the three dimensional matrix to allow optimal solutions in terms of minimum resource utilisation to be objectively sought.

The SfB System, as shown above, can be adapted to provide communication efficiency with regard to the physical entity of the project by allowing the three dimensional matrix to be expressed in arrangements appropriate to the requirements of the analytic design process and the synthetic construction process.

A METHODOLOGY

A solution has been developed[9] at an operational level whereby project information can be produced at different levels of detail, in compliance with the requirements of the three dimensional matrix, in order to reflect the state that the design and construction solution has reached during the realisation of a project.

Further, by using the SfB System, as the basic syntax for co-ordinating the project information, all the various manifestations of the project information are fully co-ordinated to each other, notwithstanding their purpose or level of detail.

It is not possible, in this paper, to describe fully and illustrate the system   that has been developed.   Figs. 1 and 2 show a part of the bill of quantities representing the completion of the analytic design process and the commencement of the synthetic construction process and could thus form the basis of a lump sum tendering strategy.

Fig. 1 shows the separate identification of the labour resource and the use of Table 1 and 2 combinations of the SfB System together with locational information, as a means of presenting the stochastic labour resource in order to facilitate activity derivation.   Fig. 2 shows the presentation of the commodity resources independent of Table 1 and 2 SfB combinations or locations as their cost to purchase is independent of these factors.

Using the above information, the construction management groups concerned can, using the labour resource description, establish a logic for constructing the project such as a network and the bill of quantities, as shown in Figs. 1 and 2, is then re-sorted to accord with the established project logic.   Figs. 3 and 4 illustrate how the resources needed to achieve the activities can be assembled.

| OFFICE BLOCK 0001 | ELEMENT: EXTERNAL WALLS (21) | CONSTRUCTION : BLOCKWORK F |

Item	Locat	Code	Description		Qty	Unit	Hours	Rate	Total Time (Hours)	Total Cost £	
	FLOORS	30 F30	Block Walls Blockwork	B/F					355.31	975	91
		F350	Walls								
		F260	Normal strength blocks in CLM (1.1.6)								
16	00	F1125	100 mm thick; fair face one side; flush pointing		6	M²	1.94	2.17	11.64	25	26
17	00	F1250	Fair cutting against soffits		8	M	1.00	2.17	8.00	17	36
		F270	High Strength Blocks in CLM (1.1.6)								
18	01	F1150	200 mm thick		58	M²	1.94	2.17	112.52	244	17
19	02	F1150	200 mm thick		75	M²	2.13	2.17	159.25	346	66
20	03	F1150	200 mm thick		92	M²	2.13	2.17	195.98	405	23
21	04	F1150	200 mm thick		73	M²	2.32	2.17	169.36	330	25
22	01	F1152	Rough cutting		3	M	0.81	2.17	2.43	5	07
23	02	F1152	Rough cutting		8	M	0.81	2.17	6.48	14	06
24	03	F1152	Rough cutting		8	M	0.81	2.17	6.48	14	06
25	04	F1152	Rough cutting		8	M	0.81	2.17	6.48	14	06
26	01	F1153	200 mm thick; against existing brickwork		27	M²	1.94	2.17	52.38	113	66
27	02	F1155	200mm thick; against existing brickwork		17	M²	2.13	2.17	36.21	78	58
28	00	F1185	200 mm thick; fair face one side, flush pointing		73	M²	2.58	2.17	188.34	408	70
29	00	F1260	Fair cutting against soffits		4	M	0.60	2.17	2.40	5	21
		F400	Piers								
		F280	High strength blocks in CLM (1.1.6)								
30	01	F2000	400 mm thick		1	M²	4.03	2.17	4.03	8	75
31	02	F2000	400 mm thick		1	M²	4.44	2.17	4.44	9	63
32	03	F2000	400 mm thick		1	M²	4.44	2.17	4.44	9	63
33	04	F2000	400 mm thick		1	M²	4.84	2.17	4.84	10	50
34	01	F2010	450 mm thick		1	M²	4.44	2.17	4.44	9	63
35	02	F2010	450 mm thick		1	M²	4.88	2.17	4.88	10	59
36	03	F2010	450 mm thick		1	M²	4.88	2.17	4.88	10	59
37	04	F2010	450 mm thick		1	M²	5.32	2.17	5.32	11	54
38	02	F2020	500 mm thick		2	M²	5.32	2.17	10.64	23	09
39	03	F2020	500 mm thick		4	M²	5.32	2.17	21.28	46	18
40	04	F2020	500 mm thick		3	M²	5.65	2.17	16.96	36	78
41	01	F2025	500 mm thick; against existing brickwork		4	M²	4.61	2.17	18.44	40	01
42	02	F2025	500 mm thick; against existing brickwork		2	M²	5.32	2.17	10.64	23	09
		F90	Sundries								
		F450	Damp proof courses								
		F210	Horizontal								
				C/F					1428.49	3248	45

FIGURE 1

Item	Locat	Code	Description	Qty	Unit	Hours	Rate	Total Time (Hours)	Total Cost £
		e550	Slates						
1		e5100	400 x 200 mm	100	NR	–	0.17	–	17 00
		f200	Pre-cast concrete						
2		f2100	Mix D; Copings 400 mm x average 100 mm thick x 1.00 m long. Weathered once, throated twice.	28	NR	–	5.25	–	147 00
3		f2200	Mix D; Lintols 100 x 112 x 1050 mm long	1	NR	–	2.63	–	2 63
4		f2200	Mix D; Lintols 100 x 225 x 1125 mm long	1	NR	–	3.15	–	3 15
5		f2210	Mix D; Lintols 200 x 225 x 925 mm long	2	NP	–	3.06	–	6 12
6		f2230	Mix D; Lintols 200 x 225 x 1275 mm	4	NR	–	3.41	–	13 64
		f710	Gypsum						
7			12 mm thick	450	M²	–	0.88	–	396 00
8		f7115	Gyptex cove cornice; 100 mm girth	182	M	–	0.30	–	54 60
			Normal strength blocks						
9		f2050	450 x 225 x 100 mm	836	NR	–	0.17	–	142 12
10		f2060	450 x 225 x 100 mm; fair faced one side	60	NR	–	0.23	–	13 80
		f230	High Strength Blocks						
11		f2050	450 x 225 x 100 mm	18275	NR	–	0.17	–	3106 75
12		f2060	450 x 225 x 100 mm; fair faced one side	1475	NR	–	0.23	–	339 25
		g240	Facing Bricks						
13		g2010	Type A	19440	NR	–	0.05	–	972 00
		g250	Vitrified salt glazed clay pipes and fittings						
14		g2200	100 mm diameter pipes	42	M	–	1.14	–	47 88
15		g2210	150 mm diameter pipes	26	M	–	1.84	–	47 84
16		g2250	Bends; 100 mm	7	NR	–	0.96	–	6 72
17		g2255	Bends; 150 mm	2	NR	–	1.58	–	3 16
18		g2300	trapped gullies 150 x 150 mm top; 100 mm outlet and inlet	6	NR	–	2.19	–	13 14
19		g2310	Raising pieces 300 mm high	10	NR	–	1.31	–	13 10
20		g2350	Channels, white glazed; half section; 100 mm curved, 750 mm long	1	NR	–	4.37	–	4 37
21		g2360	Channels; white glazed; half section; 150 mm straight; 750 mm long	2	NR	–	4.37	–	8 74
22		g2370	Three-quarter section 100 mm branch bends	7	NR	–	1.31	–	9 17
		g340	Ceramic Tiles						
23		g3120	100 x 100 x 4 mm; Eggshell	32	M²	–	2.05	–	65 60
			C/F						5433 78

FIGURE 2

| STAGE 5 p.107 | ACTIVITY : BRICK AND BLOCK EXTERNAL WALL | RESOURCE : LABOUR c |
| OFFICE BLOCK 0001 | ELEMENT: EXTERNAL WALLS (21) | CONSTRUCTION : BLOCKWORK F |

Item	Locat	Code	Description	Qty	Unit	Hours	Rate	Total Time (Hours)	Total Cost £	
	FLOORS	60	Block Walls							
		F30	Blockwork							
		F350	Walls							
		F280	High Strength Blocks in CLM (1:1:6)							
1	03	F1150	200 mm thick	92	M²	2.13	2.17	195.96	405	23
2	03	F1152	Rough cutting	8	M	0.81	2.17	6.48	14	06
		F400	Piers							
		F280	High Strength Blocks in CLM (1:1:6)							
3	03	F2000	400 mm thick	1	M²	4.44	2.17	4.44	9	63
4	03	F2010	450 mm thick	1	M²	4.88	2.17	4.88	10	59
		F400	Piers							
		F280	High Strength Blocks in CLM (1:1:6)							
5	03	F2020	500 mm thick	4	M²	5.32	2.17	21.28	46	18
		F80	Sundries							
		F450	Damp Proof Courses							
		F210	Horizontal							
6	03	F1150	200 mm wide	71	M	0.19	2.17	13.49	29	27
7	03	F2000	over 225 mm wide	2	M²	0.40	2.17	0.80	1	74
			C/F	-				247.33	516	70

FIGURE 3

									Total Time (Hours)	Total Cost £	

ACTIVITY : BRICK AND BLOCK EXTERNAL WALLS   **RESOURCE** : COMMODITIES y

OFFICE BLOCK C001   **ELEMENT** : EXTERNAL WALLS (21)   **CONSTRUCTION** : BLOCKWORK F

Item	Locat	Code	Description	Qty	Unit	Hours	Rate	Total Time (Hours)	Total Cost £	
	FLOORS	60	Block Walls B/F					247.33	516	70
		f230	High Strength Blocks							
8	03	f2050	450 x 225 x 100 mm	2219	NR	–	0.17	–	377	23
		n250	Damp Proof Courses							
9	03	n2020	200 mm wide	10	Rolls	–	5.25	–	52	50
10	03	n2025	350 mm wide	0.50	Rolls	–	8.05	–	4	03
11	03	n2030	400 mm wide	0.50	Rolls	–	10.06	–	5	03
12	03	n2035	450 mm wide	0.50	Rolls	–	10.50	–	5	25
13	03	n2040	500 mm wide	0.50	Rolls	–	13.13	–	6	57
		q430	Cement, Lime and Sand Mortar (1:1:6)							
14	03	q4120	for 200 mm block walls	92	M²	–	1.12	–	103	04
15	03	q4130	for 400 mm block piers	1	M²	–	2.24	–	2	24
16	03	q4140	for 450 mm block piers	1	M²	–	2.52	–	2	52
17	03	q4150	for 500 mm block piers	4	M²	–	2.80	–	11	20
			C/F					247.33	1086	31

FIGURE 4

105

Thus, the bill of quantities, once re-sorted as above, forms the basis for planning and controlling the task of construction. Further, all the project information is fully cross-referenced by the co-ordinating syntax   provided by the SfB System.

CONCLUSIONS

The paper has shown that it is feasible to produce a bill of quantities as a fully co-ordinated sub-set of the project information which allows the identification of the resources needed together with their costs in order to establish a project. The paper also shows that it is possible to provide bills of quantities which can provide a dynamic tool in the planning and controlling of the task of construction.

It is suggested that unless bills can be produced which achieve the above requirements, then objective appraisals of design and construction solutions cannot be reliably executed, despite recent advances in cost modelling techniques. It is important that reliable cost data bases can be established and it is hoped that this paper has provided a direction to pursue in their establishment.

REFERENCES

1 FLETCHER & MOORE   "Standard Phraseology for Bills of Quantities, 4th Edition" 1979 George Godwin Ltd.

2 LOCAL AUTHORITY   "LAMSAC Coded Library of Descriptions
  MANAGEMENT SERVICES   for Bills of Quantities (SMM6)
  AND COMPUTER   LAMSAC 1979
  COMMITTEE

3 SEELEY, I.H.   "Building Economics" Macmillan 1972

4 SALZMAN, L.F.   "Building in England Down to 1540" Oxford Press 1952

5 R.I.C.S. & N.F.B.T.E.   "Standard Method of Measurement of Building Works 6th Edition" NFBTE & RICS 1979.

6 R.I.C.S. & N.F.B.T.E.   "Practice Manual SMM6" NFBTE & RICS 1979.

7 JAGGAR, D.M.   "The Quantification of Building Work at Resource Level" Ph.D. Thesis 1982 (In preparation)

8 CIB REPORT NO. 40   "An Introductory Guide to the use of SfB" An Foras Forbartha 1977.

9 JAGGAR, D.M.   op cit.

RESEARCH BASED INFORMATION: DEVELOPMENTS IN INFORMATION TECHNOLOGY

DOUGLAS ROBERTSON, Building Cost Information Service

1. Introduction

Information, whether it is cost or price, quantitive or qualitative,
contractual or descriptive, is a requirement common to all research
activities whether they are being conducted academically or for
direct application to a live project.
    The Building Cost Information Service has always aimed to cater
mainly for the second of these activities namely the practitioner
grappling to solve day-to-day problems. However since such a large
proportion of research is information collection and analysis BCIS
data are frequently called upon to assist with students' dissertations,
further degree studies and substantial research projects. BCIS
collects data within strictures laid down by the Quantity Surveyors
Divisional Council of the RICS, each subscriber having to clear with
the contracting parties, the client and the contractor, that it is
permissable to submit data to the Service. Permission is sought and
usually given on the understanding that the data has a limited
circulation and defined use. Data are normally made available for
research purposes when such requests are made to the BCIS Management
Committee. They have to be very careful about any application which
could lead to commercial exploitation of information submitted by
BCIS subscribers.

2. Research based information

2.1 Research, development and information go hand-in-hand: research
depends on good information; research has to be undertaken to
produce the information that is required; development is continuous.
Information collected and published in BCIS must surely be very
familiar to all researchers into cost and price information. Its
familiarity comes from its consistency, regularity and quantity.

2.2 The research and development which went into the preparation of
the Standard Form of Cost Analysis was spread over six years between
1963 and 1969 but having obtained the agreement of the profession
and the Government Departments it has ever since provided a sensible
and acceptable basis for collecting data in various degrees of detail.

All data collection is expensive and subscribers are unlikely to embark upon expenditure of about £500 to prepare a detailed cost analysis unless it is clearly going to be of use for their own internal office purposes. BCIS has recently undertaken further studies into ways of simplifying the classifications and instructions of the SFCA with a view to reducing the cost of preparing analyses. However, it has been decided to defer the adoption of these suggestions because they have been overtaken by the very rapid developments in communications technology which will be discussed later.

2.3 The widely used BCIS Tender Price Index also stemmed from careful research to produce a statistical method which, whilst undergoing continual refinement, has been rigorous enough to operate consistently since its base year of 1974. Like most good information, it is expensive to calculate depending as it does upon a statistical analysis of 25% by value of items in Bills of Quantities and requiring, for statistical acceptability, a sample of about 300 Bills each year. The reciprocal nature of BCIS is maintained through the TPI: each subscriber who submits a priced Bill of Quantities for indexing receives a brief report on the job's individual index and how it compares with the quarterly average. This is an important piece of ongoing statistical work and serves as a continuous reminder of how cautiously one should treat the rates which contractors place against Bill items.

2.4 The price information collected during the calculation of individual tender price indices is a valuable contribution to the BCIS price files which are systematically being put on computer. Once these files are complete, BCIS will be able to make its own research studies of the inter-relationship of many variables such as comparisons of contract size, contract period and tender level.

2.5 BCIS has not exclusively concentrated on information derived from Bills of Quantities and the "Cost File" published in the magazine "Building" has allowed the Service to build up its experience of labour and material costs. In the future BCIS will seek to bring together its cost information and its BQ price information so that there can be a better understanding of the determinants of costs and prices.

2.6 The British Insurance Association asked BCIS to research into house rebuilding costs so that it could improve the general guidance it gave to policy holders. The research identified the variations in house type, age, size, quality and location and built up models capable of embracing 456 permutations. The models are repriced every month to produce the index used for insurance index-linking and the annual updates of the published guides have made a major impact upon household insurance throughout the country.

2.7   The data collected by BCIS each year is of sufficient quality and quantity to merit regular statistical analysis. Information, derived from the concise and detailed analyses, the Tender Price Index record cards and the recently introduced £/m^2 questionnaire, forms the basis of the price files which for the past two years have been processed by a North Star Horizon micro-computer to produce a number of cost studies. The Location Study identifies different pricing levels throughout the country, the Preliminaries Survey analyses the trend in preliminary percentage additions of all the projects submitted to BCIS and the £/m^2 study produces ranges of current costs of most building types.

2.8   Quantity surveyors, and for that matter building contractors, will always find it extremely difficult to judge market conditions and the tender climate. The Tender Price Index and BCIS Cost Index have been combined to produce a measure of market conditions which give a fair but retrospective statement of events. Recently BCIS has become more involved in providing information to help its members to forecast tender prices and building costs over the coming 24 months. To this end models have been prepared which will calculate projections in response to variations of labour and material costs assumed over the coming period. Making assumptions about cost trends is difficult enough and this is not made easier when assessments also have to be made on the keeness of competition and other market condition factors. BCIS considers it important when making these forecasts that all subscribers have before them a clear statement of the assumptions that have been used so that they can make their own judgement of trends.

3.   Developments in Information Technology

3.1   A great deal of research and development work has been done and is being done which will help quantity surveyors make good use of new communications technology. BCIS has benefited enormously in its understanding of the constraints and the opportunities presented by these developments from significant research undertaken under Professor John Bennett at Reading University and Ron McCaffer at Loughborough University. BCIS were very pleased to receive a commission from the Property Services Agency to "design, develop and test a computer based data bank, storing and providing access to data for use by quantity surveyors for providing estimates of capital costs of individual building projects and for statistical analysis on aspects of elemental prices relevant to cost advice in connection with the Department's construction programme". Don Green's paper describes this work. At the time of writing, this project is nearing completion and the experience gained from it and the programs which it will provide are fundamental to the way the BCIS Management Committee are thinking about the future developments of the Service.

3.2   The quantity surveying profession is, as every profession has to be, dynamic: some of its ideas for its own development are shown by the accompanying contributions. Neither are other professions standing still. Quantity surveyors ignore at their peril the developments in the field of computer aided design (CAD) and even computer aided manufacturer (CAM). Both have to be looked at very carefully since they are both capable of devouring some of the quantity surveyors' activities.

3.3   No research and development moves faster than the current rate of progress being made in communication technology; the proliferation of new micro-computers almost daily shows the growth in this market. The BCIS experience using the North Star Horizon shows the considerable scope for micro-computers in quantity surveyors' offices. BCIS participation on "Prestel" has also shown how communications are going to be simplified, nationally and internationally. Micro-computers talking to main frame or mini computers holding centralised data is very much with us. The BCIS micro through a direct GPO telephone line, an acoustic coupler and an emulator (costing about £1,000) has been adapted to receive teletype information from a main frame computer. Direct access and off-line editing for Prestel is just as easy.

3.4   BCIS is heavily and very actively committed to research and development into various aspects of communications technology. Bearing in mind the approach adopted by BCIS towards research as indicated earlier in the paper, the Management Committee are also taking a cool look at the commercial aspects of major developments. Timing and cost are important and whilst it will still need an act of faith, the decisions are being made only after thorough researching. Not all BCIS subscribers are going to purchase micro-computers in the next 3 years or so and there will be a period where both a hardcopy (paper) service and electronic communications will have to be run in parallel. Costs have got to be exposed and considered and these are certainly going to be high for BCIS centrally and for the BCIS subscriber using the data base. The benefits have therefore to be clearly seen to be worthwhile.

3.5   Research is well underway within the following framework:

    (1)   Work closely with PSA (DQSS) to develop a cost data bank to allow computer access to existing data as a necessary first step which cannot be missed.
    (2)   Make available manipulative programs to allow the best use of existing data.
    (3)   Enhance the North Star Horizon micro-computer to give teletype reception of data from the PSA Bureau in the short term and simulate one of the approaches which could be adopted by quantity surveying practices.
    (4)   Develop a BCIS service which would permit micro-computers to communicate with the central data bank and also allow data to be manipulated locally, i.e. on the subscriber's micro not just on the main frame bureau.

(5) Store on the BCIS data bank a) all detailed cost analyses b) ancillary information i.e. indices and location factors and c) data from concise analyses.

(6) Develop a data bank to take into account new thinking on elements, quantity factors, approximate quantities and unit rates and develop manipulative programs to make use of this new data bank to assist in the more detailed levels of cost planning.

(7) Take advice to establish the correct hardware and related software for a BCIS commercially orientated service including research into the use of a) Gateway arrangements on Prestel b) direct links between micro and main frame or mini computers c) the use of a Bureau and d) use of a computer specifically for BCIS purposes.

3.6 The results of the researches at Reading and Loughborough Universities, along with other relevant research will be very care-fully considered before BCIS decides on the nature and detail of the data to be held for the later stages of cost planning (Item 6 above). Joint research with help from a computer software house has already been undertaken into alternative computer and communications techniques. BCIS has also sought advice from its subscribers on the use they already make of computers and the equipment they may already have or are contemplating. A register has been built up of existing software which would be of interest to BCIS subscribers. BCIS is also aware that in developing its plans for a central computerised cost data bank it should consider all of the professions potential computer uses as individual firms may be helped to invest in their own computers if they see a wide scope of applications. Suggestions are being sought on appropriate applications in four main areas a) using the central data bank of price information b) using computerised cost data c) other QS professional requirements and d) QS office management applications.

4. Conclusion

At the beginning of this paper it was suggested that a great deal of research time is devoted to information collection, sortation and analysis and that a fair amount of this comes through BCIS. The BCIS research and development programme described above is aimed at improving the supply, selection, matching and manipulation of data and it is intended that this will help both quantity surveying research and practice.

SECTION IV

COST PLANNING

VALUE ANALYSIS IN EARLY BUILDING DESIGN

JOHN R. KELLY, Heriot-Watt University

1. Introduction

Current research suggests that a large proportion of the cost of a
building is committed by the sketch design stage and there is
evidence to confirm that at this early design stage little cost
advice is offered to the designer. The designer alone determines a
cost effective solution from the infinite number of options
available by the comparison of the relative values of the
alternative possibilities. This exercise, termed value analysis and
illustrated in figure 1, draws upon a data base of intuition,
experience and 'rules of thumb'.

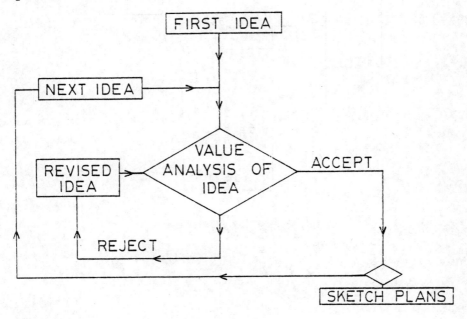

FIGURE 1   Value analysis in design

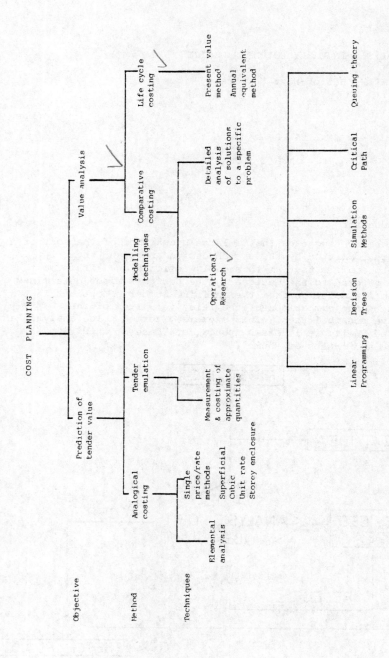

FIGURE 2   The objectives, methods and techniques of cost planning

Bathurst and Bulter (1) define cost planning as "the term used to describe any system of bringing cost advice to bear upon the design process". Thus "cost planning" is used as the generic term to cover the various techniques which have been derived for:

1) the prediction of tender value
2) the cost control of the design such that the anticipated tender value does not exceed the budget figure.
3) the comparative costing of differing solutions for components or elements of construction.
4) the prediction of the total ownership cost of a building, its subsystems, components and/or parts over its productive lifetime.

The relationship of the various objectives, methods and types of cost planning techniques are presented in a schematic manner in figure 2.

Methods are currently available which permit the prediction of tender value within known limits and considerable work is being undertaken to make this process more accurate. Specific methodologies for the application of value analysis to, and the realisation of a suitable data base for, building construction is an area within which little structured work has been done.

This paper will outline, firstly

a data base format which is useful to the designer at sketch design stage and secondly

a means of establishing a data base in the specified format.

2. An appreciation of building design

Quotations abound which relate architecture to art, however, were architecture to be purely art then any attempt to quantify, in mathematical terms, the value, the function and the benefits to the user would be in vain. The subjective analysis of building design as an art form certainly takes place and the "Beaux Arts" method of artistic appreciation was the only form of criticism used in architectural schools up until the post war period. Contemporary authors writing on the subject of architecture recognise the antipathy between artistic creativity and functional values. Broadbent (2) recognising this in design states, "design is primarily a means of resolving a conflict which exists between logical analysis and creative thought".

The inextricability of the cognition and mathematical evaluation of the artistic creativity of a particular designer is acknowledged, however, work has been undertaken to understand the process of design. Through the centuraries several have sought to influence this process, Vitruvius' 'De Architectura', Serlio's 'The Orders' (1537), and Durand's 'Precis' (1802) all lay down design rules. Lang (3) sets down a process of design which has a good correlation with the views of other authors and is reproduced in figure 3.

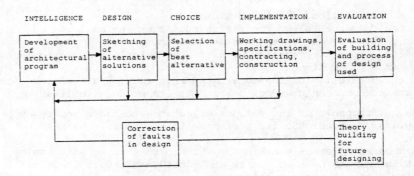

Figure 3   The design process

Levin (4) considers that the designer relates to three operations. The first is the identification of design parameters which are understood to be the factual limitations upon the building such as the shape of the site, the location, building regulations and the like.   The second is the identification of dependent variables which are those factors described or implied in the brief and are defined in terms of a performance specification.   These factors include the required space, accommodation, environment and quality. The third is the identification of independent variables which are not specifically specified but are to be considered in the design i.e. future fuel costs, future requirements for services, future styles of workplace configuration etc.

It is possible to suppose that the designer will follow the process outlined by Lang and will consider the parameters identified by Levin and bring to bear either written design rules subscribed to or rules which the designer has empirically derived from experience.

At the end of the nineteenth century an Italian economist Vilfredo Pareto developed a curve known as Pareto's law of distribution (Figure 4).   The law states that in a situation where a significant number of elements are involved, a small number of elements (20%) contain the greater percentage of costs (80%).

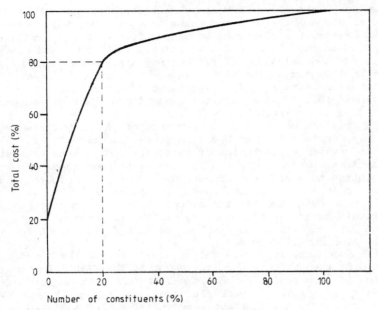

Figure 4   Pareto's Law of Distribution

This principle has been applied to the design process by several authors who point out that by the completion of 20% of the design (sketch design stage), 80% of the costs are committed.   This statement, whilst not being proved, attracts sufficient support from widespread authorities for it to be accepted as a base for further work.

Currently the designer carries out the evaluation of an idea in the development of the sketch design from a data base of intuition, experience and 'rules of thumb'.   If a data base were to be derived for components of construction in a format which is useful to the designer at this stage in the design then the result of the designers evaluation may be more cost effective.

3.   A study of existing value analysis techniques

Figure 2 identifies two procedures for comparative costing, namely the detailed analysis of solutions to a particular problem and operational research techniques.

Those procedures which are used to derive a solution to a specific problem rely upon comparative costing as an objective and a frame of reference against which comparisons may be made.   The following are examples.

1. Elemental analysis.   In the use of the accepted method of cost planning by elemental analysis comparative costing is triggered by an overspend on one particular element.   Alternative design solutions are sought which bring the predicted tender value within the target cost.   An inherent weakness in this method lies in the fact that

firstly, quality, appearance and the client's wishes are often sacrificed in the search for a reduction in cost. Secondly, the frame of reference, the elemental breakdown of the base analysis is comprised of data usually obtained in competitive tender and thereby the balance of the analysis will reflect any distortions made by the contractor in the structuring of the tender. Finally the search for an economic alternative is only triggered by an overspend, there is no provision in the technique for the economic analysis of the whole design or of elements which have met their targets.

2. Early Cost Advice, DOE/PSA Directorate of Quantity Surveying Services (5). For use at sketch design stage, the aim of this method is to give an appreciation of the cost consequences of various design solutions. Rates are given for the major components of each element relating to a base specification which may be adjusted and repriced by applying alternative rates from given tables. The technique may be criticised on the grounds that it is too complex for use at a time when the architect is developing the design very rapidly and also that since the tables are in a fixed format an optimum solution could be found.

3. Functional Analysis System Technique. Described by Macedo, Dobrow and O'Rourke (6) the technique promotes the identification of the function of each part of each element of construction which is then expressed in terms of a verb plus noun i.e. support floor. All possible solutions to the function are determined and analysed to achieve a cost effective design.

The above examples are methods of the detailed analysis of solutions based upon comparative costing, the commencement of the exercise being dependent upon a state of inconsistency with the frame of reference. Characteristic problems of the techniques relate to the number of alternative solutions to be considered, which must be limited and the considerable effort involved in evaluating each solution. Operational research techniques tend to overcome these problems by logically examining all alternatives within defined parameters.

Operational research evolved in the late 1930's as a scientific method of solution applied to practical problems. Originally the approach was more important than the method although the subsequent appearance of texts detailing specific techniques tended to result in the concept of operational research being understood as a set of techniques. Notwithstanding this the operational research objective is one which has been used by several researchers into building costs although a particular technique has rarely been used. Operational research demands the construction of a mathematical model to represent reality. The model may be statistically derived from finite examples of the element or may be configured to permit a simulation such that it yields a motion picture of the reality of the subject to be modelled.

Precedent to the construction of the model is the determination of the study objectives and the parameters within which those objectives may be met. The parameters form the boundary of the solution space. Brandon (7) illustrates the solution space (figure 5) within which an element or a building may be designed as an irregular circumscription

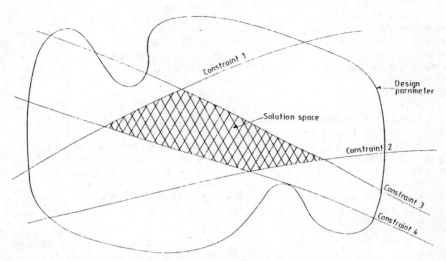

Figure 5   Diagramatic representation of the solution space

and the constraints imposed by the designer as lines over the space. It may be assumed that the boundary line identifies the design parameters and the constraint lines the dependent and independent variables as defined earlier. Authors accepting this basic method refer now to the search for the optimum, however, it is suggested that the attainment of the optimum is possible only where the following can be met:

1. Constraints must be objective. Constraints must represent an exact boundary outside of which the search may not continue. Subjective concepts are not valid as constraints with the present level of mathematics, however, advances in fuzzy set theory may soon change this situation

2. Constraints must be correctly defined. The constraints, within which the problem is to be solved, must include the optimum.

3. An infinite search is necessary. Within the constraints a test of each of the infinite possibilities must be made.

As a consequence of the latter requirement most models tend towards a sub-optimal solution, i.e. a solution realised within the rules laid down for the search. These rules must be carefully considered since the application of a particular rule of search may have a cost consequence which is a function of the rule and not the model.

In a study of operational research models of elements of construction the following display interesting characteristics.

1. The Wilderness Report (1964). This study is based upon the analysis of 1195 worked examples of hypothetical steel framed

buildings. The analysis is presented in the form of charts which permit a cost to be realised given, storey height, column spacing in the direction of the slab span, column spacing across the slab span and the loading of floors. The interesting aspects of the study lie in the facts that output is for comparative purposes only, that the model represents hypothetical rather than real data and that trends and indeed an optimum could be identified by an iterative study of the charts.

2. A cost model for lift installations. In developing the model Wiles (8) investigated the factors which determined the cost of a lift. The model is presented in the form of a series of linear regression equations and a cost may be realised by the application of, the number of floors served, number of persons accommodated in the car and the speed of the lift (high or low). The equations, however, are not conducive to the rapid realisation of costs. It is suggested that were the model to be run 324 times then all results for high and low speed lifts transporting between 8 and 16 people over between 3 and 20 floors could be tabulated on a single A4 sheet.

The above review of operational research techniques leads to the suggestion that the model itself is an inappropriate aid in the rapid realisation of the cost significance of design decisions if that model were capable of being repeatedly run between wide parameters to identify the cost trends associated with changing design. The repeated running of operational research models to identify trends may lead to the establishment of a data base which is useful to the designer at sketch design stage.

4. A methodology for the development of a data base in a format useful to the designer at sketch design stage.

In the first part of this paper it was established that designers draw upon general data in order to carry out the value analysis required by the development of the sketch design. In the second part various techniques have been described which are too specific or complex for the above evaluation, however, it is suggested that the analysis of the results of a repeatedly run model may lead to a data base of general rules.

The general rules derived from an analysis of the results of the lift model above show that, inter alia, significant cost increases occur at the 5th, 8th and 11th floor level. The lift model has an inherent advantage in terms of repeated running since the number of floors served and number of people to be accommodated are finite. In an element whose design is infinitely variable a problem occurs in the repeated running of the model.

The work of Goodacre, Kelly and Cornick (9) into the cost factors of dimensional co-ordination proved that the application of the rules of dimensional co-ordination to design had no measureable effect upon the cost of construction. The application of the rules of dimensional co-ordination to a model of an infinitely variable element would reduce the number of options to be considered from the infinite to the finite, albeit a large number.

From the above the following methodology was derived:

1. Determine study objectives and the parameters within which those objectives may be met. In design terms this entails the identification of the solution space.

2. Construct a simulation model for the analysis of any option within the solution space.

3. Modify the model to analyse a representative proportion of all options within the solution space. This may be achieved by a logical progression stepping in line with the rules of dimensional co-ordination.

4. Produce results in a form which is conducive to rapid analysis and draw general conclusions.

5. Validate the conclusions by re-running the same model with different parameters.

Experiments were conducted on a model of an in-situ reinforced concrete frame and upper floors designed in accordance with BSCP 110. The model designs the beams and floors, although the reinforcement calculations are radically simplified and produces approximate quantities which are priced at rates in Spons Architects and Builders Price Book. During the repeated running the model changes shape in step with the rules of dimensional co-ordination. The last experiment in the first series, carried out using a micro computer, took approximately 8 hours to calculate and print the 7992 priced options which fell within the parameters chosen. It is estimated that the approximate total number of options considered was 110,000. Later experiments were conducted using a Burroughs B6930 mainframe computer which carried out the same basic operation, but wrote the data to disk and used 5 minutes of CPU time. This data was subsequently analysed with the aid of a utility program, the Statistical Package for the Social Sciences. A high correlation was noted between total cost and area and total cost per square metre and floor span. It is concluded that in comparing the costs of reinforced concrete frames for buildings of differing areas area is a dominant factor, however, when a building of one area is being considered, which is normally the case, then the costs of reinforced concrete frames vary based on other factors of which floor span is dominant.

The general rules which may be interpreted from the various experiments concerned with a reinforced concrete frame are as follows:

1. Floor span is the dominant factor in the consideration of the total cost of an in-situ reinforced concrete frame.

2. With a simply supported floor cost increases as floor span increases above 3 to 3.6 metres and also cost increases as floor span decreases below 3 metres.

3. With a trough floor cost decreases as span increases with a constant cost for spans of between 9 and 20 metres (See figure 6).

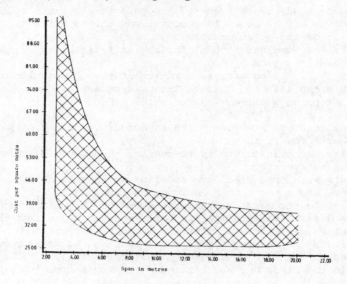

Figure 6   Area of 2664 results for frame costs

The above proves the applicability of the methodology to both a
lift installation and to a reinforced concrete frame. It is
suggested that the same methodology may be applied to other elements
of construction, however, further work is required in the analysis of
the inter-relationships between elements and how these may affect the
general rules. For example it is shown above that the cost per square
metre for a frame with a trough floor remains constant for a floor
span of between 9 and 20 metres, however, as span increases the floor
zone becomes deeper and the area of external wall becomes greater.

5. Conclusion

Levin (4) states "One major task facing building research workers is
the analysis of cost data to discover systematic cost variation with
changes in design parameters. Without this cost analysis will remain
a checking tool and real design with costs cannot take place".
    This paper has outlined an investigation into design methods and
determined that at the commencement of the design the parameters
within which the design shall develop will be identified. There is
a consensus of opinion relating to the development of the design
which confirms the iterative and empiric nature of that development.
Analyses of the value of alternative design options are taken by the
designer founded upon a data base of intuition experience and 'rules
of thumb'. It has been established that a large proportion of the
cost of a building is committed by the completion of the sketch
design and yet methods of cost planning most commonly used, which

involve the detailed study of alternative solutions, follow this critical stage. Several cost models have been developed which provide answers to 'what if' questions quickly,however it is suggested that the formulation of a model is an inappropriate place to stop in the provision of cost advice.

It has been proved that the repeated running of cost models provides data from which to form general rules which are of use to the designer at the sketch design stage.

6. References

1. Bathurst, P. E. and Butler, D. A. 'Building Cost Control Techniques and Economics' Heinemann 1977
2. Broadbent, G. H. 'Design method in architecture' Architects Journal 1966 September 14 Pages 679-684
3. Lang, J. 'A model of the designing process' from 'Designing for human behaviour' Lang, J. Et al Halstead Press 1974
4. Levin, P. H. 'Decision making in urban design' BRS current paper design series 49
5. DOE/PSA 'Early cost advice (B & CE elements) - offices, sleeping quarters' March 1977
6. Macedo, Jr. M. C., Dobrow, P. V. and O'Rourke, J. J. 'Value management for construction' Wiley 1978
7. Brandon, P. 'A framework for cost exploration and strategic cost planning in design' Chartered Surveyor Building and Quantity Surveying Quarterly Volume 5 No.4 Summer 1978
8. Wiles, R. M. 'A cost model for lift installations' The Quantity Surveyor May 1976
9. Goodacre, P. E., Kelly, J. R., and Cornick, T. C. 'Cost factors of dimensional co-ordination' Spon 1981

ALGORITHMS IN PRICE MODELLING THE EFFECT OF INCORPORATING PROVISION
FOR PHYSICALLY HANDICAPPED PEOPLE INTO EXISTING HOUSEPLANS

ALAN MORRIS, Polytechnic Of The South Bank

Summary

An explanation is given of the relationship between algorithms and
price models drawn from experience in incorporating provision for
physically handicapped people into existing houseplans, over a range
of different circumstances.
    The research method adopted was to build up information from a
simple base, classifying response patterns and developing algorithms
as a preliminary to the models.
    The use of specially generated data in the work avoided problems
which are often met when data from a sample of different building
projects are the basis in modelling.

1. Introduction

1.1 The purpose of this paper is to recommend the use of algorithms
(3.1)in price modelling, in certain circumstances. The particular
circumstances are those in which an additional design requirement is
imposed upon an existing solution. The studies underlying the paper
are in the field of housing for physically handicapped people. The
algorithms clarify what the issues are, suggesting variables for the
model (p.3).

1.2 Data generated for the purpose (p.8) were used in the study, as
opposed to data from a range of building projects. This source was
necessary in order for the work to be sustainable.

1.3 Beyond the immediate recommendation in this paper, there would
seem a range of circumstances in which algorithms can be useful in
price modelling.

1.4 The paper is phrased within the context of price prediction. In
addition to price prediction, the models were for identifying features
to look for [1]in picking out houseplans for adaptation to the special
needs of handicapped people.

1.5 The paper gives a brief introduction to the relationship between prediction, algorithms and price models; then an example from the studies underlying the paper.

2. Prediction

2.1 Times past saw an age of giants. Surveyors, giants at their art, gave expert witness to values without one scrap of back-up being proferred. And they got away with it.

2.2 In modern days that is changed. There must be assembled data, ready to hand behind the expert view. But plenty of room still remains for judgemental excursions: implicit to professional work, there is a bridging out into the uncertain.

3. Professional judgement, algorithms, models

3.1 Parallel to a strong tradition of professional judgement in prediction there is a move everywhere towards routine systems. The link between one thing and another is expressed in an <u>algorithm</u>: this term is used where any set of operations is reduced to a uniform procedure[2]. A familiar example is that computers solve problems by tracing through algorithms.

The algorithm may replace what was already dullest routine or a matter at first sight requiring complex judgement. The function of the algorithm is to define the steps in arriving at an answer and the relationship of one step to another, supplanting any subjectivity there might have been.

3.2 Also parallel to professional judgement is a move to statistical analysis: in which uncertainties may be pinned down to statistical measures and data in the prediction the better organised and understood in consequence.

So there has come into use <u>stochastic</u> models[3]. The term stochastic implies uncertainty, or that there is a limitation, expressed statistically, upon the reliability of the model. In the context of prediction the stochastic model may be used either where time or where understanding discounts a detailed approach: the purpose of the model may be to cut corners, or it may be to avoid problems. The model is essentially a simple expression picking out main influences.

The simplicity of the stochastic model may be its virtue: its sought-for end. Or the model may be only an interim measure until surer, less variable relationships are identified.

A price model describes the relationship between the price of something and factors that affect the price. A stochastic price model describing the relationship between price y and k variables $x_1, x_2, \ldots, x_k$ may be in the form of a linear regression equation:-

$$y_i = B_o + B_1 x_{i1} + B_2 x_{i2} + \ldots + B_k x_{ik} + e_i$$

(In the expression the B's are constants (parameters) and the e's random errors[4].)

   A model must have an acceptable reliability in order to give
credence. Broadly, the less the scope of the model then the greater
is its reliability. So that the level of reliability that is required
places a limitation upon the scope of the model.

3.3 Beyond the stochastic model, or suite of models, the rest is for
professional judgement in the prediction.

4. Relationship between algorithms and models

4.1 Algorithms stamp out an inexorable procedure. Stochastic models
make out a reflection of the broad position. An algorithm may be a
mass of detail whereas a model could be appreciated quickly.
Algorithm or model may be the basis for a prediction - each the
preferred method of approach in different circumstances.

4.2 In the majority of cases, the use of algorithms can extend to
only a part of the prediction. A stochastic model may penetrate
further, to where details are insufficient for an algorithm. The
scope and simplicity of the model are compensations for its variability.

4.3 In developing a model, algorithms may be an assistance: the
sequence in the algorithms - primarily the 'IF' statements (p.7) -
may suggest variables for the model. To take an example, the
Constituents of an algorithm in Figure 1 p.7 suggest the variable
'Partitions' in developing a model to predict the price of incorporating facilities for handicapped people in houseplans.
   A model which has been already set up may be improved through the
medium of algorithms. New or improved terms (see 5.8 p.7) or even a
reduction in the existing terms may be suggested: the purpose - toward
essentials in a simple well-aligned expression.
   What has been said of algorithms assisting in developing models can
also be true of models assisting in developing algorithms. If the
purpose is to go for predictions based upon algorithms then the terms
of a preliminary model will provide the focus.

5. Application

5.1 The recommendation here, to utilise algorithms in price modelling, comes from experience in adjusting houseplans to map out the
price of providing suitable housing for handicapped people[5].

5.2 An important element in adjusting the houseplans consists in
widening corridors and door openings at the expense of living and
storage space. In turn there may have to be floor area added to make
up the living and storage space that is lost to the corridor.
   Typically, there are adjustments in which a requirement for additional corridor space or door width has repercussions reaching out to
the perimeter of the houseplan.

5.3 Over a period, algorithms were developed reducing to a uniform procedure the different sequences of adjustments entailed by such repercussions.

5.4 Figure 1 (p.7) gives the constituents of an abbreviated algorithm for arriving at the length of the extra floor space required in sub-stituting a 900mm for a 800mm wide doorset (5.2). This serves the purpose of an example to show the nature of the algorithms described. In the study there are several alternative approaches included in mapping out the prices, of which making up space lost to the corridor is one. Further detail is available in the paper 'Algorithms and price models in Mobility Housing'[6].

5.5 There are four constituents of the algorithm in Figure 1: the originating adjustment; progress to the perimeter; the perimeter adjustment; and adjustments for furniture.

5.6 The following gives a guide to the chief of these constituents, and the general nature of the terms found in the algorithms.

OA - the originating adjustment

> This constituent defines the length over which living and storage space is transferred to corridor space at the immediate location of the doorset.

CL - a confined length

> This is a length of 800mm wide corridor, to be extended to 900mm wide:-

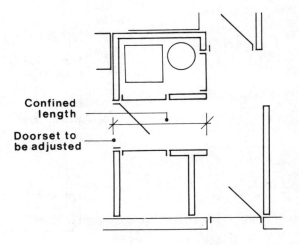

Confined
length

Doorset to
be adjusted

SL - a side length: a length of partition extending beyond one
side of the confined length.

Adjustment at the side length is
necessary in order to facilitate wheelchair access to the confined length:-

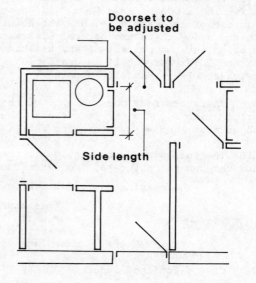

ML - a manouvre length: a length of partition that would obstruct
wheelchair turning in to the 900mm wide doorset.

When the doorset width is increased
the approach must be extended across the
extra width:-

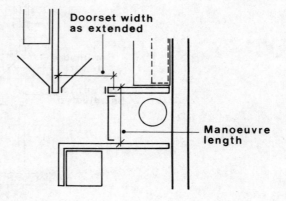

PP - progress towards the perimeter

This constituent comprises all the
alterations to the originating length
that occur in extending out to the
perimeter:-

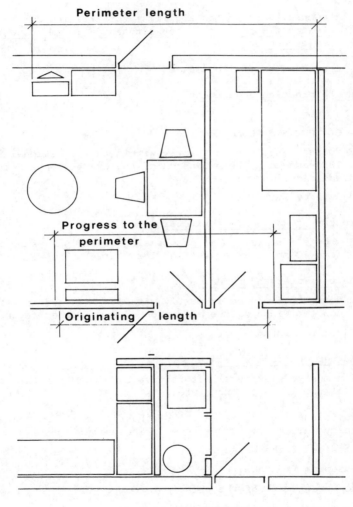

BASIS OF SKETCHES:
SINGLE STOREY HOUSING
DESIGN GUIDE
NATIONAL BUILDING AGENCY 1971

P - partition

PA - the perimeter adjustment

> This constituent gives the total length over which an extension of the floor area is required in consequence of substituting the 900mm wide doorset.

Figure 1. <u>Constituents of an algorithm for arriving at the total extra floor space in substituting a 900mm for a 800mm wide doorset</u>

x = length of extra floorspace

. <u>The originating adjustment</u>

x = CL

IF SL THEN add 900mm to x

IF doorset adjoining or partly within this additional 900mm THEN add to x the extension of the doorset beyond the additional 900mm.

Adjust for furniture*.

. <u>Progress towards the perimeter</u>

IF x cuts a partition in progressing to the perimeter THEN (add 100mm to x; adjust for furniture*).

. <u>The perimeter adjustment</u>

IF doorset adjoining or partly within x THEN add to x the doorset width or the extension of the doorset beyond x.

Adjust for furniture*.

. <u>*Adjustment for furniture</u>

IF x must cut a piece of furniture THEN add to x the extension of the piece of furniture.

5.7 Provisional variables in the stochastic model had been identified already (4.3 p.3). These were, principally:

. the size of the houseplan

. houseplan frontage

. the arrangement of the principal rooms.

5.8 In consequence of the algorithms these variables were considerably changed. For example, the location of the various cupboards and the heater compartment in the houseplan had an influence upon the adjustments not at first recognised. Conversely, the arrangement of the principal rooms within the houseplan was not a consistent influence throughout but only within certain conditions.

The influence of the algorithms upon the models is, as described (see 4.3 p.3), through the individual terms of the algorithms. The algorithm in Figure 1 identifies the confined length as an element in the originating length (p.7). A confined length is generally found in the context of access to one or more stores, cupboards or compartments. The term confined length in the algorithm led to the significance of the positioning of minor rooms referred to.

Familiarity with the adjustments, built up in the course of devising the algorithms, also played a part.

5.9 So the variables were altered, reflecting a more developed view.

6. Data

6.1 Price data from a range of building projects have generally been the basis of building price modelling in the past. Often models derived from this basis have not proved useful[7]. Experience shows that many building contract data have too much unexplained for use in modelling. And often individual data are missing. The model is pitched at the level of generality of the available data with little of a say in the make-up of variables. There is scant question of substituting new variables in order to improve the model.

In the face of these impediments what is desirable may give place to what is available. The screening of variables may have to be in the sole discretion of the statistical procedure[8].

6.2 To find a suitable alternative basis it is necessary to generate data in a controlled exercise, the models defined by reference to the control conditions. That is the approach taken in the studies which form the basis of this paper.

6.3 The method of using algorithms in price modelling recommended in the paper is likely to be best adapted to studies in which purpose-generated data are used. Without such data there may not be sufficient information or control for the method to be appropriate.

7. Conclusions

7.1 The terms in predictive price models should reflect, as closely as possible the factors that are of predominant effect upon the price.

7.2 In the building context the factors that are of predominant effect upon the price are those which, at the level of generality of the model, immediately govern the amount and nature of the construction work.

7.3 In circumstances in which alterations to existing houseplans are made there may be required a series of adjustments extending out to the perimeter.

7.4 When these adjustments are reduced to a uniform process - an algorithm - the IF statements (p.7) give factors governing the construction work (7.2) resulting from the alteration: the IF statements may suggest variables for the required model. The purpose in preparing the model is for its simplicity, and scope extending beyond the algorithm.

7.5 Considerable information and control is necessary for modelling at the level of detail described[9]. For the purpose, specially prepared data are likely to be required, accepting the limitations that this implies.

8. Acknowledgement

8.1 Although I remain responsible for opinions expressed in this paper I am indebted to Dr. Derek Fisher, Polytechnic of the South Bank, for his advice in its preparation.

References

1. See Draper, N.R. and Smith, H., Applied Regression Analysis, 1966, p.234.

2. Encyclopaedia Britannica, Algorithm, 1972, Vol.1. p.630.

3. See Johnston, J., Econometric Methods, 1972, p.8. et seq.

4. Agha, M., Multiple Regression Analysis, course notes, Thames Polytechnic, 1973.

5. See Morris, A.S., Price of Mobility Housing in Single Storey Accommodation, 1981.

6. See Morris, A.S., Algorithms and Price Models in Mobility Housing, 1982.

7. See Morris, A.S., Regression Analysis in the Early Forecasting of Construction Prices, M.Sc (Research) thesis, Longborough University of Technology, 1976.

8. Draper and Smith, op cit, p.237.

9. Morris, A.S., Assessing Cost-effectiveness in House Designs for Physically Handicapped People, CIB Symposium Quality and Cost in Building, Lausanne 1980, Vol.2. p.128.

THE ESTABLISHMENT AND WORK OF A COST RESEARCH DEPARTMENT IN A PROFESSIONAL OFFICE

ROB SMITH, Davis Belfield and Everest

Unlike the academic world there are no grants or funds
available for research and development work carried out
in private practice, so that the professional practitioner must finance such activity from fee income.
However, this constraint does at least provide the
necessary pressure to ensure that the research focuses
on a definable end product. Such an approach may involve
compromises that the academic purist would find unacceptable. However, the commercial needs of private practice
are such that research and development work must satisfy
the prorities of the day. It must also be acknowledged
that it is the business of private practitioners to
provide advice to their clients which, as far as cost
advice is concerned, must facilitate real commercial
decision making. Although such advice may be founded
on extensive statistical appraisals, at the end of the
day it must be presented in absolute terms, which
precludes the academic security of means, modes and
standard errors.
    As far as DB&E are concerned we are a professional
firm of chartered quantity surveyors with our head
office in London employing approximately 200 staff. We
have a further 11 regional offices and total staff
numbers of in excess of 350 people. Our workload is
derived from both private and public sectors so that our
experience encompases almost every conceivable type of
building, which in terms of cost advice provides a
broad cost data base. However, as with any large
organisation with considerable diversity of activity,
it is of paramount importance that the available cost
information is centrally collated if the feedback from
pockets of knowledge and expertise is to be disseminated
throughout the firm.

In order to initiate this very important function the
partnership established a cost information section.
That was over 15 years ago and at that time demanded
the part time attention of an associate. A library
was established and the foundations laid for our tender
index. Since those early days, the role and importance
of the department has grown, as can be witnessed by its
current size and responsibility. The Data department
which is our modern equivalent of the cost information
section, now requires the full time effort of 5
surveyors, a librarian and is supported by a mini
computer.

The department's areas of activity continue to involve
the processing of the cost information generated by the
firm, while the establishment of a nucleus of expertise
in this field has also resulted in research consultancy
work.

From the very early days of the cost information
section the research work undertaken by the department
concentrated on the areas that formed the focus for
private sector investment. This resulted in cost studies
into car parking, hotels and offices, which in the early
70's were building types that formed the backbone of
the property boom. Apart from the cost information
generated from within the firm, contact was made with
the industry, investors and operators, in order that
the financial implications of design, construction,
investment and occupation could be fully understood.

It was during this period that a change in inflationary trends occurred which highlighted the need to
reappraise the use of the two Spon's titles that the
firm edited at that time, Architects' and Builders'
and Mechanical and Electrical Services Price Books. The
department commenced the indexing of bills of quantities
and initiated the Davis Belfield and Everest tender
index, which was subsequently published for the first
time in the AJ, 12 November 1975. The index has enabled
the firm to use the Price Books as extremely flexible
pricing schedules to the extent that the rates can be
adjusted to reflect trade, regional, value and complexity
price variations. With inflation becoming an
increasingly significant element in building investment
decision making it was apparent that our clients
required a cost and tender forecast. Since the tender
index was first published in the AJ we have carried out
a considerable amount of research to identify construction indicators that could be used to forecast future
trends. Some of the indicators we have monitored were
illustrated in the AJ, 25 February 1981, Example A.

Unfortunately despite the advance in cost planning techniques in the late 50's and early 60's, the change in pace of development in the 70's was not accompanied by a significant improvement in the techniques deployed by the professional quantity surveyor.

Recognising the need for advice that would enable clients to make decisions as to investment options, we decided to develop a rational approach to estimating. Our early building studies had laid down the ground rules for this type of advice at the feasibility stage. During the development of the concepts embodied in these studies we prepared a series of articles for the AJ entitled Initial Cost Estimating, that considered the following building types:

Buildings	Published in AJ
Warehouses	1.6.77
Factories	14.9.77
Housing	8.2.78
Flats and maisonettes	31.5.78
Shops	15.11.79
Hotels	21.3.79
Offices	11.7.79

While these articles were not intended to provide absolute answers, they did assist in the evaluation of building options in real terms. The facility for such evaluation is illustrated in the extract from the Warehouse article, Example B.

This method of estimating has proved extremely successful. In order to enhance the technique, the use of a computer was essential, so early in 1981, we purchased a Data General mini computer. Initially, our work on the computer concentrated on the establishment of a cost data base using data from our own projects and published cost analyses. Example C is a printout of a sample of the industrial buildings held on the computer. The programme contains the facility to define a sub-set and examine the cost significance of a selected design factor.

The Initial Cost Estimating approach has also been developed on the computer in the form of a cost model. Example D is a printout of an estimate for an office block.

In order to complete the financial appraisal of a capital investment, we have also prepared a computer programme to calculate investment and development yields as well as establishing residual values. Example E illustrates the calculation of site value.

In addition to investment appraisals, many of our clients
require cash flows. In this respect we have adapted
the DHSS research work in this field to suit our
specific requirements. Two examples of the modifications
we have made are the ability to impose a contract
completion date and the inclusion of fluctuations in
the cash flow calculations. Example F is a printout
from this programme that illustrates the various monthly
and cumulative values we are able to provide.

We have various areas of ongoing research, some of
which are unique to DB&E, others have already been
considered by researchers in other organisations. These
include:

1.   The relationship of preliminaries with the prevailing
tender market conditions, Example G.

2.   The spread of tenders, Example H.

3.   The distribution of job tender indices, Example J.

together with trade price levels, levels of fluctuations
recovery, and the relationship of labour and sundry
additions to unit prices.

All of the research work undertaken by our Data
department is of direct relevance to the operation of
our practice in that it must service our day to day needs.
As such it could be said that our approach lacks some
of the advantages attributed to pure research to the
extent that we limit both time and staff resources. In
answer to this criticism we are able to reply that our
clients provide an immediate feed back from our applied
research in that they form a far more rigourous testing
ground than could be created by any amount of simulation
or sensitivity analysis.

Building costs

Forecast of building cost and tenders

CI/SfB | | | | (Y2)

Cost prices
CI/SfB (1976 revised) (Y2)

Prospects for employment in 1981 are not good, indeed the amount of work available in the foreseeable future may be less than even the most pessimistic envisaged. As well as updating the last forecast (AJ 1.10.80), more general indicators are offered here, similar to those last year (AJ 16.1.80) which suggest general trends. The indicators are the market factor, contractors' new orders, AJ classified advertising, and company liquidations. The forecasts are by DAVIS BELFIELD and EVEREST, chartered quantity surveyors.

Work available?

A year ago, (AJ 16.1.80 p151) we predicted that tender prices would begin to level out and suggested that the average tender index for the third quarter of 1980 would be 192 (1976 = 100) or 462 (1970 = 100). In the event this forecast turned out to be absolutely correct. We had our doubts in the two intervening articles when tender levels continued to rise unabated.

However, there is now every sign that the volume of work available to the industry in the foreseeable future will be less than even the most pessimistic envisaged, and this appears to be reflected in current tenders. In the last few months there has been a dramatic downturn in tender levels. This has resulted in a much lower index for the third quarter of the year, lower than had been anticipated in July 1980, which in turn necessitates revisions to our forecasts for 1981 and 1982 made at that time. The situation augurs well for potential clients, if they have confidence in their own futures.

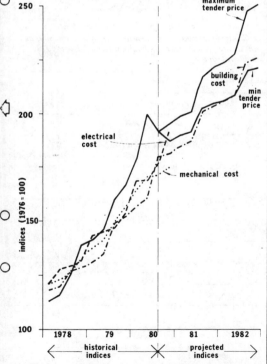

1 *The four graphs previously presented separately are here combined for easier comparison. The projected indices show some slow down in the rate of inflation.*

Table I Building costs* and tender price† indices

Year	Quarter	Building cost index (1976 = 100)	Tender price index (1976 = 100)	
1979	1	131	142	
	2	135	146	
	3	149	160	
	4	153	167	
1980	1	157	179	
	2	161	200	
	3	180	192	
			Min	*Max*
	4	182	188	195
1981	1	185	190	199
	2	187	192	201
	3	201	203	217
	4	204	205	221
1982	1	206	206	224
	2	209	209	228
	3	224	220	248
	4	226	221	251

Table II Mechanical and electrical cost indices

Year	Quarter	Mechanical cost index (1976 = 100)	Electrical cost index (1976 = 100)
1978	1	121	121
	2	123	128
	3	126	129
	4	133	132
1979	1	138	143
	2	141	145
	3	150	147
	4	157	154
1980	1	165	169
	2	171	169
	3	172	177
	4	175	192

* 'Building costs' are the costs actually incurred by the builder mainly in respect of labour and materials. The index reflects the variation in these major elements assessed from a contract wages sheet revalued for each variation in labour costs and the indices prepared by the Department of Trade and Industry for material prices.
† 'Tender price' is the price the building owner has to pay and the index reflects the level of pricing contained in the lowest tenders on a fluctuating basis for new work in the London area and takes into account market conditions as well as building costs.

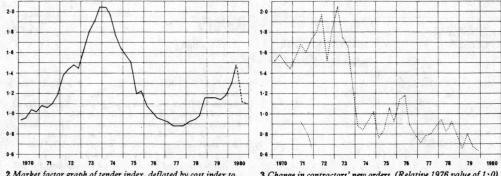

2 *Market factor graph of tender index, deflated by cost index to indicate movement of tender price relative to building costs. (Relative to 1976 value of 1·0).*

3 *Change in contractors' new orders. (Relative 1976 value of 1·0).*

4 *Change in AJ classifieds. (Relative to 1976 value of 1·0).*

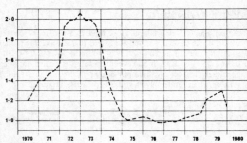

5 *Change in contractors' insolvencies. The higher the number on the vertical scale the fewer the insolvencies. (Relative to 1976 value of 1·0).*

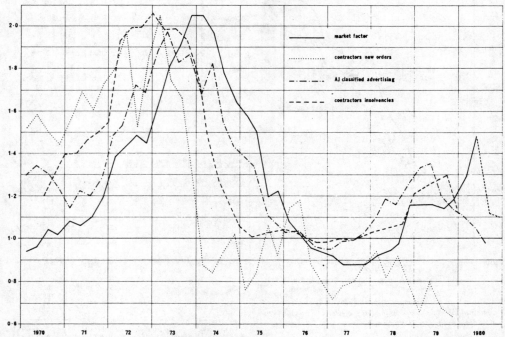

— market factor
.... contractors new orders
—·—· AJ classified advertising
– – – contractors insolvencies

6 *Combination of graphs 2 to 5. New orders are followed by changes in AJ classified and insolvencies. Market factor indicating tender price follows on 6-12 months later.*

DESIGN/COST SELECTION CHART   WAREHOUSES

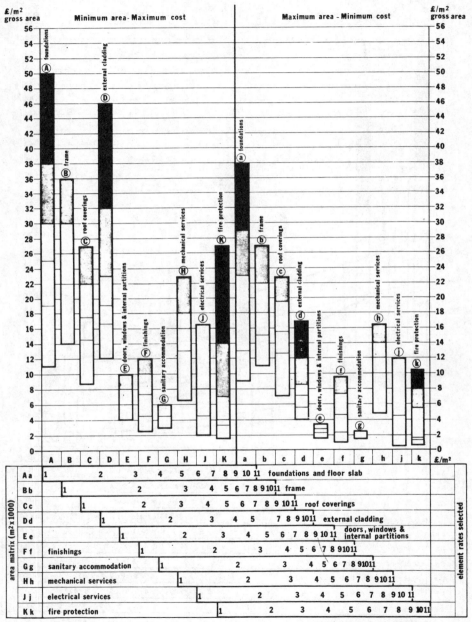

£/m² gross area — Minimum area - Maximum cost   Maximum area - Minimum cost — £/m² gross area

		A	B	C	D	E	F	G	H	J	K	a	b	c	d	e	f	g	h	j	k	£/m²
	A a	1			2		3	4	5	6	7	8	9	10	11	foundations and floor slab					
	B b		1			2		3		4	5	6	7	8	9	10	11	frame			
	C c			1		2		3		4	5	6	7	8	9	10	11	roof coverings			
area matrix (m² x 1000)	D d				1		2		3	4	5		7	8	9	10	11	external cladding			
	E e					1		2		3	4	5	6	7	8	9	10	11	doors, windows & internal partitions		
	F f	finishings					1		2		3	4	5	6	7	8	9	10	11		
	G g	sanitary accommodation					1		2		3		4	5	6	7	8	9	10	11	
	H h	mechanical services						1		2		3		4	5	6	7	8	9	10	11
	J j	electrical services							1		2		3		4	5	6	7	8	9	10	11
	K k	fire protection								1		2		3		4	5	6	7	8	9	10	11

element rates selected

Job name = _____
Site location = _____
Gross floor area = _____
Final estimate = £ _____
(excluding external works)

rate/m² for selected warehouse = £ _____
extra for additional specification = £ _____

adjusted by site location factor ⊗   = £ _____
update by × estimate date A J tender index _____
            June 1977 AJ tender index

∴ final estimated rate/m² for building = £ _____

DESIGN/COST SELECTION CHART   *worked example*   WAREHOUSES

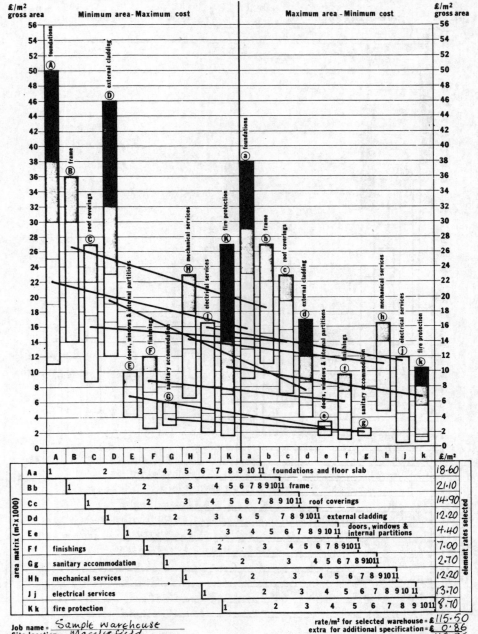

£/m² gross area — Minimum area - Maximum cost | Maximum area - Minimum cost — £/m² gross area

		A	B	C	D	E	F	G	H	J	K	a	b	c	d	e	f	g	h	j	k	£/m²
A a		1		2		3	4	5	6	7	8	9	10	11		foundations and floor slab						18.60
B b			1		2		3	4	5	6	7	8	9	10	11	frame						21.10
C c				1		2		3	4	5	6	7	8	9	10	11	roof coverings					14.90
D d					1		2		3	4	5		7	8	9	10	11	external cladding				12.20
E e					1		2		3	4	5	6	7	8	9	10	11	doors, windows & internal partitions				4.40
F f		finishings		1		2		3		4	5	6	7	8	9	10	11					7.00
G g		sanitary accommodation		1		2		3		4	5	6	7	8	9	10	11					2.70
H h		mechanical services		1		2		3		4	5	6	7	8	9	10	11					12.20
J j		electrical services		1		2		3		4	5	6	7	8	9	10	11					13.70
K k		fire protection		1		2		3		4	5	6	7	8	9	10	11					8.70

area matrix (m²x1000) — element rates selected

Job name = Sample Warehouse
Site location = Macclesfield
Gross floor area = 3,500 m²
Final estimate = £383,000
(excluding external works)

rate/m² for selected warehouse = £115.50
extra for additional specification = £ 0.86
= 116.36
adjusted by site location factor × 0.89 = £103.56

update by × $\frac{\text{estimate date AJ tender index } 265}{\text{June 1977 AJ tender index } 251}$

∴ final estimated rate/m² for building = £109.33

DB&E Ref	Building Type	Tender Date	Contract Value	GFA sq.m	Cost/ sq.m	No. of Storeys	Location	Tender Method	Fix / Fluct	Data Source	Functional Unit	
I/156	Industrial Building	08/80	409702	1617	181	1	Lothian Region	Select	Fix	BCIS (D)	1577	Usable
I/148	Industrial Building	11/79	305270	1479	173	1	Hampshire	Select	Fix	BCIS (D)	1271	Usable
I/145	Industrial Building	09/79	1072777	5838	149	1	Buckinghamshire	Select	Fluct	BCIS (D)	N/A	
I/134	Industrial Building	05/79	203896	1362	122	1	Nottinghamshire	Select	Fix	BCIS (D)	1320	Usable
I/143	Industrial Building	04/79	785046	2469	183	1	Buckinghamshire	Select	Fix	BCIS (D)	2343	Usable
I/160	Industrial Building	02/79	275999	716	322	1	Strathclyde	Select	Fix	BCIS (D)	706	Usable
I/159	Industrial Building	08/78	397220	1290	249	1	Strathclyde	Select	Fix	BCIS (D)	1240	Usable
I/137	Industrial Building	08/78	382603	1874	126	1	Nottinghamshire	Select	Fix	BCIS (D)	1788	Usable
I/130	Industrial Building	07/78	729062	4652	120	1	Nottinghamshire	Select	Fix	BCIS (D)	4523	Usable
I/124	Industrial Building	06/78	1155496	5670	164	1	Strathclyde	Select	Fix	BCIS (D)	5198	Usable
I/123	Industrial Building	04/78	1213846	5312	185	1/2	Buckinghamshire	Select	Fix	BCIS (D)	N/A	
I/158	Industrial Building	03/78	91048	201	386	1	Orkney Islands	Select	Fix	BCIS (D)	163	Usable
I/129	Industrial Building	11/77	381021	1083	155	1	Nottinghamshire	Select	Fix	BCIS (D)	1012	Usable
I/110	Industrial Building	07/77	582304	3116	131	1	Sussex	Select	Fluct	BCIS (D)	2833	Usable
I/111	Industrial Building	05/77	197156	1048	148	1	Durham	Select	Fix	BCIS (D)	994	Usable
I/157	Industrial Building	03/77	106365	243	368	1	Western Isles	Select	Fix	BCIS (D)	196	Usable
I/121	Industrial Building	02/77	912421	3450	230	1	Warwickshire	Select	Fix	BCIS (D)	3224	Usable
I/117	Industrial Building	11/76	301803	1208	200	1	Buckinghamshire	Select	Fix	BCIS (D)	N/A	
I/109	Industrial Building	07/76	208751	879	209	1	Tyne and Wear	Select	Fix	BCIS (D)	N/A	
I/120	Industrial Building	07/76	266028	956	213	1	Buckinghamshire	Select	Fix	BCIS (D)	881	Usable
I/102	Industrial Building	06/76	966058	5563	149	1/2	Buckinghamshire	Select	Fluct	BCIS (D)	5563	Usable
I/119	Industrial Building	06/76	252368	1079	191	1	Lothian Region	Select	Fix	BCIS (D)	984	Usable
I/101	Industrial Building	05/76	367238	1923	168	1	Derbyshire	Select	Fix	BCIS (D)	1814	Usable
I/100	Industrial Building	03/76	711976	107	370	1	Hampshire	Select	Fluct	BCIS (D)	N/A	
I/104	Industrial Building	12/75	462213	1843	189	1	Tyne and Wear	Select	Fix	BCIS (D)	1775	Usable
I/089	Industrial Building	12/73	453626	2510	126	1	Lothian Region	Select	Fluct	BCIS (D)	2510	Usable
I/084	Industrial Building	12/73	137881	685	169	1	Clwyd	Negot	Fix	BCIS (D)	N/A	
I/088	Industrial Building	11/73	449874	2852	120	1	Cambridgeshire	Select	Fix	BCIS (D)	2714	Usable
I/093	Industrial Building	04/72	447325	3171	119	1	Strathclyde	Negot	Fix	BCIS (D)	3136	Usable
I/092	Industrial Building	12/71	688911	5592	108	1	Strathclyde	Select	Fix	BCIS (D)	5176	Usable
I/080	Industrial Building	09/71	564709	4735	108	1	Strathclyde	Select	Fix	BCIS (D)	4528	Usable
I/079	Industrial Building	08/71	190092	566	302	1/2	South Glamorgan	Select	Fix	BCIS (D)	458	Usable
I/081	Industrial Building	08/71	675680	3880	140	1/2	Buckinghamshire	Select	Fix	BCIS (D)	3340	Usable
I/076	Industrial Building	05/70	393341	2239	146	1	Strathclyde	Negot	Fix	BCIS (D)	2115	Usable

* All costs are updated to 6/1982 - Tender Price Index 193
  Costs per square metre include preliminaries and contingencies, but exclude external works

EXAMPLE C

INITIAL COST ESTIMATE
==================

Job Name :- Office Building Example
Job No. :- n/a
Gross Floor Area :- 5000 m2
No. of Storeys :- 4 no.
Plan Area :- 1250 m2

O.L. Tender Index :- 204
Regional Variation :- -7 %

Element	Unit Quant M2IND	Av.Unit Rate £	Cost/m2 GFA £	Total Cost £
Substructure	1250	100.00	25.00	125000
Frame to roof	1250	13.70	3.43	17125
Upper Floors & Supporting Frame	3750	85.00	63.75	318750
Roof	1250	65.00	16.25	81250
External Walls	2762	225.00	124.25	621450
Windows & External Doors	1488	225.00	66.95	334800
Internal Walls & Partitions	1750	55.00	19.25	96384
Wall Finishes	6262	9.10	11.40	55984
Floor Finishes	5000	34.00	34.00	170000
Ceiling Finishes	5000	17.00	17.00	85000
Fittings & Furnishings	5000	8.00	8.00	40000
Sanitary & Disposal Installation	5000	7.00	7.00	35000
Heating & Water Installations	5000	6.00	6.00	30000
Ventilation & Air Conditioning	5000	105.00	105.00	525000
Electrical Installation	5000	80.00	80.00	400000
Lift Installation	2	32000.00	12.80	64000
Special Services	5000	4.00	4.00	20000
EXTERNAL WORKS	5.00%	-	30.20	151030
PRELIMINARIES	15.00%	-	95.15	475740
TOTALS			729.47	3647379

SPECIFICATIONS

Substructure	- Good ground - bases/strips up to 12 storeys
Frame to roof	- RC for open plan offices from 7 to 12 storeys
Upper Floors & Supporting Frame	- RC for cellular offices from 13 to 18 storeys
Roof	- Lightweight - wood on corrug. decking + insulation
External Walls	- Hand made facing brick & block cavity walls
Windows & External Doors	- Double glazed purpose made hardwood doors
Internal Walls & Partitions	- Blockwork or stud & hardwood doors
Wall Finishes	- Plaster & vinyl throughout
Floor Finishes	- Screed & vinyl tiles throughout
Ceiling Finishes	- Plaster & emulsion to soffits
Fittings & Furnishings	- Medium quality - reception desk/shelves/cupbds etc
Sanitary & Disposal Installation	- Normal services for medium rise building
Heating & Water Installations	- Only water supplies
Ventilation & Air Conditioning	- Variable air volume system to small office
Electrical Installation	- Open plan owner occ. - high qual. light & power
Lift Installation	- Passenger lift for 3/5 storeys
Special Services	- Firefighting - hosereels etc.
External Works & Drainage	- Restricted urban site

STAFF NO. 402
DATE 7/6/1982
DAVIS BELFIELD & EVEREST

EXAMPLE D

DEVELOPERS BUDGET — DATA

Prime Offices
Office Building Example

DEVELOPMENT COSTS

CONSTRUCTION:-
```
Demolitions                                              50000
Advance Works                                           100000
Buildings:  5000 m2 x   660 =  3300000        0
               0 m2 x     0 =        0        0
               0 m2 x     0 =        0
                                                       3300000
External Works                                          200000
                                                       3650000
                                                        511000
Professional Fees     14.0 %                           4161000
                                                        104025
Developers Contingency  2.5 %                          4265025
                                                        682404
Finance 16.0 % over 24 x 1/2 months                    4947429
                                                       =======
```

SITE:-
```
Site Purchase                                              0        0
Acquisition fees & Expenses  4.0 %                         0        0
                                                       -------
                                                           0
                                                       -------
Finance 15.0 % over 36 months
Prior to construction       6
Construction period        24
Letting period              6
                          ----
                          36 months                        0
                                                       -------
```

MARKETING:-
```
Promotion/advertising    5.0 %                          28632
  of first years rental
Letting fees 10.0 % of first years rental               57265
                                                        85897
                                                       -------
```

DEVELOPERS PROFIT:-
```
Construction                    4947429
Site                                  0
Marketing                         85897
                                5033326
Profit @ 20.0 % of development cost                   1006665
                                                      -------
```

DEVELOPMENT VALUE

** Building 1 **
```
Gross Floor Area   53820 sq.ft      Annual Rent/sq.ft  14.00
LESS  20.0 %       10764sq.ft       LESS   5.0 %        0.70
Lettable Area      43056 sq.ft  @   Net rent/sq.ft     13.30

Annual Income                        572645
LESS Ground Rent                      10000
                                    -------
Net Annual Income                    562645
@ Yield 5.00 %                    x   20.00
Gross Development Value            11252900
                                  --------
```

RESIDUAL SITE VALUE

```
Development Value                             11252900

LESS Development Costs
Construction             4947429
Marketing                  85897
Developers Profit        1006665              6039991
                                             -------
LESS Profit div.(1 +20.0 /100 %)             5212909
                                              868818
                                             -------
                                             4344091
LESS Finance - interest @15.0 % pa
     over   36 months div.(1 + 52.1 /100 %)  1488015
                                             -------
                                             2856076
LESS Acquisition Fees & Expenses
     div.(1 + 4.0 /100 %)                     103843
                                             -------
SITE MARKET VALUE                            2745227
                                             =======
```

EXAMPLE E

CASH FLOW FORECAST & PREDICTION OF CONTRACT PERIOD

Job Name	OFFICE BUILDING EXAMPLE
Job Number	NOT AVAILABLE
Tender Date	NOT APPLICABLE
Contract Value	£ 3647379
Retention Percentage	3 %
Last Valuation: No.	9
Date	1/6/82
Value of Work	£ 750000
Value of Flucts.	£ 15000
Materials off-site	£ 0
Fluctuating Contract?	YES
Base Month	8/81 (INDEX 195)
Start on Site	1/9/81
DB & E Indices ?	YES
Clause 38/39/40(31B,31A,31F)	CLAUSE 40 (31F)
Abatement/Non-adj. Element	10 %
Prediction of Contract Period	30.7 MONTHS
Imposed Contract Period	26.0 MONTHS

(0)	(1)	(2)	(3)	(4)	(5)	(6)	(7)	(8)	(9)	(10)	(11)	(12)
VALT'N NO.	MONTHLY VALUE WORK(G) £	CUMLTVE VALUE WORK(G) £	MONTHLY VALUE FLUC(G) £	CUMLTVE VALUE FLUC(G) £	CUMLTVE VALUE TOTAL (G) £	CUMLTVE RETNT/N £	CUMLTVE TOTAL (N) £	MONTHLY TOTAL £	MONTHLY RETNT/N £	MONTHLY TOTAL (N) £	CUMLTVE WORK (N) £	CUMLTVE FLUCTS (N) £
					(2)+(4)		(5)-(6)	(1)+(3)		(8)-(9)	(2)-(6*)	(4)-(6*)
10	151800	901800	7700	22700	924500	27800	896700	159500	5300	154200	874700	22000
11	162700	1064500	14300	37000	1101500	33000	1068500	177000	5200	171800	1032600	35900
12	172500	1237000	15100	52100	1289100	38700	1250400	187600	5700	181900	1199900	50500
13	180200	1417200	16600	68700	1485900	44600	1441300	196800	5900	190900	1374700	66600
14	186300	1603500	17200	85900	1689400	50700	1638700	203500	6100	197400	1555400	83300
15	190500	1794000	18500	104400	1898400	56900	1841500	209000	6200	202800	1740200	101300
16	192800	1986800	19600	124000	2110800	63300	2047500	212400	6400	206000	1927200	120300
17	193400	2180200	19600	143600	2323800	69700	2254100	213000	6400	206600	2114800	139300
18	192100	2372300	20400	164000	2536300	76100	2460200	212500	6400	206100	2301100	159100
19	189200	2561500	21000	185000	2746500	82400	2664100	210200	6300	203900	2484700	179400
20	184400	2745900	21300	206300	2952200	88600	2863600	205700	6200	199500	2663500	200100
21	177400	2923300	21300	227600	3150900	94500	3056400	198700	5900	192800	2835600	220800
22	169100	3092400	20300	247900	3340300	100200	3240100	189400	5700	183700	2999600	240500
23	158600	3251000	26400	274300	3525300	105700	3419600	185000	5500	179500	3153500	266100
24	146700	3397700	24400	298700	3696400	110900	3585500	171100	5200	165900	3295800	289700
25	132500	3530200	22600	321300	3851500	115500	3736000	155100	4600	150500	3424300	311700
26	117200	3647400	20500	341800	3989200	119700	3869500	137700	4200	133500	3538000	331700
– RELEASE OF HALF OF RETENTION AT COMPLETION						59850	3929350	59850	59850	0	3592700	336650
– FINAL RELEASE OF RETENTION AFTER M.G. DEFECTS						0	3989200	59850	0	59850	3647400	341800

*RETENTION SPLIT DEPENDENT ON FLUCTUATIONS CLAUSE

STAFF NO. 402
DATE 7/6/1982
DAVIS BELFIELD & EVEREST

COST INDICES

10 - 206	11 - 214	12 - 214	16 - 217	17 - 217	18 - 218
13 - 215	14 - 215	15 - 216	22 - 221	23 - 231	24 - 231
19 - 219	20 - 220	21 - 221			
25 - 232	26 - 233				

EXAMPLE F

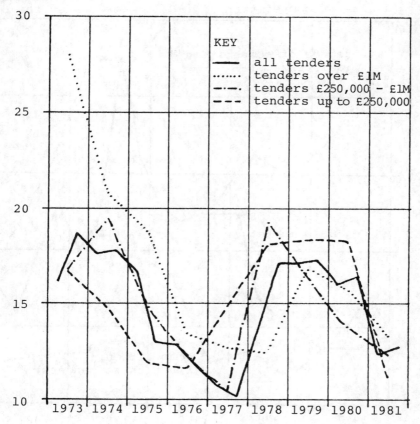

percent

KEY
——— all tenders
.... tenders over £1M
—·—· tenders £250,000 - £1M
— — tenders up to £250,000

EXAMPLE G : PRELIMINARIES PERCENTAGES

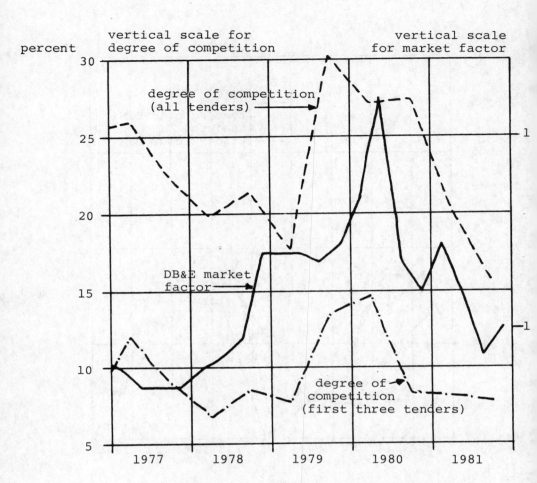

percent

vertical scale for
degree of competition

vertical scale
for market factor

degree of competition
(all tenders)

DB&E market
factor

degree of
competition
(first three tenders)

EXAMPLE H : DEGREE OF COMPETITION AND
THE DB&E MARKET FACTOR

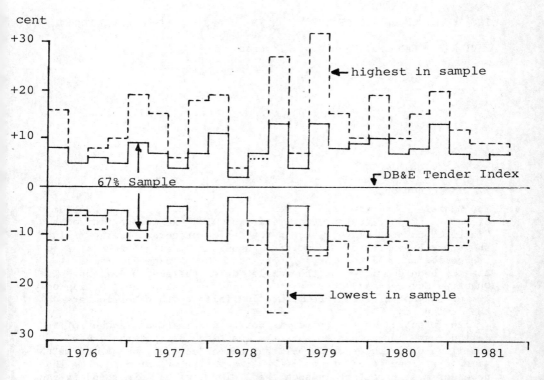

EXAMPLE J : DISTRIBUTION OF SUCCESSFUL TENDERS
AROUND THE DB&E TENDER INDEX

THE EFFECT OF DESIGN DECISIONS ON THE COSTS OF OFFICE DEVELOPMENT

PAUL R.F. TOWNSEND, Turner and Townsend

1. Introduction

The Quantity Surveyor is, at present, often faced with the problem
of being asked to control cost, starting with the preparation of an
"initial" cost estimate, late in the design process. Brandon (1)
shows how expenditure is committed through the design process up to
the period of tenders. He suggests that 80% of expenditure has
already been committed by the completion of the sketch designs. The
Quantity Surveyor may then be faced with the task of controlling only
20% of the anticipated expenditure in liaison with other members of
the design team.

The intention in this paper is not to castigate any member of the
design team, but rather to show the Quantity Surveyor that cost
modelling by computer is a tool that can effectively be used from the
time of a client's brief until the time for return of tenders. The
power of most computers is such that, with limited resources, it can
be used at the very earliest stages of design to make cost
comparisons between the Architect's initial design ideas. The
necessary changes in design made throughout the design period can be
readily costed using the computer and even major changes,
e.g. raising the building on piloti to allow more car parking on the
site, can be considered and accepted or rejected without involving
the Quantity Surveyor in the usual flurry of activity to revise his
earlier estimates.

2. The Definition of Cost

Cost is a term which is basic to the Quantity Surveyor. His
definition of cost is usually the tender figure. However the cost to
the client is that which is crucial and must be recognised as such by
the Quantity Surveyor. The cost to the client may be required in any
form and it is apparent that the Quantity Surveyor must be able to
give the anticipated cost in any form required.

The term "cost" in this article will be any of the following:-
(a) total cost (b) cost per square metre (c) cost per square metre of
net usable floor area (d) annual cost-in-use or (e) cost per work
unit.

3. *The Shape and Form of the Proposed Development*

The shape of each floor of the proposed development is defined by six
points defined in the datafile used by the program. The six points
are shown in Figure 1.

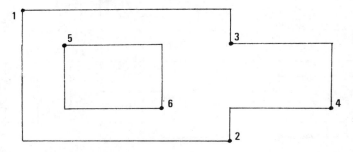

Fig. 1. The determination of shape

As can be seen Points 1 and 2, 3 and 4 and 5 and 6 each define
rectangular spaces. Points 1 and 2 define the initial rectangular
form, Points 3 and 4 define an addition to the initial rectangle and
Points 5 and 6 define a court area either totally within the initial
rectangle defined by Points 1 and 2 or partly or fully opening the
court area.

Hence an "L" shaped development can be modelled in two ways as
shown in Figure 2. Method A uses the addition of the further
rectangle shown by Points 3 and 4. Method B uses the court area
positioned at the corner of the development, positioned at Points 5
and 6.

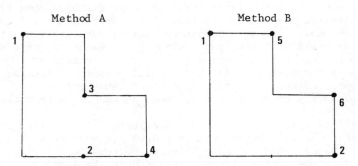

Fig. 2. The alternatives for an "L" shape

As can be seen from Figure 3 the number of shapes which can be
modelled from the three rectangular areas defined will represent the
majority of office developments.

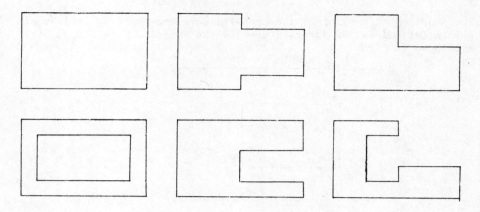

Fig. 3. The available shapes

The shape and form of the development is expressed in the three
forms:- (a) wall:floor ratio (b) plan/shape ratio, as defined by
Banks (2) and (c) POP ratio, as defined by Building Performance
Research Unit (3).

4. *The Cost Database*

The cost database used in any cost model is of crucial importance to
the results obtained from the model. If any doubt about the
validity of the cost database is apparent then the results may be
meaningless.

A cost model can be one of two kinds:- a comparative model or a
predictive model. In the latter case the cost database must be more
rigorously tested than in the former. The comparative model can
function when costs in the database are out of date, provided that
the relative costs of items have not changed. Hence, for a cost
database with concrete at £30/m^3 and steel reinforcement at
£300/tonne, a comparative model could still be used if the current,
or anticipated future, cost of concrete was £40/m^3 and of steel was
£400/tonne, as the relative cost between the two items remains at
1:10.

The predictive model can, it may be argued, only function for a
limited period of time before the cost database must be updated.
This limit to the period of time between the updating of costs may
tend to reduce the number of costs included in the database, so as to
speed the updating process.

The cost database for the capital costs used in this model was a
standard building "price book" and for the running costs were a
number of different sources.

The running costs allowed by the model are lighting, heating,
air-conditioning (if specified), insurance, maintenance and cleaning.
A further running cost of the building might be the annual cost of

staffing the building. This cost when considered against the annual
cost-in-use of the building will show the limited cost of the
building and will stimulate the design team to attempt to provide an
environment which will produce the maximum efficiency from the
employees. It is for this reason that some attempt has been made to
allow for the objective factors in the environment e.g. temperature,
lighting level and noise level. No attempt has been made to include
subjective factors in the analysis e.g. "niceness of the workplace".

Algorithms have been developed to show the efficiency of an
employee based on data given by Vernon (4), Kerslake (5) and
Chrenko (6). These algorithms represent the efficiency index of
staff under different temperatures, lighting levels and noise levels.
The number of staff employed with the building multiplied by the
efficiency index gives the number of work units supported by the
proposed development.

The capital and running costs have been combined using the annual
equivalent value of the building costs, the land costs and the fees
and adding these to the annual running costs to produce an annual
cost-in-use for the proposed development.

5. *The use of the model*

The model is, in its present form, relevant at all stages of the
design process. For a particular proposed development the model can
be constantly updated as design decisions are made. At the earliest
stages of design the design data used by the model can be average
values e.g. a lighting level of 500 lux, but as design progresses
the average values will be "firmed up" to allow for greater
accuracy when making cost comparisons.

6. *The results*

In addition to the use of the model on particular projects, the
model can be used to investigate the costs of office buildings in
general. Some examples of these general investigations are detailed
in the following graphs.

6.1 *The optimum window area*

The proposed development under consideration is one of 12 metre
depth. The three graphs following show the cost per square metre,
the cost per square metre of daylit area and the annual cost-in-use
per square metre related to the percentage of the external wall area
which is glazed.

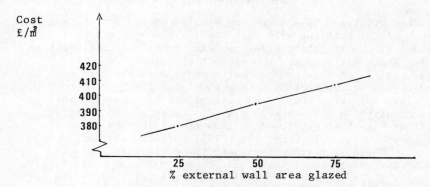

Graph 1.   Capital Cost per Square Metre of Gross Floor Area

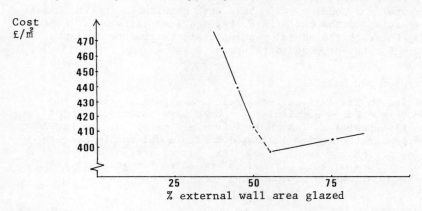

Graph 2.   Capital Cost per Square Metre of Daylit Floor Area

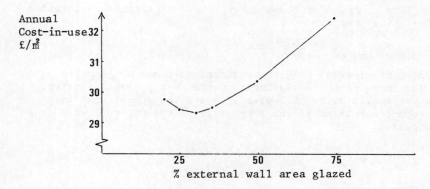

Graph 3.   Annual Cost-in-Use per Square Metre of Gross Floor Area

Graphs 1, 2 and 3 show how, with a different definition of cost,
different optima are produced for the percentage of the external
wall which should be glazed.

Each client will have a different cost limit or limits to apply
to the project. A speculative developer will probably be primarily
concerned with Graph 1, allowing the window area to be the minimum
percentage at which he feels he can let the office. The owner
occupier has a considerable interest in the running costs of the
building so may consider Graph 3 to be the one with which he is most
concerned.

6.2 *The optimum spacing of floor trunking*

Dickens and Hawkes (7) considered the provision of telephone
installations by means of trunking laid in the floor screed. The
conclusion of the report was that to allow for flexibility in the
use of the building, the best system was the spur system, Figure 4.

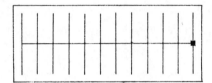

Fig. 4.   The Spur System of Trunking

The three variables of the spur system are (a) the distance between
spurs (b) the distance between outlet points along each spur and
(c) the distance at which each spur terminates from the internal
face of the external wall. In addition to these three variables the
outlets can be staggered, Figure 5.

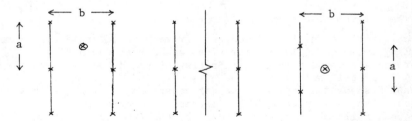

Fig. 5.   The Comparison between Staggered and Non-Staggered Outlets

The maximum distance from an outlet, neglecting the variable denoting
the distance at which the spur terminates from the external wall, for

the non-staggered outlet is

$$\frac{a^2 + b^2}{4}$$

while that of the staggered outlet is

$$\frac{a^2 + 4b^2}{8b}$$

This, with a equal to 3 metres and b equal to 4 metres, gives the maximum distance from a non-staggered outlet to be 2.5 metres while from a staggered outlet is 2.28 metres.

   Graph 4 shows the effect of altering the three variables, using staggered outlets. The line drawn on the graph represents the approximate lowest cost at which a maximum distance from an outlet can be achieved. It can be seen that the curve reaches a lowest cost at about £11 per square metre and for maximum distances from an outlet of less than 1 metre the cost becomes prohibitive.

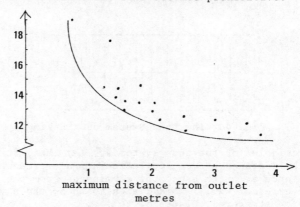

Cost
£/m²

maximum distance from outlet
metres

Graph 4.   The Cost of Providing Electric Outlets in Floor Trunking

6.3   *The optimum slab span*

The design of the reinforced concrete frame and floors is calculated according to Morgan (8), minimum sizes are set to reach fire resistance standards detailed in British Standard Codes of Practice (BSCP) 3, part IV and 114.

   Graph 5 shows the capital cost per square metre as the slab span changes. The rise in cost from 6 metres to 7 metres and the subsequent fall to 8 metres is caused by the 7 metre span being the same depth, but with a reduction in the number of columns, causing a fall in the cost per square metre.

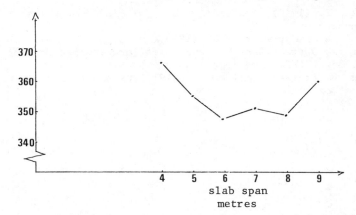

Graph 5.   The Effect on Cost of Different Slab Spans

7. *Conclusions*

The increase in the power and the reduction in the cost of computers over the last decade has brought the processing capacity necessary for Quantity Surveyors within the finance available to most Quantity Surveying practices. The subsequent increase in the use of computers has led other members of the design team and the client to be given information of a more detailed nature and faster then was previously the case.

If, as seems likely, the growth in the use of computers continues in the next decade, it can be expected that the construction industry and its clients will become accustomed to the better service provided by computers. Those practices providing this service will find themselves more advantageously placed to operate in an increasingly competitive market.

A computer will, in the near future, become an invaluable tool to the Quantity Surveyor. The use of a computer for simple but repetitive tasks such as cash flow forecasts or NEDO formula price adjustments will soon be joined by sophisticated cost models, computer aided taking off with sophisticated post-contract aids and other tasks, hitherto not considered.

References

1.   Brandon, P.S. (1978), A Framework for Cost Exploration and
       Strategic Cost Planning in Design, *Building and Quantity
       Surveying Quarterly Vol. 5 Nr. 4.*
2.   Banks, D.G. (1974), Uses of a Length/Breadth Index and a Plan
       Shape Ratio, *Building and Quantity Surveying Quarterly Vol. 2
       Nr. 1.*
3.   Building Performance Research Unit (1972), *Building Performance.*
4.   Vernon, H.M. (1921), *Industrial Fatigue and Efficiency.*
5.   Kerslake, D.Mc.K. (1972), *The Stress of Hot Environments.*

6. Chrenko, P. (1974), *Bedford's Basic Principles of Heating and Ventilation.*
7. Dickens, A. and Hawkes, D.U. (1970), *A Computer Simulation of a Service System,* Land Use and Built Form Studies Working Paper Nr. 36, LUBFS, Cambridge.
8. Morgan, W. revised Hamilton, I. (1973), *Students Structural Handbook.*

PROBLEMS OF LOCALITY IN CONSTRUCTION COST FORECASTING AND
CONTROL

D. W. AVERY, Glasgow College of Building and Printing

1. Introduction

This paper is intended to concentrate attention on differences in construction costs which are attributable to
locality. It will attempt to identify the main causes
of difference, consider their potential importance and
suggest implications for the work of cost forecasting,
cost planning and formal cost control. Features normally
associated with individual sites, such as slope and ground
conditions are not considered. Whilst the paper is
largely based on the study of local prices and costs in
Scotland with particular emphasis on remoteness the principles discussed have wider application and might be
especially relevant to overseas conditions.
    With the production of tender price indices in recent
years(1)it has been possible to detect and measure
regional or local influence on tender levels(2). As a
result those with responsibility for setting cost limits
on public sector building, most notably in the field of
housing, have been able to apply effective regional allowances and revise them on occasion to reflect changing
regional conditions. Regional allowances calculated for
local authority housing in England and Wales have shown
that expensive areas produce tender levels in the order
of 20% above the lowest(3). Mainland Scotland excluding
Orkney and Shetland and the Western Isles shows a similar
range(4)but whereas in England tenders for housing appear
to be highest in the major urban areas, in Scotland they
are highest in parts which are remote from the main areas
of activity. In general terms that means the West High-
lands of Scotland but the highest levels of all are found
in the islands where they are shown to be on average more
than 40% above the lowest.
    Although the allowances calculated for the purposes of
housing cost control are often adopted as a useful general
guide it cannot be assumed that the forces which determine
the level of housing tenders will influence tenders for

other types of building work in exactly the same way. In
view of the obvious importance of local cost influence
uncertainty and anxiety often surround the cost control
of work in unfamiliar locations especially if they are
remote where there is little experience of similar work.
   Locational influences can be considered in two groups.
The first group contains factors which might have a bearing on the cost of executing work to any given design and
they consequently concern tenderers as well as those engaged in forecasting or controlling tender levels.
Factors in this group are found also to have a potential
for influencing design itself on economic grounds. In
the second group are factors which influence design
directly and are therefore of interest more specifically
to those taking architectural decisions or engaged in cost
forecasting and control.

2.   Effects of location when working to given design

2.1   Remoteness from source of material supply
Remoteness is, of course, an imprecise term and its effect
on the cost of materials needed for a project is by no
means constant. Apart from the fact that various materials are likely to come from different places freight
charges will be influenced by volume or weight and size
of order. Further costs are likely if materials are
dangerous, fragile or difficult to handle. Heavy materials with low initial cost such as sand and gravel tend
to be worst affected by distance and difficult transport
conditions. Journeys to remote country areas and over
ferry crossings can double or treble their cost per cubic
metre. Expensive manufactured goods tend to be affected
much less and as a matter of policy some manufacturers.
offer delivery anywhere in Britain without special charge.

2.2   Labour cost and productivity
The determinants of labour cost exhibit local variation
although detailed local investigation may be needed to
assess the total effect. Nationally agreed hourly rates
which themselves show locational differences are often
increased to meet conditions in the local labour market.
The market is influenced both by the availability of construction work and activity in other local industries.
Estimators also judge the level of productivity to be
expected in a locality, the quality of industrial relations and the effects of local working arrangements such
as 'labour only' sub contracting. Travelling and lodging
allowances may have to be paid and have been shown to
increase the hourly labour rate by around 35%. Such costs
might attach to a large or complicated project in an area
where a small simple job could be handled by the local

labour force. Exceptionally high cost is also incurred
when a specialist has to travel a long way to carry out
a small quantity of work as he might on a small project.
Size and character of project have been found to be very
significant when attempting to evaluate the effect of
locality on labour cost.

2.3 Water, power and sewage connections
Reasonably easy access is commonly available to water
and electricity supplies and to sewers. Consequently
connections normally make up a very small proportion of
the total project cost. Although there will always be
exceptions high costs can be expected in sparsely pop-
ulated places and services might even have to be provided independently. The impact on the cost of a small
project is sometimes dramatic.

2.4 Water and power for the works and mechanical plant
Unless expensive temporary connections have to be made
the cost of providing water and power for carrying out
the works is unlikely to be a significant local factor
in preparing cost forecasts.
    Contractors introduce mechanical plant to save labour
and to improve speed of production. Although plant costs
are normally only a small proportion of total expenditure
the cost of transport to and from site reduces its potential advantage. Such loss might have to be weighed
against the cost of importing labour.

2.5 Security
Security is an increasingly important overhead cost in
many urban areas but one which hardly ever features
significantly in tenders for work away from the major
connurbations.

2.6 Climate, weather and hours of daylight
Climatic conditions throughout Britain are well document-
ed. Some of the worst conditions occur in Scotland but
there is much local variation. The problem is not confined to winter building for vulnerable work is frequently
scheduled to be done during the summer, especially in
latitudes where winter days are very short. Lacy,
writing a few years ago(5)reported that 'Clearly it is
not yet possible to formulate any general rules about the
extent of hindrance caused by bad weather'.
    Rainfall and high winds during the hours of daylight
are serious features and meteorological records show
average annual percentage of wet hours by day in the
United Kingdom ranging from under 4% to more than 14%.
In Scotland the wettest areas are in the West and the
severest gales occur in the islands to the North and West
during winter. There is room for much more study of this

subject but the coincidence of unfavourable climate and
high cost in the West could be misleading. Bad weather
by itself, especially in summer, seems less guilty than
other factors of adding massively to total project cost.
Of course an estimating error could very seriously affect
a contractor's profit and unpredictable weather as distinct from recorded climate constitutes a disconcerting
risk. Perhaps it will never be possible to measure the
effect of a district's reputation for bad weather on
contractors' willingness to tender and hence on the level
of competition.

2.7 Market conditions or climate of tendering

Changes in the level of competitiveness over a period of
time are indicated by comparing changes in tender levels
with changes in the level of contractors' costs. During
the years from 1970 to 1982 nationally produced tender
price indices have shown dramatic moves both upwards and
downwards against a pattern of relatively steady cost
increases. Tender price indices for specific types of
construction in different places normally show deviations
from the national pattern in response to local changes in
market conditions.

Small projects in remote areas pose a particularly
difficult problem for the cost forecaster, firstly because the smallness itself is likely to enhance unit
rates and secondly because distance from the main centres
of activity tends to produce separate market characteristics. While large and attractive projects in remote
areas often create their own climate of competitiveness
among major contractors, the costs and risks associated
with small projects frequently make them unattractive to
outside firms. It does not follow that prices from local
contractors will necessarily be higher on that account
but observations of tenders for small buildings prepared
by local contractors in different parts of Scotland do
suggest at least partial immunity from the influence of
outside competition.

2.8 Local tendering customs

Contractors accustomed to tendering for small projects in
remote areas often work without bills of quantities and
it is conceivable, although difficult to demonstrate, that
different methods of tendering produce different levels
of pricing.

3. Effects of location on building design

3.1 Tradition and policy

Traditionally architectural practice has responded to
local requirements producing buildings appropriate to
the local climate and topography and suited to withstand

other natural conditions such as risk of insect attack
or dry rot. Local traditions also acknowledge economic
forces and were developed around the availability of
materials and other resources. Following a trend during
the last three decades towards uniformity of design there
are now signs of renewed attention to the physical
criteria which buildings have to satisfy and to the
strength of local traditions which grew up in response
to them(6).

Building regulations and local planning requirements
are devised to ensure physical suitability but much detailed attention is now being directed by architects to
the most extreme conditions which may occur during a
building's life and to the peculiarities of micro-climate.

3.2 Climatic conditions

Lacy(5)refers to many climatic features capable of influencing building design. Some of them such as salt spray
and atmospheric pollution deserve special attention only
in certain places. Others, such as wind, rain and low
temperatures occur everywhere but in some places they
are severe enough to warrant special attention.

In Scotland many of the worst conditions are apt to
occur in remote areas. Measures designed to combat
severe conditions include additional strength, thermal
insulation and weather tightness of walls, roofs, windows
and external doors. Problems like insect or fungal
attack are countered by careful design, preservative
treatments and use of appropriate materials. The cost of
such features tends to be hidden among the general costs
of a building and local differences are hard to isolate.
Sidwell(7)referred to measurable differences between
housebuilding traditions in Scotland and England and
assessed average costs.

3.3 Compliance with local practice

In recent years interest in environmental and conservation matters has reinforced the desirability of harmon-
ising design with existing traditional buildings even
beyond legal planning requirements. Functional requirements may be served at the same time.

The cost of deliberate response to local requirements
can of course be evaluated at the time of design. For
example a typical feature of Scottish architecture is
the strong steeply pitched slated and gable ended roof.
Such a roof might be designed for a new building to
satisfy both local climatic conditions and architectural
tradition. The cost per square metre can be calculated
and compared with that of a light low pitched roof with
corrugated sheet covering which might have been acceptable in another region.

3.4 Response to vandalism

Vandalism is not entirely confined to known locations but
risks in some areas are clearly higher than in others.
Appropriate precautions can have a fundamental effect on
layout and specification, generally in the interests of
impregnability, and they are somewhat akin to the precautions against hostile weather. Building elements most
affected are again the walls, windows, doors and roof but
in this case fencing, landscaping and lighting costs are
also likely to be higher than usual.

Protection against vandalism is more likely to be needed in certain urban areas than remote country ones and
costs can again be quantified.

4. Summary of observations and their implications

Location is clearly capable of exerting a powerful influence on the major components of construction cost and
also on the way in which design criteria are satisfied.
The influence of a locality falls unevenly on the costs
of the various materials needed for construction and also
on labour according to project circumstances. Market conditions, varying as they do from place to place and over
a period of time, are often difficult to judge. They call
for attention, with special consideration being given to
the size and nature of projects in areas which are remote
from the full range of construction activity. Even when
local climatic conditions are known their effect on tender levels is difficult to judge. Other potentially
expensive features such as service connections can only
be forecast with any certainty after specific detailed
investigation.

One of the difficulties encountered in early cost forecasting and cost control is that features of local importance can occur separately or in various combinations. The
problem tends to be most acute when considering remote
places about which little information is available from
previous projects and possibly none at all from projects
of similar type and size.

5. An approach to the exploration of location costs

It has become clear over a period of time that several of
the factors described above deserve more detailed study.
One approach currently being made in Glasgow is by studying tenders for work of broadly similar function but
differing in size, specification and, of course, locality.

At present there are three strands to the study.

5.1 Cost analysis

The BCIS Form of Detailed Analysis is being used to yield information about the effects of design, size of unit, detailed functional requirements, specification, quality and contract particulars. The analyses provide useful material for the study of locational costs by showing architectural response to local requirements, describing project conditions and indicating market conditions.

5.2 Comparison of unit rates

As a more direct approach rates for measured items are being compared using the PSA Schedule of Rates for Building Work 1980 for a common base. The procedure is based on that used by government agencies and BCIS for tender price index calculations. Elements are being used for sectional detail and weighting purposes. In the search for as much detailed information as possible the selection of items within each section is not being limited to 25% of the total value.

5.3 Material costs

Since rates in bills of quantities do not separate material and labour costs further enquiries are being conducted into the cost, delivered to site, of major materials.

5.4 Development

A difficulty inherant in remote localities is that of finding a sufficient number of similar projects to establish reliable local indices but the three strands of detailed information presently being investigated are starting to provide insights to individual projects which would not be provided by any one line of study alone. The work described above is at any early stage but there are indications that the systamatic accumulation of such details could provide an increasingly useful bank of information in elemental form.

6. Potential application of study

It is hoped that more detailed understanding of locational differences in tender levels will have practical applications.

6.1 Forecasting

Detailed assessment of locational influence would help to explain the underlying reasons for allowances calculated by government departments or others for defined types of building. Early cost forecasts, whether for broadly similar work or work with distinctly different characteristics could then be made with greater confidence.

6.2   Cost planning
It seems that there will always be a need for more information about the influence of locality on individual
elements and specific forms of construction for cost
planning. More information should make it easier to
compare design options and possibly lead to the development of guidelines for economical construction in various
locations.

6.3   Control
If the magnitude of locational differences and inevitable
uncertainty about market conditions in remote areas
prevent the most rigid forms of cost control from being
applied the need for thorough and effective cost planning
becomes all the greater.

References

1.   Azzaro, D.W. (1976), Measuring the level of tender
        prices. Chartered Surveyor Building and Quantity
        Surveying Quarterly. Vol 4 No 2 Winter 1976.

2.   Beeston, D.T. (1975), One statistician's view of
        estimating. Chartered Surveyor Building and
        Quantity Surveying Quarterly. Vol 2 No 4
        Summer 1975.

3.   Department of the Environment and Welsh Office (1972),
        The Housing Cost Yardstick. Joint Circular 45/72,
        95/72.

4.   Scottish Development Department (1979), Housing Cost
        Indicators - Appendix B to SDD Memorandum No 32/
        1979.

5.   Lacy, R.E. (1977), Climate and building in Britain.
        HMSO 1977.

6.   Aspinall, A. (1977), Building materials and planning
        control. The Architects' Journal. 7 Dec. 1977.
        1153 - 1167.

7.   Sidwell, N. (1970), The cost of private house build-
        ing in Scotland. HMSO 1970.

SECTION V

COST MODELLING

EXPERIMENTS IN PROBABILISTIC COST MODELLING

ALAN WILSON, Liverpool Polytechnic

Introduction

This paper discusses some preliminary results from a major research
project into the design, development and testing of models for
predicting the economic consequences of building design decisions.
Such economic response models generally have one of the following
purposes.

(i)   To predict the total price which the client will have to
        pay for the building. (Tender Price models)

(ii)   To allow the selection, from a range of possible design
        alternatives, at any stage in the design evolution, of the
        optimum design according to some predefined criterion of
        economic performance.

(iii)   To predict the economic effects upon society of changes in
        design codes and regulations.

It has been found that the purpose of the model has a considerable
influence upon the most appropriate approach to model formulation.
It will prove useful to characterise models according to purpose and
the classification used here is to identify those under (i) above
as macro models, and those under (ii) and (iii) as micro models.
This paper is essentially concerned with micro models.

It is possible to propose a great many economic criteria of performance which could be adopted in the economic modelling of building
design decisions.   That adopted here is monetary cost to the client.
We are thus dealing with micro cost models.

Approaches to Modelling

Architecture is concerned with the creation of space.   A good building design will capture and articulate space to satisfy both
quantitatively and qualitatively the demands of the processes to be
accommodated.   The total number of design decisions which must be
taken is enormous.   They vary, for example, from the choice of
structural frame type to the position of light switches, from the
number of storeys to the type of window fastening.   It is convenient

to regard the decisions as design variables which simply take different values in different buildings. Since it is these decisions which alone determine the nature of the building, it is they which give rise to the cost of the building.

Thus,   Cost, $C = f_1 (V_1, V_2, V_3 \ldots\ldots V_N)$ (1)

where V_1, V_2, etc. are the design variables.

However, the cost of building work is actually incurred and price is usually expressed, not in terms of design variables, but in terms of the resources of all kinds which the design decisions commit.

Thus, $C = f_2 (\Sigma Rj)$ (2)

where Rj are the resources committed.

The central task of cost modelling is the reconciliation of equations (1) and (2).

It is possible to recognise two different, although not mutually exclusive, approaches to the construction of cost models.

        deduction
        induction

Deductive methods involve the analysis of cost data over whichever design variables are being considered, with the objective of deriving formal mathematical expressions which succinctly relate a wide range of design variable values to cost. This approach draws heavily upon the techniques of statistics; correlation and least squares regression, in particular. Disadvantages of this approach arise from the not inconsiderable limitations of these statistical techniques, and on the total dependence upon the suitability of the cost data used.

Inductive methods, on the other hand, involve, not analysis of a set of given cost data, but rather the synthesis of cost of individual discrete design solutions from the constituent components of the design. Whilst deductive methods are, perhaps, more important in the early stages of design, inductive methods are more important in the later stages. Deductive methods arise largely from equation (1), inductive methods arise largely from equation (2). Inductive methods require the summation of cost over some suitably defined set of subsystems appropriate to the design. The most detailed level of subsystem definition would be the individual resources themselves, but several other levels of aggregation are in common use, e.g. operational activities and constructional elements.

This paper is primarily concerned with inductive, micro, cost models. Such concentration upon taxonomy may be thought excessive, and even pedantic, but it is suggested that it may be conducive to the continued development of economic modelling in building design.

Uncertainty

Whilst the uncertainties implicit in any industrial cost estimating have long been recognised at least qualitatively, it is only in the fairly recent past, with the stimulus of computer based cost modelling that there has been real movement towards attempting to quantify them. In his seminal paper on probabilistic estimating[1] Spooner commented upon the lack of data on uncertainty, in 1974. Unfortunately, little has been added in the intervening years.

A major problem is the unsuitable form of existing cost data. In particular, there appears to be widespread confusion in the building industry between cost and price. It is a recognised phenomenon in economics that the selling price of any product is determined by the market for the product, and not by the manufacturer's input costs. Costs are certainly important, since they determine profitability, but their relationship to selling price is often extremely tenuous. Thus, an analysis of tender prices does not reflect the variability of costs, but rather reflects the exigencies of the market place. Further, bills of quantity rates, insofar as they represent anything, reflect this market abstraction.

In the absence of objective data some writers in the area of probabilistic estimating have made assumptions which are unproven but convenient. Prime amongst these, for example, is the assumption that cost uncertainties are always symmetrical, i.e. that they can be accounted for by an expression of the form $£X \pm a\%$. It was the intuitive feeling that this was an oversimplification which, amongst other things, prompted the research described in this paper.

Much of the author's experience of modelling has been devoted to micro models, i.e. cost models for fairly tightly defined design optimisation problems[2,3], and these are the types of cost model primarily under consideration in the present research. Such models have two important characteristics.

Firstly, inductive micro cost models do not include the large number of subsystems which is the case in macro (tender price) models. One excuse widely used by analysts for ignoring cost uncertainty in macro models is the "swings and roundabouts" argument. More formally, this is expressed by the Central Limit Theorem of statistics which suggests that the summation of the mean costs of each subsystem will tend towards the mean cost of the total system and no matter how assymmetrical the uncertainty in each subsystem, the final composite uncertainty in total cost will tend to the symmetry of the Normal Distribution. Whilst such assumptions may be valid in macro models with large numbers of constituent subsystems, to assume that it is the case for micro models with only a few subsystems seems rather foolhardy.

The second characteristic of micro models is that since, in the search for optimal or at least improved design solutions, we are comparing alternative solutions to the same design problem, we are interested only in those components which are likely to differ. This allows us to avoid incorporating some of the more troublesome items of any estimate such as profit, overheads and tactical marketing considerations which have significance only in the case of macro cost models.   They are likely to be constant for each design alternative, provided we take care in our model formulation.

Experimental Method

The objective then is to obtain data indicating the distribution of uncertainty around subsystem cost estimates, and to determine the effects of such uncertainties upon the evaluation of alternative design solutions.   It is a vast task.   It will take a great deal of time and manpower to achieve fully.   This paper describes the strategy adopted and presents early results.

Controlled experiments of the kind possible in the scientific world are generally impractical in economics and impossible in building economics.   So, unfortunately, there appeared no way of obtaining unbiased objective prime source data on cost uncertainty.   Further, no secondary sources of data, i.e. data acquired for other purposes, seemed suitable, largely for the reasons discussed above.

The experimental approach adopted to the acquisition of data was the Delphi method which is often used in the field of technological forecasting.[4]   Essentially, it comprises a structured approach to exploiting the specialised knowledge and judgement of experts in a particular field.   In a commercial environment, such as the one we are dealing with, it is difficult to define "an expert".   Eventually, though, a panel of eight competent contractor's estimators was constituted, where competence was judged on the basis of performance, position, experience and standing.   It is difficult to be dogmatic about the optimum size of such a panel.   Certainly the statistical view of "the more the better" seems appropriate, to be tempered only by administrative convenience and compatibility of experience.   To call the assembled estimators a panel is, perhaps, misleading, since at no time did they meet each other, or indeed, know each other's identity.   The anonymity is an advantage of this experimental approach.   The weakness, of course, of the Delphi method is its reliance upon subjective judgement, but if the experiment is carefully controlled, this can become a virtue.

The constructional element chosen as the vehicle for the first experiment was a 150mm thick, suspended, cast in-situ, reinforced concrete, floor slab.   This may be unexciting, but it is an important, common item in modern construction and it does require a number of resource types.

Round 1 in the experiment consisted of identifying how the estimators themselves went about pricing such a floor slab.   The differences in approach are considerable.   Whilst all the estimators worked at a detailed level, they often identified different constituent items as significant.   After some feedback between experimenter and estimators, a pro forma of 17 items necessary to the pricing of the slab was adopted.   (There still remained discrepancies in interpret- ation, especially with regard to the need for skilled and unskilled labour in certain of the constituent tasks.)

The second round was the lynchpin of the experiment in that its purpose was to obtain the uncertainty data.   The problems facing the experimenter are considerable since he must design a questionnaire which will elicit the required information without leading the participant.

It is important to realise that the way the questions are framed very much influences the response of the participants.   It soon became apparent that it was necessary to make assumptions about the shape of the uncertainty probability distribution in order that the appropriate questions could be asked, and the correct parameters sought.   The distribution function will, in practice, be continuous, but it seemed unlikely that the estimators would be able to associate quantitative levels of probability with their estimates. (This is a common criticism of the PERT procedure)   Thus, it was necessary to choose a distribution function which could be characterised by a few significant parameters.   The four distributions most often proposed for this type of problem are the uniform, normal, triangular and beta distributions.   That selected was the triangular distribution.   It requires only three point estimates to define it; the probable value, and the limiting minimum and maximum values.   In addition to having the support of other workers in the field[1], it has the considerable merit of simplicity.

The estimators were then asked to fill in a questionnaire which, for each of the seventeen identified constituent items in the cost synthesis of the concrete floor slab, sought their estimates of minimum, probable and maximum values.   The questionnaire reminded the participants that the values sought were to be net of overheads, profit and tactical considerations.

The third round of the Delphi experimental method consists, in this case, of an analysis of the results of round 2, the presentation of those results in a suitable format, and their return to the participating estimators for comment and amendment.   At this stage the estimators see the estimates of their peers for the first time. (Although it is not possible for them to associate names and values) The purpose here is a movement towards consensus, to give them an

opportunity to change their minds in the face of the collected body
of results, and, if necessary, to make apparent their reasons for
dissent.   This feedback is essential to guard against the
experimenter influencing the analysis of results.

Results

A study of this kind necessarily generates a large number of results,
and it is inappropriate in this paper to attempt to publish them in
full.   Thus a few typical results have been selected for discussion.

Figure 1 illustrates the results of Round 2 for seven of the subsystems.   In Figure 2 these same results have been processed using
a very simple statistical analysis.   There are, of course, always
dangers in reading too much into what is necessarily a very small
sample, but it is suggested that the results do allow a few
general conclusions.

The variability in pricing all of the items is considerable, but it
does appear that estimators are constrained by the traditional
practice of single figure estimating, and when the opportunity is
presented to them expressing the uncertainties in their estimates
they rise to the occasion.   A more complex statistical analysis
upon the total set of results revealed no consistent bias amongst
individual estimators, and so the results for each item can be
averaged across the estimators with some confidence.   Two phenomena
are displayed in Table 2 and it is important that they are
distinguished.   Firstly, there is the variability in each of the
point estimates measured by the coefficient of variation, and
secondly, there is the perceived range of uncertainty around each
item indicated by the difference between the minima and maxima.

Looking firstly at the variability of the three point estimates in
each category, some general conclusions are possible.   As might be
suspected, the variability in estimating materials prices is
consistently less than that for labour prices.   Perhaps more
surprisingly, the variability in estimating the maximum and minimum
values is not always greater than that for the probable value.
It does not seem that the opportunity to express the range of
uncertainty reduces the variability of the single point (probable)
estimate from the levels which are generally accepted by the
industry.

Perhaps the most striking feature of the results emerges when we
look at the distribution of the maxima about the probable estimates.
There is a pronounced and consistent skewness to the right in all
cases, for both labour and material items.   Whilst the range of
the uncertainty about the probable value (a_1 to a_2) is considerable
(> 100% in one case), the assymmetry (a_2/a_1) averages at 2.6.   The
skewness appears greater in the case of the labour items as indeed

Figure 1

SUBSYSTEM	ESTIMATOR	A	B	C	D	E	F	G	H
READY MIXED CONCRETE ($£/m^3$)	Min.	24.34	24.00	32.00	28.00	25.36	26.50	26.75	27.00
	Prob.	25.86	26.00	34.00	29.50	28.58	28.60	27.35	28.00
	Max.	27.96	30.00	34.00	35.00	35.60	30.60	28.50	32.00
LABOUR; PLACING & FINISHING CONCRETE ($manhrs/m^3$)	Min.	0.46	0.50	0.38	0.36	0.45	0.67	0.50	0.40
	Prob.	0.51	0.75	0.45	0.40	0.56	0.74	0.50	0.45
	Max.	0.66	1.125	0.75	0.45	0.90	0.80	0.60	0.60
BAR REINFORCEMENT ($£/Tonne$)	Min.	245.40	240.00	290.00	243.25	243.25	235.00	247.66	238.00
	Prob.	247.66	300.00	320.00	275.68	275.68	240.00	247.66	262.00
	Max.	286.18	440.00	380.00	303.54	300.00	400.00	247.66	304.00
LABOUR - STEEL FIXING ($man.hrs/Tonne$)	Min.	17.00	37.00	22.00	26.00	16.00	25.00	32.00	35.00
	Prob.	25.00	43.00	26.00	37.00	30.00	32.00	32.00	53.00
	Max.	37.00	54.00	34.00	73.00	66.00	52.00	38.00	143.00
PLYWOOD FORMWORK ($£/m^2$)	Min.	3.11	5.30	4.00	3.40	2.90	3.10	3.50	3.90
	Prob.	3.49	5.55	6.00	3.50	3.49	3.42	3.50	3.90
	Max.	4.52	6.50	6.00	4.50	3.72	3.90	3.85	9.68
LABOUR - MAKING, FIXING & STRIKING FORMWORK ($man.hrs/m^2$)	Min.	1.00	2.00	3.60	2.48	1.50	2.90	2.00	1.39
	Prob.	1.65	2.50	4.00	2.48	1.60	3.10	2.25	1.57
	Max.	2.27	3.50	5.00	2.48	2.10	3.20	3.00	2.34

Figure 2

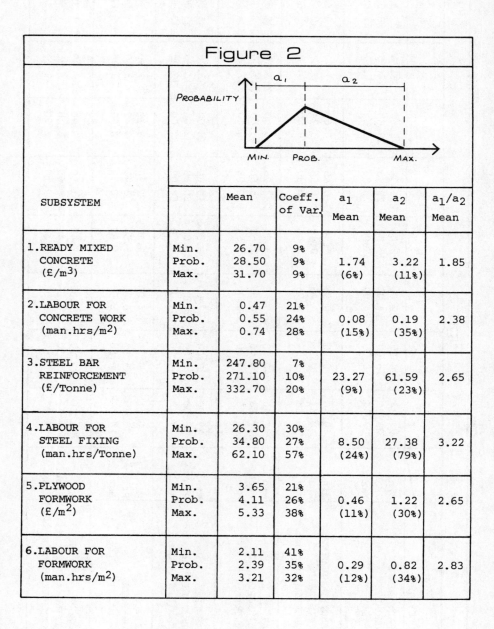

SUBSYSTEM		Mean	Coeff. of Var.	a_1 Mean	a_2 Mean	a_1/a_2 Mean
1.READY MIXED CONCRETE ($£/m^3$)	Min. Prob. Max.	26.70 28.50 31.70	9% 9% 9%	1.74 (6%)	3.22 (11%)	1.85
2.LABOUR FOR CONCRETE WORK (man.hrs/m^2)	Min. Prob. Max.	0.47 0.55 0.74	21% 24% 28%	0.08 (15%)	0.19 (35%)	2.38
3.STEEL BAR REINFORCEMENT ($£$/Tonne)	Min. Prob. Max.	247.80 271.10 332.70	7% 10% 20%	23.27 (9%)	61.59 (23%)	2.65
4.LABOUR FOR STEEL FIXING (man.hrs/Tonne)	Min. Prob. Max.	26.30 34.80 62.10	30% 27% 57%	8.50 (24%)	27.38 (79%)	3.22
5.PLYWOOD FORMWORK ($£/m^2$)	Min. Prob. Max.	3.65 4.11 5.33	21% 26% 38%	0.46 (11%)	1.22 (30%)	2.65
6.LABOUR FOR FORMWORK (man.hrs/m^2)	Min. Prob. Max.	2.11 2.39 3.21	41% 35% 32%	0.29 (12%)	0.82 (34%)	2.83

does the uncertainty range, an unsurprising result perhaps, but the extent of the skewness in material prices is, perhaps, surprising.

Monte Carlo Simulation

Having examined the uncertainty around the individual subsystem costs we must now investigate the way in which these individual uncertainties combine to influence the uncertainty surrounding the total cost of our concrete slab. The pronounced assymmetry of the uncertainties seemed to suggest that the often used, simple analysis of variance was inappropriate. The technique adopted was Monte Carlo Simulation.

A micro computer program was written which would randomly select costs for each subsystem in accordance with the triangular probability distribution obtained experimentally for each subsystem. These values were then combined in a simple inductive cost model aggregating the subsystems of Figure 2, with suitable coefficients for hourly wage rates, etc., incorporated, to give a single deterministic cost for $100m^2$ of suspended concrete slab. This whole process is then repeated a large number of times to enable a frequency diagram to be plotted (Figure 3a). When using Monte Carlo simulation there is often some doubt as to the "randomness" of the computer generated random numbers. In our trials we have found those produced by a typical micro computer standard function to be more than adequate. Another issue facing the investigator is the number of iterations necessary for a reliable result. The error of convergence reduces in proportion to the square root of the number of iterations, so the more iterations, the better, but there is a diminishing return on accuracy. The number of iterations carried out in this case and illustrated in Figure 3 was 500.

Because of the use of random numbers, each 500 simulations will result in a different relative probability distribution. This non-repeatability tends to disturb traditional cost estimators, and care is necessary in interpreting such probabilistic cost distributions. Of most value perhaps to the estimator is the cumulative probability diagram of Figure 3b, but the relative diagram of Figure 3a is useful as an indication of the resulting skewness of the total cost profile. Both Figures 3a and 3b are, in reality, continuous curves, but it is computationally more convenient to divide the output into equal band widths and this accounts for the stepwise appearance.

Diagrams such as that of Figure 3b bring a new dimension to estimating. For each proposed cost value of our concrete slab, we can now associate a probability of occurrence. The likelihood that the cost will be less than £5,500 is zero, the likelihood that the cost will be between £5,500 and £7,092 is 100%. (The monetary amounts shown in the figures are not significant since any monetary value has a transient validity, but they may be regarded as indices.) The median cost of our slab is £6,062 - this is the value which has a 50% chance of being exceeded.

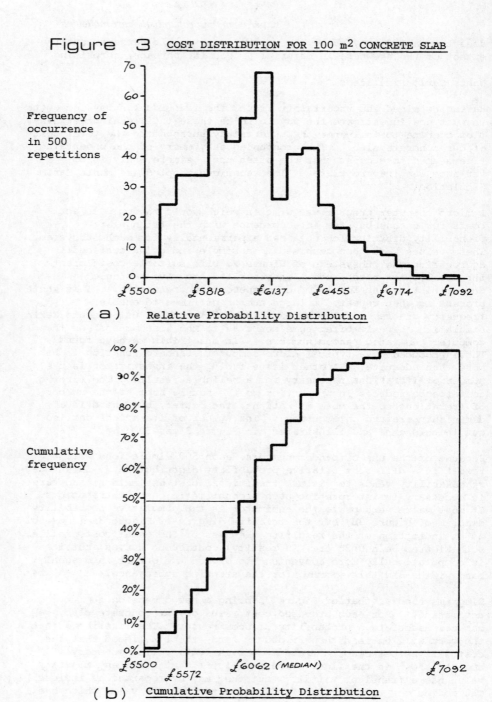

Figure 3 COST DISTRIBUTION FOR 100 m² CONCRETE SLAB

Frequency of occurrence in 500 repetitions

(a) Relative Probability Distribution

Cumulative frequency

(b) Cumulative Probability Distribution

178

It is illuminating to interpret the estimators' most probable values
in terms of Figure 3b.   By substituting the average probable
estimates for each subsystem in the inductive cost model used, the
value of £5,500 was obtained.   Inspection of the graph shows that
this has an 87% chance of being exceeded.   A strange interpretation
of most probable!

There is one major omission in the analysis described so far and in
the results illustrated in Figure 3.   Each subsystem has been
regarded as an independent entity with respect to the choice of
random values.   No account has been taken of the obvious inter-
dependence of some of the items.   For example, bad weather would
adversely affect all three labour items, to continue to regard them
as randomly unrelated would be inaccurate.   The omission of this
phenomenon, (known statistically as covariance), is partly because
its treatment is more appropriate to a further paper, and partly
because experiments have not yet revealed the correct way of
dealing with it.

Conclusions

It is hoped that this modest paper has emphasised the need for the
collection of more data revealing the nature and distribution of
uncertainties in the cost estimation of building work.   The paper
has suggested one experimental approach to the collection of such
data, and has shown how the data can be incorporated in the context
of the expanding discipline of design cost modelling.   It is hoped
that the inappropriateness of some analyses of cost uncertainty
have been highlighted, particularly those which fail to make
allowance for the extent of the skewness of uncertainty distribution.

The work is continuing, primarily in the general area of data
acquisition for probabilistic design cost models, and in the
specific area of the treatment of covariance.

Acknowledgements

The author is indebted to the contractors' estimators who freely
gave of their time and expertise to take part in this study, and
to John Raftery for his computational assistance.

The financial assistance of the Nuffield Foundation in the ongoing
research is gratefully acknowledged.

References

1   SPOONER, J.E.   "Probabilistic Estimating" J.Con. Divn.
                                     A.S.C.E., March, p.65. (1974)

2   WILSON, A.J.,   "The Optimal Design of Drainage
     TEMPLEMAN, A.B. &   Systems", Engineering Optimisation,
     BRITCH, A.L.   1, p.111. (1974)

3   WILSON, A.J. &   "An approach to the Optimal Thermal
     TEMPLEMAN, A.B.   Design of Office Buildings", Building
                                     Science, Spring. (1976)

4   JONES, H. & TWISS, B.C. "Forecasting Technology for Planning
                                     Decisions", Macmillan Press Ltd.
                                     (1978)

A PROBABILISTIC PLANNING MODEL

KRISHAN MATHUR, Dundee College of Technology

Introduction

Cost planning and cost control are at the heart of the services of a cost adviser. But to set a "single figure" cost limit for the whole building or for its elements based on the cost analysis of a single previous building is not only to progressively stifle innovatory solutions, but also to misjudge and misrepresent the true nature of cost which has not only a mean value but also variance.

1. The Basic Model - A Deterministic Model

The basic model for the prediction of cost in a cost plan is

$$[\text{Element Cost}] = [\text{Element Unit Quantity}] \times [\text{Element Unit Rate}]$$
or $\qquad C = Q \times R \qquad\qquad\qquad (1)$

Then the total cost of 'n' elements,

$$TC = C_1 + C_2 + C_3 + \ldots + C_n$$
or $\qquad TC = Q_1 R_1 + Q_2 R_2 + \ldots + Q_n R_n$
or $\qquad TC = \sum_n Q_i R_i \qquad\qquad\qquad (2)$

Traditionally in equations (1) and (2), the element unit rate, R, is taken from the analysis of a previous comparable design, adjusted for quality and price.

This approach has several difficulties.

(a) Selecting a comparable design: What are the criteria of this eclectic selection and how exhaustive is the search? To what attributes of the design are the criteria applied? If there are two equally comparable designs, it will be interesting to compare the two cost plans obtained independently from the two designs.

The question of an exhaustive search is, of course, limited by the limitations of human memory, but can be overcome by machine storage and retrieval. A prerequisite for machine retrieval will be a well defined set of attributes (cost, size, specification, or whatever they may be) and criteria.

How much can the selected attributes and criteria be able to reflect the performance of the comparable design in relation to the desired performance of the proposed design?

(b)   Adjustment for quality and quantity: This adjustment to obtain the cost of an element in the proposed design is a linear model. That is to say, it assumes that cost is linearly related to size and quality. There is, unfortunately, little evidence to support the use of this model. However, this problem cannot be overcome until there is an adequate information base to develop reliable models for these adjustments.

If the cost analyses of a large number of previous designs are available, then an arithmetic mean of the unit rates of an element from these analyses can be calculated. The mean value so obtained may be more reliable than the value from a single comparable design because errors of the element priced too high or too low are likely to cancel out.

Whether the cost from a single comparable design is used, or a mean from several designs; it is taken to be a unique or fixed value for the proposed design. This, however, does not reflect the true nature of cost. Costs fluctuate from one project to another. While one might be able to identify some of the factors that affect variation in cost, it will be quite difficult to identify all the variables affecting cost and establish relationships for accurate prediction of cost for a new project. Cost is, therefore, not deterministic. Its variation from one project to the next can at best be ascribed to an element of chance - or probability.

2.   Representation of Cost Data

The most convenient and useful method of classifying historical cost data is a frequency distribution. This is illustrated in Figure 1(a) which shows that the element cost for 15% of the sample was under £15; between £15 and £20 for 20% of the sample; between £20 and £25 for 30% of the sample; between £25 and £30 for 25% of the sample; and over £30 for the remaining 10% sample. This is also known as a relative frequency distribution, or a probability distribution if the frequencies are expressed as a fraction of unity. In terms of probability, it means that if an observation is taken from a large sample then there is a 25% chance, or probability of .25, that the cost will be in the range £25 to £30.

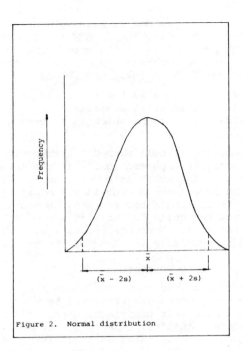

Figure 2. Normal distribution

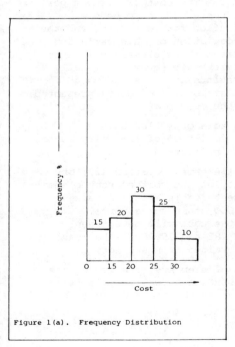

Figure 1(a). Frequency Distribution

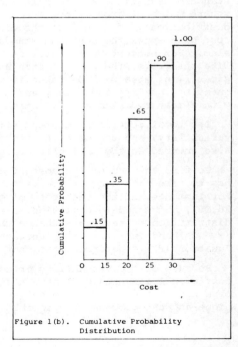

Figure 1(b). Cumulative Probability Distribution

If these probabilities are added successively we get what is called the Cumulative Probability function shown in Figure 1(b). This implies that the probability of the cost being £27.5 (median of £25 and £30) <u>or less</u> is .90; probability of .65 for the cost being £22.5 (median of £20 and £25) <u>or less</u>; and so on.

If sufficient historical data is not available to formulate a frequency distribution, it is still possible to get a rough idea of the size of the standard deviation, which is a measure of the variability of the data.

For a normal (symmetrical, bell shaped) distribution, Figure 2, 95% of the observations are expected to lie in the interval $(\bar{x} - 2s)$ and $(\bar{x} + 2s)$, where $\bar{x}$ is the arithmetic mean, and s the standard deviation. If it can be estimated that in 95% of all projects the element cost would lie somewhere between say, £5 and £30, then the length of the interval 4s equals £25 or that s = 6.25. An estimate of the standard deviation is therefore £6.25, and, crude that it may seem, it enables us to acquire a feeling for the variability of data.

3.   A proposed probabilistic approach

As suggested in the previous section, the element unit rate taken from a large number of previous projects will be a probability distribution; in obtaining this distribution each project rate will be adjusted for price and quality. The cost plan will, therefore, be presented with both the mean value of this distribution and its standard deviation, as shown in Figure 3. The cost of element per square metre of gross floor area will, therefore, also have a distribution - identical to the distribution for unit rate, and the cost plan shows the mean and standard deviation of this variation also. For elements which are not represented as element unit rate, like the frame, the cost of such an element per square metre of gross floor area will be taken from previous projects, adjusted for quantity, quality and price, and the distribution of costs presented with a mean and standard deviation in the cost plan.

The total cost of an element will, therefore, also have variability, and the mean and standard deviation of the distribution also entered in the cost plan as shown in Figure 3.

To find the sum of the cost of all elements, Equation (2), one has to take into account the variability of the cost of each element. For instance, if the mean cost of element 'i' in Figure 3 is $\bar{x}$ = 26,082, and standard deviation, s = 1,650, then for a normal distribution there is 95% chance that the cost will be between $(\bar{x} + 2s)$ and $(\bar{x} - 2s)$, i.e. between 22,782 and 29,382. The model in Equation (2) is deterministic, but the values of R are not.

Because of the variability of costs of elements, the sum of the cost of all elements will have a distribution as well. This is obtained by the use of a technique known as the Monte Carlo technique and the process is described below.

Element	Element Unit Quantity	Element Unit Rate		Element cost per m^2 of gross floor area		Total cost of element	
		Mean £	Standard Deviation £	Mean £	Standard Deviation £	Mean £	Standard Deviation £
1							
2							
3							
.							
.							
.							
i							
.							
.							
.							
n							
Total Cost							

Figure 3.   The Cost Plan

3.1   Total-Cost Frequency Distribution

It has been argued that the actual cost of an element in the proposed project cannot be predicted in a determinable way. The actual value that the cost will take from the whole range of values is probabilistic. In the absence of any other reliable information, a random number can be used to pick a value. The cumulative probability distribution, described in the previous section, for the element cost is obtained. The cost corresponding to a random number between 0 and 100 from the cumulative probability distribution is as likely a cost as any other in the distribution of cost, and therefore taken to be the predicted cost of the element in one event.

Costs of other elements are also obtained in this way. The sum of all these costs, then, gives one 'total cost' in the whole range of total cost. Other values are obtained by taking more random numbers and obtaining cost of elements from their respective cumulative frequency distributions. It should be noted that a new random number will give different costs of elements than in the first round, and therefore a different sum. If this process is carried on for a sufficiently large number of times, a distribution of total cost can be obtained. The mean and standard deviation of this distribution are then entered at the bottom of the table in Figure 3.

To obtain a distribution and a cumulative probability distribution for each one of the elements; and using 500 or more random numbers to obtain the distribution for total cost may seem a daunting task done manually, but it can be handled and processed quite conveniently on a digital computer.

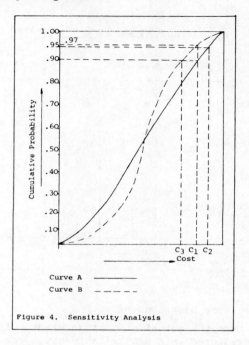

Figure 4.   Sensitivity Analysis

3.2   Sensitivity Analysis

The total cost can now be stated with some confidence. From the cumulative probability distribution, Curve A of Figure 4, one can state that there is 90% chance that the total cost will not exceed C_1. For a greater degree of confidence, say for 95% probability of not exceeding the total cost, the cost in Figure 4 is C_2. In other words, a quotation smaller than C_2 for the total cost will have a relatively smaller chance of success.

The variability of the costs of individual elements contributes directly to the variability of the total cost of all elements. The greater the variability of an individual element cost, i.e. the greater the value of s, the standard deviation, the greater it will contribute to the variability of the total cost. An element, or elements with the highest value of s in the cost plan are, therefore, the "cost significant" elements.

Greater attention is required to be paid to a cost significant element to obtain greater reliability of data and reduce the value of s. A new distribution of total cost can then be obtained using the method described with a revised value of s for the cost significant element. Comparing this revised distribution (Curve B) with the original distribution (Curve A), one can interpret the result in two ways:

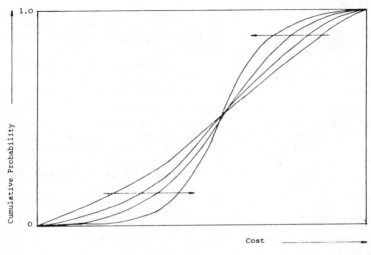

Figure 5.   Increasing reliability of cost

(a)   For the same degree of confidence (90% chance of success), the total cost expected is reduced from C_1 to C_3;   or

(b)   For the total cost of C_1, one can put in greater degree of confidence in the revised distribution - an increase from 90% to close to 97% chance of success.

This sensitivity analysis can be carried out any number of times with all the cost significant elements, or for all elements taken in turn.

3.3 From Outline-Proposals Cost-Plan to Scheme-Design Cost-Plan

As the design progresses from Outline Proposals (RIBA PoW Stage C) to scheme design (Stage D), some of the element costs (like M. & E. Services) may be obtained with greater confidence. For example, if an element is now quoted to cost in a range between £x and £y, and if the revised distribution can be taken to be a normal distribution (shown in Figure 2) then as explained in section 2, the standard deviation will be $(x - y)/4$. With this revised distribution, and other distributions remaining unchanged, a new total cost can be obtained.

As each of the elements, for which a revision is possible in the scheme design, is revised in turn in the cost plan, the increasing reliability of the total cost is illustrated in Figure 5 by the cumulative probability curves shifting gradually from right to left at the top. In the revised cost plan one would observe smaller values of s for the elements which have been revised, and also for the total cost. As explained earlier, it may be possible now to quote a smaller total cost or quote the same cost as in outline proposals but with greater confidence.

4. Application to allocation problem

When an upper limit on the total cost of the proposed project is fixed, the problem in traditional cost planning approach is to allocate this cost among various elements. To illustrate the application of the probabilistic approach, take the elements shown in Figure 6. The problem is to allocate C to C_1 and C_2; C_1 to C_{11}, C_{12}, C_{13} and C_{14}; C_2 to C_{21}, C_{22}, C_{23}; and so on.

From the cost analyses of a large number of previous designs one would get the frequency distribution of the cost of building section expressed as a percentage of the total cost. This is shown as X_1 in Figure 7. The distribution of the cost of services section as a percentage of the total cost is shown as X_2. The distribution of the cost of elements of the building section as a ratio of the cost of building section are shown as X_{11}, X_{12}, X_{13} and X_{14}. Similarly there will be distributions for other elements and sub-elements.

The steps in the allocation procedure will be as follows.

1. Establish cumulative probability distributions from the frequency distributions in Figure 7.

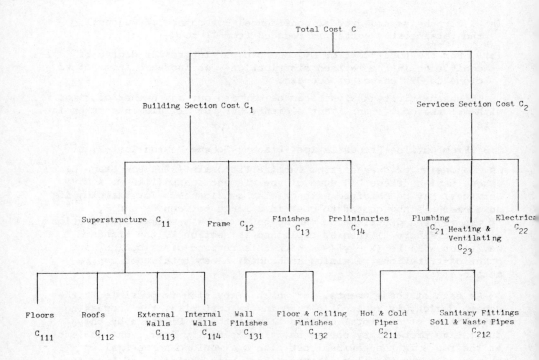

Figure 6. Allocation of Total Cost to elements

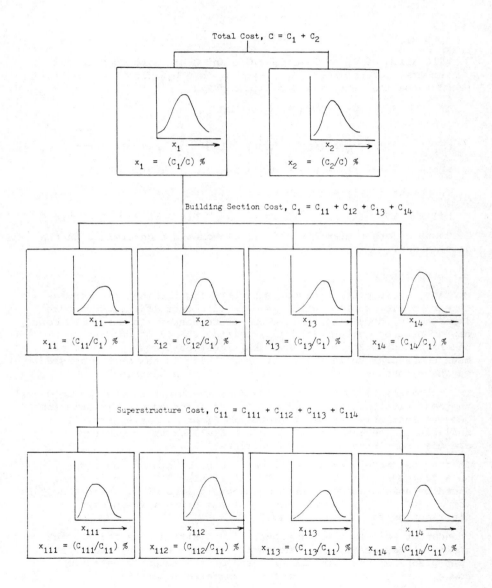

Figure 7   Distributions of cost ratios.

2. Allocation of C to C_1 and C_2: Using two random numbers obtain values W_1 and W_2 from the cumulative distributions for X_1 and X_2. The sum of W_1 and W_2 should be 100, but if this is not so, calculate values of X_1 and X_2 as follows:

$$X_1 = \frac{W_1}{W_1 + W_2} \quad \text{and} \quad X_2 = \frac{W_2}{W_1 + W_2}$$

Then $C_1 = X_1 \times C$ and $C_2 = X_2 \times C$

3. Allocation of C_1 to C_{11}, C_{12}, C_{13} and C_{14}: Using four random numbers, obtain values W_{11}, W_{12}, W_{13} and W_{14} from the cumulative distributions for X_{11}, X_{12}, X_{13} and X_{14}.

If $\quad W_{11} + W_{12} + W_{13} + W_{14} \neq 100$,

then $\quad X_{11} = \dfrac{W_{11}}{W} \ ; \ X_{12} = \dfrac{W_{12}}{W} \ ; \ X_{13} = \dfrac{W_{13}}{W} \ $ and $ \ X_{14} = \dfrac{W_{14}}{W} \ ;$

where $\quad W = W_{11} + W_{12} + W_{13} + W_{14}$

The costs of elements, then, are

$C_{11} = X_{11} \times C_1; \ C_{12} = X_{12} \times C_1; \ C_{13} = X_{13} \times C_1$ and $C_{14} = X_{14} \times C_1$

Costs of other elements and sub-elements are determined in the same manner.

5. Summary

Quoting a firm cost of elements is treating the cost planning as a deterministic problem, which is deceiving oneself and, much more importantly, deceiving those whom we are seeking to serve. A cost can at best be quoted with an associated chance of success as demonstrated in Figure 4. Has the traditional cost planning not been successful so far; one may ask. This, it has been suggested, may have been due to better cost control than planning.

The probabilistic approach proposed in the paper allows one to quote the cost with increasing degree of confidence as more information becomes available, as demonstrated in Figure 5. A sensitivity analysis can be carried out to get a feel for those elements which are most sensitive or 'cost-significant'.

In the cost model (Equation 2) each value of element unit rate, R, is a frequency distribution. The total cost, TC is, therefore, found by using random numbers to choose values of R. Repeating this for a sufficiently large number of times gives a frequency distribution for the total cost.

One cannot formulate prior probability distributions for element costs, it may be argued. One could also ask that in traditional cost planning, what are the criteria of the eclectic selection and how exhaustive is the search for a comparable design?

The proposed probabilistic approach can also be applied to the traditional elemental cost planning to allocate the total cost among various elements.

COST MODELLING: A TENTATIVE SPECIFICATION

SIDNEY NEWTON, University of Strathclyde

1. Summary

This paper looks in general at the various approaches to cost modelling.

The output required is identified, with the computer program GOAL
(General Outline Appraisal of Layouts) providing an exemplar of the
design environment.

A wide variety of cost modelling techniques are considered, and the
shortcomings of existing methods, largely attributed to the complex-
ities of cost generation (being the mechanism by which a tender price
is built-up by the contractor, and transformed into a cost to the
client). This cost generation process is shown to be structured by
context, and difficult to translate using the unit of finished work.

A specification is proposed, which provides for the development of
a cost model fully capable of producing the output required, within
the restrictions imposed by existing cost generation proceedures. The
practicability of this specification is exemplified by the computer-
based cost simulation model, ACE (Analysis of Construction Economics).

2. Cost estimating within the design activity

Like many aspects of quantity surveying, cost estimating is shrouded
in the mystique of intuition and expert judgement, and few of the good
estimators will know, even themselves, why they are consistently more
accurate than others.

The main function of a cost estimate is, usually, to predict and
translate a cost expressed in terms of operations (being a piece of
work that can be completed by a man, or a gang of men, without inter-
uption from others (1)), to a cost expressed in terms of an activity
(being an expression of the building users requirements from a space).
Thus, what is ostensibly a simple numeric process, is in reality a
highly complex mechanism for communicating cost information between
design and construction.

It is therefore unacceptable to treat cost estimating as an end in
itself; it should be considered in this role, of servicing design.

At the most general level, design is seen as making explicit proposals for how a change from some existing state to some future state

might be achieved. All facets of design are thus related either to describing a state of the system, or to the mode of its change. In the context of the built environment, it must be recognised that any particular state of the system is merely a sub-system of society at large, and that any mode of change will therefore have social, political, economic, aesthetic, and other implications.

It is a function of the designer to resolve these often conflicting aspects, and thereby determine which particular solution from the many possible solutions (the potential solution space), best suits the requirements of the user/ client. A major weakness in current design methods, is their failure to consider a large enough proportion of the potential solution space.

GOAL (2) (an acronym for General Outline Appraisal of Layouts) will serve as an example of how the ability of the computer rapidly to evaluate a variety of design alternatives, is being used to increase appreciably the scope of an iterative design method. The program represents over 10 years of experience in computer-aided design by the research group ABACUS (Architectural and Building Aids Computer Unit, of Strathclyde), and may give a clearer insight into some of the possible requirements from future cost models.

2.1 The costing process within GOAL
The basic philosophy of GOAL, as propounded by Maver (3), is illustrated in figure 1.

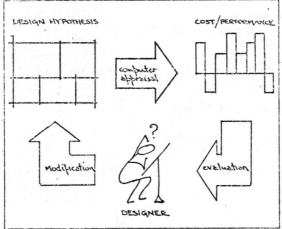

Figure 1. Concept underlying new generation models

The designer (or design team) generate a design hypothesis which is fed into the computer; GOAL then models the scheme and predicts future reality in cost and performance terms; the designer evaluates the cost/ performance profile and modifies the design hypothesis accordingly.

This general structure is shown in more detail in figure 2.

Prior to running the program, the user will construct a STANDARD DATA file for the particular building type under investigation, if a file does not already exist. The Standard Data File will contain the

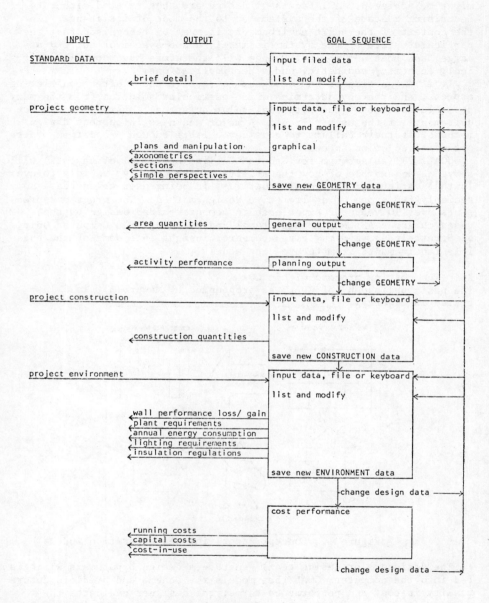

Figure 2. The general structure of GOAL

194

relative properties of alternative functional, constructional, and environmental choices, and therefore acts as a palette from which the more project specific files may be created (that is the GEOMETRY, ENVIRONMENT, and CONSTRUCTION files).

Using the design of a hotel as a typical example, the geometry of a scheme would be described as floor-by-floor sketch plans, perhaps as figure 3. Each space is uniquely identified by a series of coordinates in 3-dimensional space and labelled in respect of its functional type. Each surface of the building can then be allocated a discrete constructional type, selected from the Standard Data File. Finally, the project environment is specified in relation to the site (its location, exposure, etc.), the building (its orientation, fuel type, etc.), and financial constaints (capital cost limit, running cost limit, etc.).

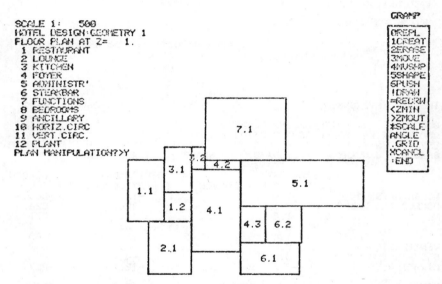

Figure 3.   Ground floor plan of proposed scheme, as output by GOAL

Several aspects of building performance can then be evaluated, typical examples being:

area quantities (see figure 4)
energy performance (see figure 5)
functional (planning) performance (see figure 6)
Energy regulations (FF3/ FF4) compliance (see figure 7)

All performance calculations, are based on well-established techniques. Capital cost performance is a simple simulation of the cost estimating process common in practice. A variety of cost structures may be described by the user in the Standard Data File, but the general form is one of a number of cost groups, each cost group having a number of elements. There is, associated with each cost element in

```
OUTPUT : GENERAL                                27-Jan-81 12:25
GEOMETRY    : HOTEL DESIGN:GEOMETRY 1

AREAS (M2)

COMPONENT           EXT.FLR  INT.FLR  TOT.FLR EXT.WALL    ROOF  VOL(M3)
    1  RESTAURANT     147.0     0.0    147.0    78.0     0.0    441.0
    2  LOUNGE         109.3     0.0    109.3    78.0   103.3    327.8
    3  KITCHEN         75.0     0.0     75.0    36.0    27.0    225.0
    4  FOYER          258.0     0.0    258.0    33.0    96.3    774.0
    5  ADMINISTR'     270.0     0.0    270.0   121.5     0.0    810.0
    6  STEAKBAR       155.0     0.0    155.0   100.5   127.0    465.0
    7  FUNCTIONS      252.0     0.0    252.0   129.0   252.0    756.0
    8  BEDROOMS       101.3  1410.8   1512.0   834.0   756.0   4536.0

TOTAL               101.3U  1410.8   2778.3  1410.0  1367.5   8334.8
                   1266.3G

WALL TO FLOOR RATIO    0.51

VOLUME COMPACTNESS     0.55
```

Figure 4.   A selection of the area quantities for a proposed
scheme, as output by GOAL.

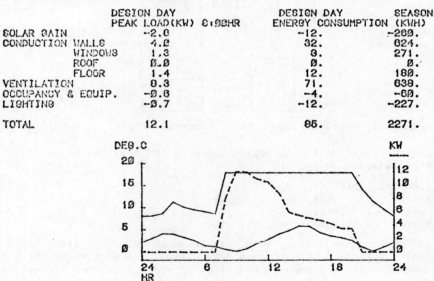

```
THERMAL ENVIRONMENT                             14-Oct-81 10:22

GEOMETRY     : HOTEL DESIGN GEOMETRY 1              GOAL V3.1
CONSTRUCTION: HOTEL CONSTRUCTION 1
ENVIRONMENT : HOTEL SITE,GLASGOW
SEASON : JANUARY   ( 31 DAYS)      SPACE :   1 RESTAURANT

                    DESIGN DAY              DESIGN DAY      SEASON
                    PEAK LOAD(KW) 8:00HR  ENERGY CONSUMPTION (KWH)
SOLAR GAIN            -2.0                   -12.          -260.
CONDUCTION WALLS       4.0                    32.           624.
           WINDOWS     1.3                     8.           271.
           ROOF        0.0                     0.             0.
           FLOOR       1.4                    12.           180.
VENTILATION            8.3                    71.           639.
OCCUPANCY & EQUIP.    -0.6                    -4.           -60.
LIGHTING              -0.7                   -12.          -227.

TOTAL                 12.1                    85.          2271.
```

Figure 5.   The evaluation of room load and energy consumption for
a proposed scheme, as output by GOAL

196

```
OUTPUT: ACTIVITY PERFORMANCE
GEOMETRY    : HOTEL DESIGN:GEOMETRY 1

TOO FAR APART
              COMPONENTS      DISTANCE  ASS.   "COST"   FACTOR

   6 STEAKBAR    3 KITCHEN      36.5    10.0    730.     2.66
   7 FUNCTIONS   1 RESTAURANT   33.9     9.0    610.     2.23
   6 STEAKBAR    1 RESTAURANT   35.3     8.0    564.     2.06
   5 ADMINISTR'  2 LOUNGE       43.5     5.0    435.     1.59
   7 FUNCTIONS   3 KITCHEN      21.7     9.0    390.     1.42
   5 ADMINISTR'  4 FOYER        22.8     7.0    319.     1.16
   7 FUNCTIONS   4 FOYER        22.3     6.0    268.     0.98
   3 KITCHEN     2 LOUNGE       21.3     6.0    256.     0.93

TOO CLOSE TOGETHER
              COMPONENTS      DISTANCE  ASS.   "COST"   FACTOR

   8 BEDROOMS    7 FUNCTIONS    24.8     1.0     50.     0.18
   8 BEDROOMS    3 KITCHEN      26.1     1.0     52.     0.19
   8 BEDROOMS    1 RESTAURANT   27.1     1.0     54.     0.20
   8 BEDROOMS    6 STEAKBAR     28.4     1.0     57.     0.21
   8 BEDROOMS    4 FOYER        15.6     2.0     62.     0.23
   6 STEAKBAR    5 ADMINISTR'   18.4     4.0    147.     0.54
   7 FUNCTIONS   5 ADMINISTR'   24.5     3.0    147.     0.54
   3 KITCHEN     1 RESTAURANT   12.2    10.0    244.     0.89

TOTAL TRAVEL COST                              8769.
STANDARD DEVIATION                                      0.77
```

Figure 6. An analysis of the activity performance of a
proposed scheme, as output by GOAL

```
ENERGY REGULATION FF3 &FF4                       27-Jan-81 12:32
ENVIRONMENT : HOTEL SITE:GLASGOW
CONSTRUCTION: HOTEL CONSTRUCTION 1
GEOMETRY    : HOTEL DESIGN:GEOMETRY 1
                                       ALLOWED  ALLOWED
                  AREA   U  HEATLOSS     AREA   HEATLOSS
GLAZED AREAS
  WINDOWS
        SINGLE  584.46  5.7 3331.42
         TOTAL  584.46      3331.42     352.50
  ROOFLIGHTS
         TOTAL    0.00         0.00     273.50

GLAZING TOTAL   584.46      3331.42             3568.20
  UNGLAZED AREAS
  WALLS
                825.54  1.7 1403.42
         TOTAL  825.54      1403.42
  ROOFS
               1367.50  0.7  957.25
         TOTAL 1367.50       957.25
  FLOORS
                101.25  1.7  172.13
         TOTAL  101.25       172.13

UNGLAZED TOTAL 2294.29      2532.79             1376.57

SCHEME SATISFIES FF4 ,PARTS:   A2

SCHEME IS NOT DEEMED TO SATISFY FF3
```

Figure 7. An appraisal of Energy regulations (parts FF3/FF4)
compliance for a proposed scheme, as output by
GOAL

each cost group;

a. an element type index which indicates if the element is non-constructional (0), a floor (1), a wall (2), glazing (3), or a roof (4),

b. a construction type number, which is an index to the construction of that type held in the Standard Data File,

c. a unit cost of the element,

d. a unit quantity code. Several different quantities are allowed, which provide a parameter more closely identifiable with the element function (see figure 8)

Allowable quantity codes are:

CODE		QUANTITY
0		unity
1	NOCC	total number of occupants of the building
11	TFA	total floor area (gross)
12	GFA	ground floor area
13	TUFA	total upper floor area
14	IUFA	internal (non-exposed) upper floor area
15	EUFA	external (exposed) upper floor area
16	TFAC	total floor area of each construction type
17	FFA	total framed ground floor area
21	TEWA	total external wall area
22	TWAC	total wall area of given wall construction type
31	TGA	total glazed window area
32	TGAC	total window area of given glazing type
33	TRLA	total rooflight area
34	TLAC	total rooflight area of given glazing type
41	TRFA	total roof area
42	TRAC	total roof area of given roof construction type

Figure 8.   The choices of quantity codes currently available in GOAL

In each case, the unit rate is input by the quantity surveyor and updated by index if necessary. The requisite quantity is then calculated and multiplied by the relevant unit rate to give an elemental cost breakdown (see figure 9).

The designer can easily and interactively modify the design data including, by graphical means, the geometry. A systematic-cycling through the program will provide the designer with a wealth of explicit information which can then be used in comparative evaluations of the differing design solutions and alternatives, in order to approach a 'better' design.

The cost model in GOAL is just a straight-forward automation of accepted costing proceedures, but its speed means that it does not interrupt the flow of the design process. It also links the user to a centralised data base, which reduces duplication of information by the various disciplines, and helps to overcome the increasing bureaucracy of statutory requirements, company strategies, etc. To a limited extent the program is able to assist in solution generation, and presents information in a familiar form.

```
COST ANALYSIS : CAPITAL COST GROUPS                27-Jan-81 12:33

ENVIRONMENT : HOTEL SITE:GLASGOW
CONSTRUCTION: HOTEL CONSTRUCTION 1
GEOMETRY    : HOTEL DESIGN:GEOMETRY 1
                                                %        /SQ.M.

A  PRELIMS & CONTINGENCIES            19253.    3.9        6.9
BC FOUNDATIONS &GROUND FLOOR          29630.    6.0       10.7
D  FRAME                              22793.    4.6        8.2
E  ROOF & ROOFLIGHTS                  29702.    6.0       10.7
F  WALLS,WINDOWS,DOORS                75194.   15.3       27.1
G  INTERNAL WALLS                     52259.   10.6       18.8
H  UPPER FLOORS                       15120.    3.1        5.4
I  STAIRCASES                         13891.    2.8        5.0
J  FIXTURES                           55565.   11.3       20.0
KM INSTALLATIONS                      93099.   18.9       33.5
NOR SERVICES                          46119.    9.4       16.6
ST EXTERNAL SERVICES                  39034.    7.9       14.1

   TOTAL                             491659.  100.0      177.0
```

Figure 9. The elemental cost breakdown for a proposed scheme, as output by GOAL

However, single elemental unit rates are only valid for a range of solutions quite close to the original, and misleading predictions may be made where the design solution is altered radically. The program also fails to give any automatic indications, in explicit terms, as to how well a proposed solution compares to the population of previous, similar projects — a design performance can only be judged as good or bad in relation to other projects. Perhaps more significantly, the costing mechanism employed is inflexible, and unsuited to the wide variety of alternative design methods adopted in practice.

Finally, it is apparent that this particular costing technique is fundamentaly incompatible with prefered techniques later in the design process. Design is a dynamic process, with decisions being made and subsequently changed. If a cost model is to mimic such a process, it too must allow previous decisions to be reviewed.

Despite these shortcomings, GOAL is used in a number of commercial architectural and engineering practices, and public bodies throughout the U.K. and Europe. Many schools of architecture are integrating the program into sketch design projects (4), and it is apparent that the basic concepts are well recognised.

2.2 The potential of alternative techniques
The very basic critique of the automated costing process in GOAL, has followed several guide lines, proposed elsewhere in more detail (5), for judging the potential of a costing system. These guide lines were also applied to an extensive survey of other existing approaches to cost estimation, and gave a matrix as figure 10.

Current practice: The only real value of current practice would appear to be that the quantity surveyor shields the designer from the true complextities of cost. Information is in a familiar form, and cost estimates can be produced for even the most innovative design solution.

Automated costing process: By automating the costing process, several of the limitations observed in current practice can be overcome,

	Current Practice	Automated Costing Process	Statistical Analysis	Parametric Study	Theoretical Analysis	Simulation	Optimisation
1.Does it minimise conflict ?	S	LS	F	F	F	S	LS
2.Does it assist in solution generation ?	F	S	S	LS	LS	LS	S
3.Does it relate project to other comparable projects ?	F	F	S	F	F	F	F
4.Does it link the user to a central data base ?	F	S	F	F	LS	LS	LS
5.Is it quick ?	F	S	S	S	F	S	S
6.Is it adaptable ?	S	F	F	F	F	F	F
7.Is it dynamic ?	F	F	F	F	F	LS	F

Where: S - satisfactory
        LS - limited satisfaction
        F - failure to satisfy

Figure 10.   The performance of existing cost modelling techniques

but at the cost of reducing flexibility, and restricted validity.

Statistic analyses: Statistic analyses have the clear advantage over all other techniques in that they can relate a proposed project to other, comparable projects. However, such techniques are often greatly restricted by the lack of suitable data, and accuracy tends to deteriorate quickly over time. Despite this, they are capable of prescribing design decisions and can greatly assist in solution generation.

Parametric studies: Parametric studies have generally failed to meet any of the requirements. The range of validity is even more restricted than statistical analyses, and they are of little use except where the level of provision is well defined and relatively static. However, given favourable conditions, they are quick and simple to use, making them quite useful predictors of cost relationships.

Theoretical analyses: To date theoretical analyses have all but completely failed to relate to the requirements of design. This, despite the fact that a theoretical model is a prerequisite of ALL practical applications of cost modelling, though not all will state the theoretical base as formally as Brandon (6).

Simulation: Simulation models would appear to have faired better than most, primarily because they are very much akin to current practice, but with the added bonus of being computer-based. By promoting the use of a prefered estimating technique earlier in the design than at present, the cost information becomes more consistent and engenders a more dynamic decision-making process. Somewhat unfortunately, the mechanisms of those models already developed are inflexible, and their empirical base dictates considerable caution in any direct application.

Optimisation: The utility of optimisation models shows a marked

down-turn as the problem definition becomes more complex, certainly
in terms of the 'dimensionality' (that is the number of disparate
criteria on which a solution is judged). The explicit expression of
an objective function, is likely to be alien to many designers and
will create conflict , most especially, when objectives change during
the course of a project. However, the concept of optimisation is a
very powerful one and the capacity to prescribe optimal, or near optimal, solutions for discrete, micro problems, should not be lightly
dismissed.

That so many attempts at cost models should fail to meet the requirements of design, suggests the cost output achievable is constrained in some respect by the way in which cost is generated, and
that these constraints are being over-simplified. A further indication is given by the fact that the use of any model is restricted by
the variability and relevance of its input data.

It is therefore intended now to investigate the mechanics of cost
generation.

3. The cost generation process

Figure 11, outlines the usual cost generation process.
Having dispatched enquiries to sub-contractors and, where necessary, visited the site, the work is arranged into work packages, or
operations. The content of each operation will be unique to the particular project, and to benifit from past experience, each should be
analysed in more detail. In theory, this gradation would be down to
a finite level of perhaps net energy, etc. In practice, the level of
detail is constrained by the increasing cost associated with identifying and controlling smaller and smaller units of measurement.
With the current standards of information processing, most contractors would appear to categorise work in terms of labour, materials,
plant, and sub-contractors work (7). But even at this level of detail, the inherent inaccuracies are quite substantial.
Pricing of the resources is generally more straight-forward, although particular conditions of contract (fixed price, etc.), an unstable client, or unusual location, etc., may severely complicate the
process.
Tender adjudication can distort the cost even further, producing a
tender price based on variable, inherently inaccurate assumptions,
and which is only notionally allocated to units of finished work. It
is apparent, however, that such inaccuracy is not not a random process
but related to the widely differing assumptions each estimator will
make regarding the labour output, material use, plant efficiency,
level of attendance on sub-contractors, firm policy, etc. Clearly,
variation in cost is structured by the <u>context</u> in which each estimate
is prepared.
In the Bills of Quantities, items are generally grouped into appropriate work sections; to be of use in comparing the cost of a given
function in one project with an equivalent function in another project, the items must be rearranged. Cost analyses are patently more
formalised than the tendering process, but the allocation of each
Bill of Quantities items to a functional element is still based on an

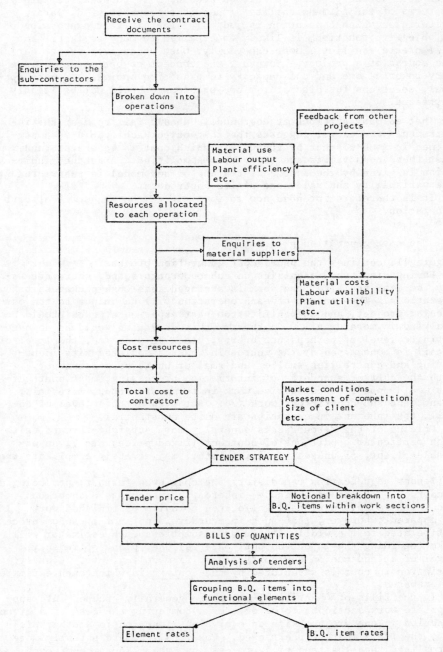

Figure 11.   The cost generation process

individuals' interpretation of general principles, and therefore remains subjective. Thus, although the cost of an activity is ultimately dependent upon the cost of an operation, the relationship can only be expressed in terms of a language (the unit of finished work) which has an imprecise syntax. This is an important concept, since it follows that, with current practice, cost relationships <u>cannot</u> be absolute.

The foregoing would suggest that the characteristics of cost generation impose quite severe limitations on the use of cost data, and gives some explanation for the paucity of success in the modelling of cost relationships. For example, if costs are context dependent, then the single rate costs used by the vast majority of existing models to relate to all contexts, are inappropriate. Further, the many cost models which assume absolute/ static cost relationships are oversimplifying the problem, and will rapidly become obsolete.

4. The specification for a new generation of cost models

Cost has been identified as a poorly understood sub-system to design. As such, existing cost models have been evaluated in relation to several guide lines which proposedly judge the potential of a costing system to meet the various requirements of the design activity.

The failure of so many, so diverse approaches, suggested that the cost output was constrained in some reapect by the way in which cost is generated. An analysis of the cost generation preocess indicated that:

(a) The cost data produced is context dependent, and inherently inaccurate.

(b) Cost relationships are never completely manifest, but implicit in the process as a whole.

The general specification for a cost model may therefore be outlined as:

- Initial formulation of the model must not require a detailed knowledge of cost.
- Cost information is to be consistent throughout the design process.
- Information is to be presented in a familiar form.
- Assistance is to be given in solution generation.
- The model should relate a proposed design to previous, comparable projects and indicate the variability of that relationship.
- The model must link the user to a central data base.
- Cost information is not to interrupt the 'flow' of design.
- The costing mechanism is to be highly flexible and adaptable to the most innovative design solution.
- The model should facilitate a dynamic decision-making process.
- The user must be allowed to interogate the results in great detail.
- The mechanics of cost generation are to be 'transparent', so that cost consequences can be traced through all inter-relationships.

4.1   ACE: Analysis of Construction Economics
In order to test the practicability of this specification, the first
version of the computer model ACE (Analysis of Construction Economics)
has been developed.

A computer-based model was considered essential in the early stages
both to link the user to a central data base, and to give the requisite speed of computation. Because the costing process is poorly understood, a simulation approach was adopted in which the model was
structured closely to reality. The prefered mode of cost estimating
was taken as using approximate quantities, and a conscious effort
made to apply this same technique to all stages of the design process.

The first step was to identify the proceedures adopted in costing
a design solution. The unit rates and corresponding quantities
necessary to cost a range of alternative designs, were determined.
For example, for the substructures, pad, pile, and raft foundations
were considered. The total cost of each alternative was analysed into
its constituent parts, and each expressed as a quantity and a unit
rate. In pad foundations, the total cost was taken as the sum of
costs for concrete, excavation, reinforcement, formwork, and earth-
work support. Concrete costs would then be a unit rate for concrete
in bases, multiplied by the total volume of base concrete. Using this
procedure for the complete range of alternatives, nearly 100 such
equations were identified.

Attempting to model reality as closely as possible, each of the 100
types of unit rates were then examined to determine what would influence the value of each. For example, the unit rate for a water pipe
was held to be dependent upon the pipe diameter, and type of jointing
used. Unit rates were built-up for every possible combination of pipe
diameter and type of joint, so that the actual rate used in any particular context, would be determined by that context. To then fully
describe the conditions of any particular design, required a total of
approximately 150 variables.

A proposed design is thus being costed using quite detailed approximate quantities and unit rates, whose values are determined by the
ambient conditions.

But how would this detailed costing process be applied to the very
early stages of design ?

With just a minimal skeleton of information describing area, shape,
quality, etc., it was possible to project a sufficiently detailed imp-
ression of a building by using statutory regulations, codes of practice, simple rules of thumb, etc. Initially this is a crude image of
a simple building, but as design proceeds, the image becomes more refined, and concrete. The 'fleshing-out' process necessitates simple
structures to be designed, heat losses to be calculated, lift strategies to be effected, etc., and a total of 250 variables are now cal-
culable

4.2 Early trials of the cost simulation model, ACE
By way of an example, the program ACE was given the following information:

- gross floor area ⎤
- plan area   ⎬ any two from these three
- number of storeys ⎦
- building type (offices)
- if building is to be framed (yes)
- tender date (8:8:1980)

By varying the number of storeys from 1 to 30, the effects of
height can be expressed against a cost per square metre of gross floor
area. Figure 12 shows how the relationship varies depending on whether it is the plan area, or the gross floor area, which is held constant, and with their values. It is apparent, that a change in context
will change the values, and therefore that the relationship is good
for only this particular instance. However, the general U-shaped
trend compares well with the hypothesis outlined by Flanagan (8);
that a cost per square metre of gross floor area will decline initially as the number of stories increases, but will eventually rise.

It is, of course, also possible to study particular changes in
more detail, as shown in figures 13, 14, and 15.
The general system is illustrated in figure 16.

5. Future prospects and problems

It should be stressed that the current version of ACE was developed
with the intention of assessing the practicability of an, as yet,
tentative specification. As such, it is based on an idiosyncratic
interpretation of cost, and is inelegant in many respects. However,
the output obtained would suggest that, even in this rickety form,
the program manages to model cost relationships with reasonable accuracy.

Two alternative strategies are now being considered for the remaining sixteen month of research.

STRATEGY A: The existing model is used to study the inherent cost
relationships in some detail. The sensitivity and stability of a
wide range of, hither to ignored, design variables may be investigated
under controlled conditions. Against this approach is the difficulty
of representing an unstable, multi-variate relationship coherently;
and the immensity of a problem structure which already exceeds 250
dimensions.
STRATEGY B: ACE, version 1, is discarded and a wholly restructured
model developed. It is expected that each quantity surveyor will
have his own methods of expressing cost allowances, and by using
rather sophisticated programming techniques, it should be possible to
allow each individual user to determine the complete mechanism by
which costs are generated.

Clearly, the latter approach would substantially reduce the time
available to investigate cost relationships in any detail.

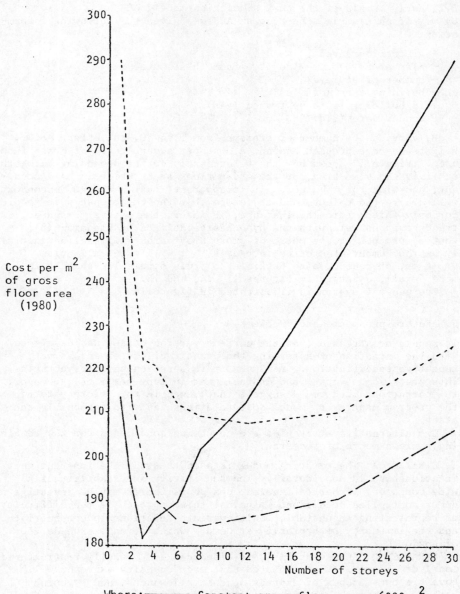

Figure 12. Selected relationships between cost per square metre of gross floor area and height, as output by ACE

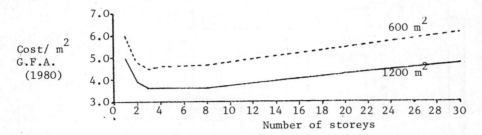

Figure 13. The effect of height on the cost of substructures, per square metre of gross floor area, for plan areas of 600 and 1200 m^2

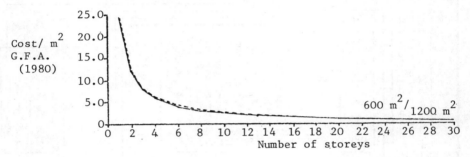

Figure 14. The effect of height on the cost of the roof, per square metre of gross floor area, for plan areas of 600 and 1200 m^2

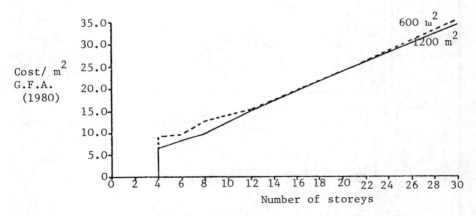

Figure 15. The effect of height on the cost of lift installations per square metre of gross floor area, for plan areas of 600 and 1200 m^2

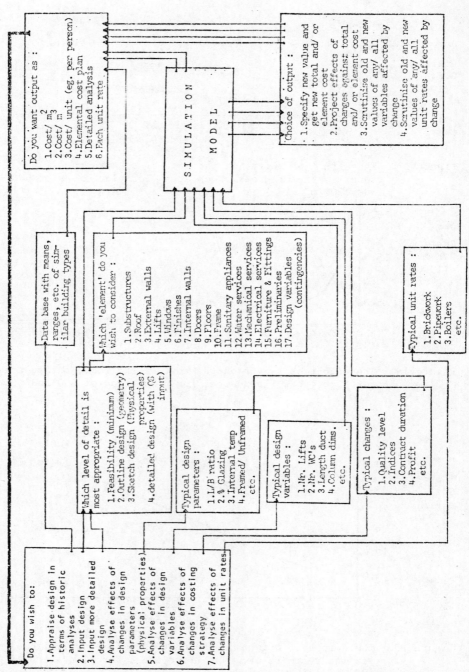

Figure 16. The general control system envisaged for ACE

208

Aknowledgements

This work has been funded by the Science and Engineering Research Council through a research studentship.

References

1. Skoyles, E.R., Introduction to Operational Bills, BRS Current
      Paper CP/ D32, 1963.
2. Sussock, H., GOAL - General Outline Appraisal of Layouts, ABACUS
      Occasional Paper no. 62 (revised), 1981.
3. Maver, T.W., A Theory of Architectural Design in which the role
      of the Computer is Identified, Building, 1970.
4. Schijf, R., Modular CAAD courses - A vehicle to discuss CAAD edu-
      cation, CAD82 Proceedings, Butterworths, 1982.
5. Newton, S., ACE: Analysis of Construction Economics, Internal Re-
      port, University of Strathclyde, April 1982.
6. Brandon, P.S., A Framework for Cost Exploration and Strategic
      Cost Planning in Design, Unpublished MSc Thesis, University of
      Bristol, October 1977.
7. Morrison, N. and Stevens, S., A Construction Cost Data Base, The
      Chartered Quantity Surveyor, June 1980.
8. Flanagan, R. and Norman, G., The Relationship Between Construct-
      ion Price and Height, B & QS Quarterly, Summer 1978.

AN ECONOMIC MODEL OF MEANS OF ESCAPE PROVISION IN
COMMERCIAL BUILDINGS

DEREK SCHOFIELD, JOHN RAFTERY, and ALAN WILSON
Liverpool Polytechnic

INTRODUCTION

This paper describes the development of a mathematical model to
investigate the economic consequences of differing strategies
employed to achieve fire safety within commercial buildings.   The
work brings together two continuing strands of research in this
Department, namely an investigation into fire defence in buildings,
and a study of economic models for use in building design.

The model is intended to aid the designer of a building in complying
with fire regulations, and to provide for him an illustration of the
cost to his client of such compliance.   Thus, the economic measure
of effectiveness used as the objective function is the discounted
life cycle cost to the client of the loss of lettable space, in
addition to the cost of the stairway itself.

Perhaps the true value of the model emerges, though, when it is used
to compare the economic consequences of complying with the various
codes and standards applying to fire defence in commercial buildings,
both in the U.K. and overseas.   Indeed, the potential of the model
to code writers in framing future fire regulations is, it is
suggested, considerable.

THE HYPOTHESIS

The need for the model arose out of research into the economic
assessment of means of escape from fire in multi-storey office
building currently being undertaken in the Department of Surveying[1,2].

The research commenced in 1979 and is based upon the hypothesis that
a systems approach to fire defence, taking account of early detection
and automatic extinction of the fire, together with pressurisation of
staircases would enable a reduction in means of escape provision and
that the resulting increase in lettable area would pay for the
additional fire defence.

A case study approach was considered to be inappropriate because of
the lack of suitable existing buildings.

THE PROBLEM

An examination of the regulations, codes of practice and statutes
relating to occupancy loads in multi-storey office buildings in the
United Kingdom shows that there is considerable variation.   Where

layouts have not been agreed at the design stage, which applies in
speculative development and also in many owner occupied buildings,
very low values of maximum floor area per person are required, which
may well result in a larger means of escape provision than will be
necessary once the building is occupied.   In recent years occupancy
loads have been falling as a result of increasing mechanisation of
office tasks.   The occupancy loads are based upon the gross floor
area, excluding staircases and lavatories, and vary in the United
Kingdom between $5m^2$ and $10m^2$ per person, as shown in Figure 1.   In
practice, however, occupancy loads may normally be as high as $25m^2$
per person.   A recent study by De Wolf and Henning[3] of 77 Canadian
Government offices disclosed an average usable area of $18.6m^2$ per
person.

In addition to variations in occupancy load, separate means of escape
legislation applies to the Inner London Boroughs, England and Wales,
except for the Inner London Boroughs and also in Scotland.   Further,
office buildings are also subject the the provisions of the Offices,
Shops and Railway Premises Act 1963, The Fire Precautions Act 1971,
and Section 78 of the Health and Safety at Work Act 1974, as outlined in Figure 2.

Apart from the different means of escape provision, recent research
undertaken by Pauls[4] at the Canadian National Research Council has
questioned the staircase requirements upon which the United Kingdom
regulations are based.

MODELLING APPROACH

The experience of previous work indicated that it would be beneficial
to ensure that the problem to be modelled was well defined from the
beginning[5,6].   Paradoxically, it was also decided that the model
should be as "open" as possible.   By this, we mean that while it is
acceptable for a user to be constrained by the parameters of the
particular problem with which he is dealing, it should not be
acceptable for him to be constrained by the model per se.   Although
most workers in the field would lay claim to this principle, it can
be argued, justifiably, that it is rarely, if ever, achieved.

The STEPS (Staircase Evaluation Program Suite) model was built and tested,
taking account of the above principles, during the first quarter of
1982.   The model considers only multi-storey office buildings and
is restricted to dog-leg staircases.   Figure 3 outlines the system
logic.   It is apparent that there are two types of input to the
model.   Firstly, there are "design decisions", these are decisions
which are under the immediate control of the designer, although they
will normally be related to the designer's solution to the problems
posed by other building subsystems.   Examples of this type of
decision are storey height and (within certain bounds) the catchment
area to be served by the staircase.   In terms of the model, these
are, of course, the independent variables.   Before any dependent

	Statutes/Guidance	Occupancy	Qualifications
Health & Safety at Work Publication No. 40	Guidance	$3.7m^2$	Gross Floor area per person, excluding staircases, corridors, lavatories, etc., for individual rooms.
Building Standards Scotland	Statutory	$3.7m^2$	As Above
Offices, Shops & Railway Premises Act 1963	Statutory	$5m^2$	Gross floor area per person, but excluding staircases, lavatories, etc.
Health & Safety at Work Publication No. 40	Guidance	$5.1m^2$	As above
Building Standards Scotland	Statutory	$5.1m^2$	As above
British Standard Code of Practice No. 3	Guidance	$7m^2$	Gross floor area per person
British Standard Code of Practice No. 3	Guidance	$9.3m^2$	Gross floor area per person, excluding staircases and lavatories
Greater London Council Code of Practice	Guidance	$10m^2$	As above

FIGURE 1 - OCCUPANCY LOADS IN STATUTES & CODES OF PRACTICE

Designated Area	Statutory Instruments		Guidance	
	Acts	Regulations	Codes of Practice	Recommendations
Inner London Boroughs	London Building Acts (Amendments) Act 1939 Sections 36 - 35 --------		GLC Code of Practice Means of Escape in Case of Fire	Health & Safety at Work Public-ation No. 40
England & Wales	Offices Shops & Railway Premises Act	Building Regs. 1972 & amends.	British Standard Code of Practice CP3	Health & Safety at Work Pub'n. No. 40
Scotland	Offices Shops & Railway Premises Act	Building Stds. (Scotland) (Consolidation) Regs. 1971		Health & Safety at Work Pub'n. No. 40

FIGURE 2 - STATUTORY INSTRUMENTS & GUIDANCE WHICH APPLIES TO MEANS OF ESCAPE IN OFFICE BUILDINGS

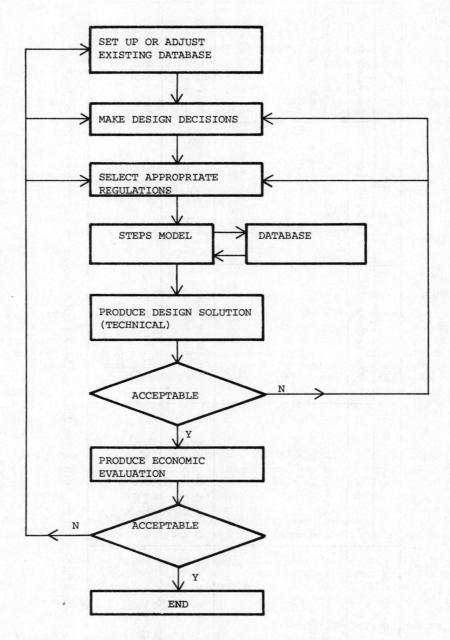

Figure 3   STEPS MODEL SYSTEM FLOWCHART

variables can be produced, it is necessary to take account of the
problem constraints, which bound the feasible solution space.   In
terms of STEPS these are known as "regulatory decisions".   These
decisions are not under the control of the designer (see Fig. 2).
Once it has been decided by the relevant planning or fire authority,
that a particular set of planning regulations apply, then the number
of possible design solutions becomes constrained.   The model
effectively becomes "closed" at this point.   The feasible space is
now defined and the solutions within it can now be evaluated.

DATABASE

The database for STEPS was designed so that it would be possible to
test proposed solutions for sensitivity, not only to design and
regulatory decisions but also to changes or relative changes in the
database.   Two types of data are held in this data file.

(i)   Items; which accurately reflect the relative costs of
         various construction components.   Price book rates are
         used, which we consider useful only for comparing solutions,
         not for cost or tender prediction.

(ii)   Relationships; functions which are used in the methods of
         economic appraisal of solutions.

Evaluation of particular solutions is carried out using a life cycle
model which incorporates the opportunity cost of the design solution.
The opportunity cost is calculated with respect to an arbitrary
datum of zero.   Although, clearly, no proposed solution can have
zero opportunity cost with respect to other solutions, the datum is
necessary so that comparisons may be made between solutions.   Its
use has a further benefit in that it precludes the temptation to use
the output for cost prediction.

THE MODEL

At an overall level STEPS seeks to model the relationships between
the design parameters and the possible solutions.   The measures
adopted are both technical and economic.   It is an interactive
model and the user is given the opportunity to explore the relationships among the following; floor area, occupancy load, width and
number of staircases, construction and/or total life cycle costs.
Clearly, each solution contains many attributes.   The user has the
choice of examining these in detail, or of reducing the problem to
a single criterion.   The latter approach is usually adopted for
sensitivity testing.   The model user should be encouraged to "play"
with the model in order to get a feel for the nature of the problem
and its solutions.

What can the user expect from the model?   Essentially, it provides
information which expresses his design options in terms of the
solution to meet a particular set of decisions.   Further, interaction with the model tells him about the sensitivity and stability of

of the solution given changes in any of the parameters on which it was produced.

At present, the model deals only with staircases designed under GLC Regulations Table 3 for commercial buildings and the work undertaken by Pauls in Canada concerned with staircase widths.   It is hoped in the very near future to incorporate GLC Regulations Table 2 and Code of Practice No 3.   Figure 4 is a schematic representation of the use of STEPS in terms of the inputs required of the user and the basic output produced by the model.   There are a variety of functional relationships embodied in STEPS, an example for the calculation of the dependent variable W (width of staircase) is as follows:

$$W = f(n)$$

where $f(n) = \dfrac{(10 \times N - 100)}{1000}$

subject to constraints

$$1.83 \geq W \geq 1.1$$

where W = width of staircase (meters)
      N = occupant load (number)

The model was tested by running a design problem in parallel with a manual problem, checking for discrepancies between the model solution and the solution and appraisal by hand.

SOME INITIAL RESULTS

Work to date enables the relationship amongst the independent and dependent variables to be analysed as illustrated in Figures 5 and 6. In Figure 5, the change in catchment area has been plotted against the total discounted cost, firstly for a staircase complying with Table 3 of the GLC Code of Practice and, secondly, for one based upon an interpretation of Pauls research work[4].   In both examples, the costs relate to a 33° slope reinforced concrete doglegged staircase, 3.6 metres floor to floor, where the occupant load is $10m^2$ per person, building life 25 years, rate of discount 5% and the rental level of £100 per m^2 per annum.

The effect on total discounted cost of a similar 33° slope stair- case complying with Table 3 of the GLC Code of Practice where different storey heights and catchment areas are analysed is shown in Figure 6.   The occupant load, building life, rate of discount and rental level being the same as in the previous example.

In addition to the two examples, it is also possible to trace the financial effect of changing the staircase slope or varying the occupant load as well as analysing the effect of different interest rates, rental levels and building lives.

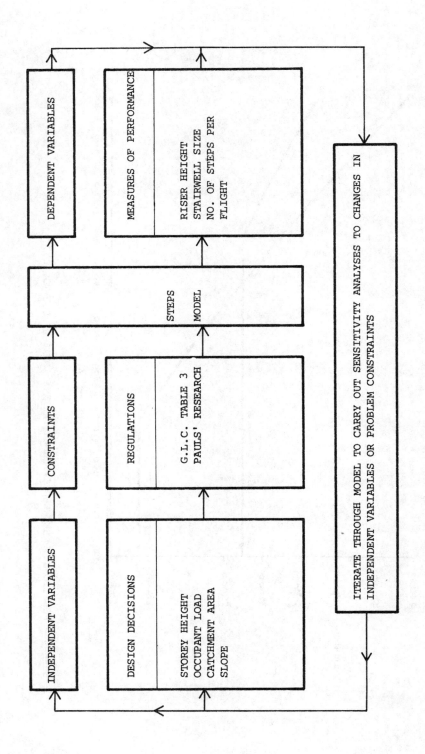

Figure 4   STEPS MODEL - SCHEMATIC PRESENTATION

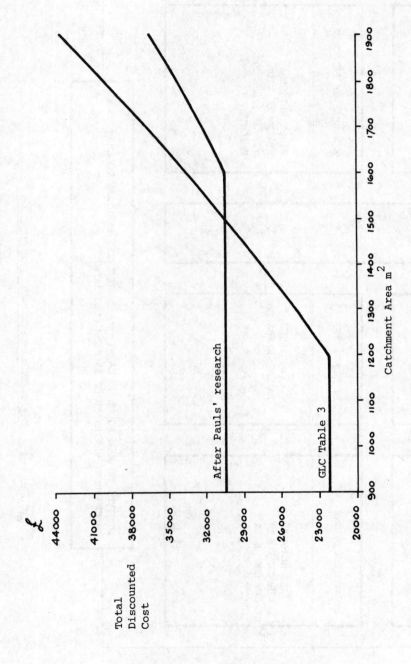

FIGURE 5 - RELATIONSHIP BETWEEN CATCHMENT AREA & TOTAL DISCOUNTED COST

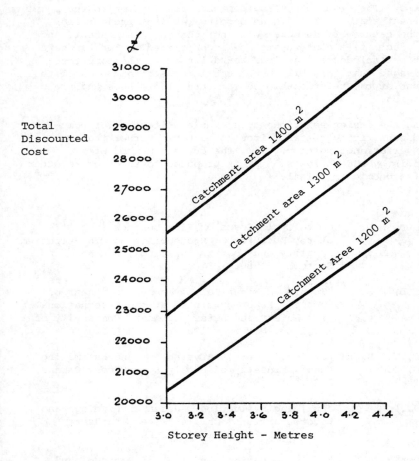

£

| Total Discounted Cost | | |

FIGURE 6 - RELATIONSHIP BETWEEN STOREY HEIGHT AND TOTAL
DISCOUNTED COST FOR CATCHMENT AREAS OF 1200,
1300, and 1400m^2.

FUTURE DEVELOPMENTS

STEPS runs on a North-Star Horizon micro computer with twin floppy
disk drives.   The description and comments above refer to the model
in its present state.   This is an ongoing research project and
although the nature of the model will not change, enhancements and
additional facilities continue to be incorporated.   The hardware
configuration includes a digitiser board linked to the computer.
It is envisaged that this will be of use in expanding the model to
take account of the relationship between travel distance and staircase provision.

In reducing the design solutions to a single criterion for comparison and sensitivity testing, from an economic rather than a
technical performance point of view, the current method of
evaluation is by DCF analysis.   It is proposed to incorporate other
methods of economic appraisal.

REFERENCES

[1]   SCHOFIELD, N.D.   "Life Cycle Costs of Fire Defence in Multi-
                             Storey Buildings" Proceedings of CIB Symposium
                             'Quality and Cost in Buildings', Lausanne,
                             September 1980.

[2]   SCHOFIELD, N.D.   "Towards a Financial Assessment of Means of
                             Escape" Paper presented at the International
                             Life Safety and Egress Seminar, University of
                             Maryland, November 1981.

[3]   DE WOLF, T.J.D. &   "Office Space: Analysing Use and Estimating
       HENNING, D.N.   Need", Ministry of Supply & Services, Canada,
                             1980.

[4]   PAULS, J.L.   "Building Evaluation: Research Findings and
                             Recommendations" Ch. 14 "Fires and Human
                             Behaviour", John Wiley & Sons, 1980.

[5]   WILSON, A.J.   "Building Design Optimisation Methodology"
                             Proceedings of CIB Symposium 'Quality and Cost
                             in Buildings', Lausanne, September 1980.

[6]   RAFTERY, J.J. &   "Some Problems of Data Collection and Model
       WILSON, A.J.   Validation" Paper for Research Seminar,
                             Liverpool Polytechnic, March 1982 (unpublished).

GENERAL PURPOSE COST MODELLING

LESLIE HOLES,   Construction Measurement Systems Ltd., and
RAY THOMAS,   Leicester Polytechnic

Introduction

Conventional building cost predictions are based on calculations
whereby one or more variable quantities, which are derived from
facts known about the project at the time, are multiplied by suitable
money rates, representing predictions of facts, as yet uncertain.

As the designing proceeds and more decisions are made, a greater
variety of variables becomes available for incorporation into the
calculations. Hence the use of, firstly, a functional unit rate,
then a rate per square metre of gross floor area or various element
unit rates and, ultimately, the unit price rates entered against the
quantities of measured items in the bill of quantities. These last
are intended to cover the cost of the resources likely to be used
during the construction of each item, and rates used at earlier
stages will probably be based on combinations of such items from
other projects. Thus, it would seem that all data on money rates
implicitly represents, or 'models', the consumption of materials and
components and the work of operatives with plant, as well as the
expenditure that will motivate these events.

Our approach is to relate building measurements to likely resource
consumption rates, and then to apply unit resource price rates that
reflect appropriate market conditions; project and general overheads
are separately considered. That is to say, we concentrate on the
need for resources and regard cost as largely derivative.

Obviously, a great deal of data is involved, and our purpose in
this paper is to outline how a computer can be used to deal with it.
Essentially, we set out to exploit the regularities to be found in
the individual facts about buildings and their construction and also
in the relations between them. Whatever the stage in the design
process, we use much the same resource data and the same computer
programs, storing within the computer the portable data that
represent these regularities.

Computer systems

There are two complementary computer-aided systems:

(1)   a system originally developed for general contractors'
      estimators and in use, as such, commercially. This can be
      used for cost modelling either on its own, provided some
      approximations of measurements are acceptable, or in
      conjunction with
(2)   a system for measurement and the production of a bill of
      quantities, which can also be used on its own.

   Both computer systems are content-free; that is, they provide a
structure for project data, whatever it is, which the computer will
relate and process under the control of the user. Data in frequent
use can be stored and recalled using simple references. The systems
are set up by creating computer files of such data, representing the
kinds of activities that are being modelled. Project input data is
prepared, using natural language for the descriptors and making
reference to entries on the files; the computer amplifies this
initial input by copying repeated data and by recalling data from
the files. This amplified input is processed in accordance with
programmed instructions and directions included in the data.

Resource and cost estimating

The Construction Industry Resource and Cost Estimating system that
we call 'CIRCE' concentrates on relating item quantities to
resources, treating the resource cost rates as simple facts which
can be stored and changed at will. Its principal characteristic is
that it will combine data on basic resources in ways that match the
ways in which the resources themselves will actually be combined on
the site; for example, to constitute a workgang, to make concrete,
to construct a concrete foundation or even the whole of a foundation
wall. These we refer to as "collectives of resources". The computer
simply responds to the data in the input, which will consist of the
details of basic resources and the specifications of the relations
between these, the collectives and the item sizes and quantities.
We have found that a ratio of no more than four constituents to one
collective works well in practice. Since each constituent can,
itself, be a collective, it is possible to relate an item quantity
to up to four, or sixteen, or sixty-four, or whatever number of
basic resources is appropriate. Each reference to a resource or
collective can be associated with at least two (and up to five)
numerical factors. The ability of the computer to store and recall
data and to implode and explode it as required presents both a
challenge and an opportunity, and CIRCE has already been used in a
variety of ways.

To start with, one must consider what regularities there are in the facts about buildings and building processes and their relationships, and seek to reflect these in project data.

Such regularities can include the following:

(1) the occupations of operatives and plant employed in the enterprise and the materials and components they habitually use,

(2) collectives of operatives and plant when working on particular processes (we use a file with about one hundred such collectives in order that all labour hours and costs can be allocated to production cost control centres),

(3) collectives of materials such as mortars and formwork,

(4) collectives of the foregoing, representing resource requirements apportioned to each measured unit of a bill item,

(5) collectives of (4) that will have the same item quantities; here it becomes necessary to consider what regularities there are in the buildings themselves.

As a result of the brief and the thoughts and calculations of the designer, the first quantitative information on the building will probably be in the form of location or sketch drawings. These will simultaneously indicate both the spaces to be provided for the occupiers and the physical structure that will enclose them.

At the risk of being obvious, it must be said that, generally, there will be horizontal constructions above and below the rooms, etc., and a vertical structure between one and the other and next to the environment. Also, that the junctions between these constructions will probably be at right-angles; that is, the building will be orthogonal. Thus, the various rooms and other 'cells' nest within the building envelope rather like a box of boxes.

We know that the finishes to the ceilings, walls and floors of the cells are made from different materials and that, although there are likely to be skirtings, and sometimes cornices, there are seldom any extra treatments at the junctions between the finishes to adjoining walls. Thus, for most projects, we shall be concerned with the surfaces at the top, at the sides, and at the bottom of each cell, and also with the junctions between the top and the sides, and between the sides and the bottom.

These three kinds of surface and two kinds of junction are also present in the building envelope in the form of the roof, the external walls, the ground floor, the eaves or verge, and the foundations to the external walls. The constructions at such surfaces have two variable dimensions and are likely to be regarded intuitively as areas. In the case of junctions, only their lengths vary, whilst the storey height is likely to be constant for each floor. These variable dimensions are, essentially, also those of the spaces they help to enclose.

Here, we begin to diverge from both CI/SfB Table 1 (1) and the Standard Form of Cost Analysis (SFCA) (2). Although in the former, elements are regarded as being "parts with particular functions", in practice, they appear to act more like classifiers of items such as those which are implied or specified by name in equivalent sections of the SFCA. However, if constructions at the surfaces of spaces are regarded as the elements proper, they become more easily recognised as entities with functions. This allows junctions to be similarly distinguished, their role being to maintain the continuity of functions between neighbouring elements.

Space-related dimensions

In view of the foregoing, we hypothesise that, in the case of orthogonal buildings, we should be able to generate the areas of elements and the lengths of junctions by copying their scalar dimensions (or 'scalars') from those of the spaces. Moreover, we should be able to generate the dimensions of measured items, representing the parts of elements and junctions, by modifying such copied dimensions if necessary, as is done manually. The computer could then generate the girth of the internal dividing walls on any one floor, as this will be, in effect:

the sum of the lengths and widths of the cells,
minus
the length and width of the building envelope,
plus
(n-1) x the thickness of the internal walls
(where n = the number of cells).

We also hypothesise that doors, windows and other openings are analogous, in that these are a combination of elements proper and junctions with other elements.

If the plan shapes of cells and the building envelope were always simple rectangles, the dimensions could be copied directly from those of the spaces, the areas of ceilings, roofs, and floors being the product of the lengths and widths, and the girths of walls being twice the sums of lengths and widths. By regarding complex orthogonal plan shapes as consisting of simple rectangular components, and applying the same rules, the total areas can be correctly obtained, but the sum of individual girths will exceed the actual perimeter by twice the dimensions of the 'inner boundaries' between the component spaces.

One apparent solution to this difficulty is to regard compensatory spaces of, initially, zero thickness as lying between component spaces. These are given a timesing factor of -1.00 as their initial purpose will be to remove the dimensions of these inner boundaries (as they have no physical existence) whilst having no effect on the areas.

When all the scalars, including the ones with negative
timesing, are modified by the addition or subtraction of, say, a
wall thickness, in order to obtain actual dimensions, these
'negative scalars' are found to continue their compensatory role,
maintaining the correctness of both the areas and the perimeters.
It would seem that 'negative spaces' can be in any plane, and the
technique has been developed so that the computer can generate the
measurements of individual cuboids.

Procedure at scheme design stage

At this stage, only the size and arrangement of the spaces, and the
general quality of the building will have been settled. Because
the details of the technical solutions which will meet the functional
requirements of the elements have yet to be decided, these can only
be represented in some way, and related to the areas of elements
and lengths of junctions, generated as described above.

An interesting possibility is to relate the various quantities
to data on the cheapest possible technical solutions, and to include
factors to represent the ratios between the likely cost of work of
an acceptable quality and these minima. Such ratios are similar to
the 'cost/worth ratios' used in Value Engineering procedures.
Suitable data models can be held on computer files, and can include
those of installations and external works. Where the file prices
of resources have become outdated, current ones can be included in
the job input.

All this can be carried out on the CIRCE system alone, although
the modification of scalars to allow for overlaps at corners, etc.
is not possible. Even so, being a single system for use within a
single organisation, it could also be a satisfactory way of
preparing tenders for design-and-build contracts, particularly as
budgets would be computed for production management.

However, if the scalars are to be modified, it is necessary to
use the bill production system to compute the quantities. In such
cases, the scalars could be retained for use later.

Procedure after the design is complete

The final stage of designing is to make decisions on the technical
solutions, that is, on the parts that will constitute the elements
and junctions, including their sizes, the materials from which they
will be made, and any constraints on the quality of the result.

There are other regularities here, as wall leaves, screeds and
other continuous parts of elements with two space-related dimensions
will only need decisions on their thicknesses, whereas joists and
other skeleton parts, and fascias, architraves and other parts of
junctions will need widths and either thicknesses or heights.

The computer-aided system we are about to outline provides for the copying of dimensions from those of spaces, and, if necessary, their subsequent modification by the addition or deduction of such widths or thicknesses, so that the actual dimensions of the planes in each of the parts are generated. No attempt is made to aggregate into girths, as the computer can deal with practically any number of individual dimension sets. Indeed, there are advantages in this naive approach, as each set can be individually retrieved for operational purposes.

This is made possible by two more regularities as follows:

(1)   the three Cartesian (scalar) dimensions of objects and spaces indicate how much there is of each. We also need to be able to state how many there are of each size,

(2)   in the case of parts that are constructed insitu, their quantities will be the product of the number and one or more dimensions.

These regularities are given effect by entering the timesing, the three dimensions and the other facts in rows in a straightforward way and by indicating to the computer which of the dimensions are to be used in calculating the quantity. Where regularities exist between spaces and parts of elements or junctions, the scalar dimensions of the spaces are entered first, and followed by the items data, but with instructions to copy and modify particular columns of the preceding scalars. In the case of regular technical solutions, these copying instructions will also be regular (e.g. a floor finish will always require lengths and widths). Both these and the quantification indicator can be attached to their part descriptors in the computer file, and recalled with the part name in project data, so that no knowledge of this aspect of the Method of Measurement would be required by the user.

We have found it beneficial to make use of the regularities of elements and junctions in order to schedule the data on technical solutions in a methodical way. This schedule is regarded as the source document and the various references to be used in the input are added to it. Given a stock of descriptors, etc., we do not see why a computer should not be programmed to produce these schedules during an orderly detail design procedure and, at the same time, generate much of its own input for bill production.

Bill production

We have hypothesised that the items in a bill of quantities are sorted chiefly on the basis of the materials being used, even though they are arranged in worksections which may be called by the names of elements or operative occupations. In our bill production system, the descriptors of these materials are placed in separate lists and numbered sequentially to indicate to the computer where their respective items are to be placed, thus giving general control over the arrangement of the output.

The computer places the items in the conventional order of 'cubes, supers, runs and numbers', by reference to the quantification indicator, and uses the widths and thicknesses to provide almost every one with a unique position.

During the first process, the computer amplifies the input by copying descriptors and items from its files. The dimensions are then copied and modified and an 'abstract' is prepared, consisting of the items in bill order, with individual dimension sets, quantities and retrieval references. This is retained in the computer and a bill of quantities is produced by copying selected portions. Where the quantity of the material is different from the item quantity, this can be calculated by the computer under a set of controls that make use of brick sizes, etc.

Within each worksection, items are arranged under headings, and the information required by estimators when dealing with an item is contained either in the item description or in the first heading above it.

Partial bills are possible, and the dimension sets can be re-expressed to suit operational needs by redefining the component spaces from which they will be copied when the program is re-run.

Estimating

Either the unit rate or the operational estimating approach is possible. Quantities from the bill are input to the CIRCE system, with references to basic resources and collectives, appropriate factors and updating details. An operation is regarded as a collective of resources, the costs of which will be apportioned to one or more bill items. In the output, item unit rates are analysed and totalled and the quantities and costs of resources are given.

Further developments are to be expected in the CIRCE system as the resources analysis can be arranged to suit the need for production information.

CIRCE seems particularly suited to the larger micro-computers and has also run on main-frame machines. A scaled-down version is available for use on a PET.

References

1. Ray-Jones, Alan and Clegg, David. CI/SfB Construction indexing manual. London: Royal Institute of British Architects Publications, Ltd., 1976.

2. Building Cost Information Service. Standard form of cost analysis. London: The Royal Institution of Chartered Surveyors, 1969.

SECTION VI

COST CONTROL

WHAT HAPPENED TO THE SCHOOL BUILDING COST LIMIT?

DR ALAN SPEDDING, Bristol Polytechnic,
                   Department of Surveying.

1. Introduction

The school building cost limit system, initiated in 1949 and aband-
oned as such in 1974, was designed to provide a framework for economy
in design and construction, within which the phenomenal post-war
school building programme could be controlled. It provided a framework for overall control of National expenditure on new school
building, but it also related the National objectives to the individual school, and, furthermore, to the level of the individual pupil.
    The cost limit, expressed in pounds sterling per "cost place"
therefore virtually demanded a method of relating "costs" to the
amount of construction work necessary to provide the accommodation
and from this need was developed the Ministry of Education cost
planning system. During a research project on the cost of school
building the author obtained data on tenders for nearly 2,000
Primary Schools. The data were analysed in various ways which
included the use of an ICL computer and the XDS statistical package
regression analysis routines. Rogue data were deleted using Quantity
Surveying experience as well as by examining deviations from plots
of ordered residual values.
    This paper touches on several of the factors which were found to
have relevance to the cost limit system, although the main thrust of
the research was a comparison of system and non-system building for
schools.

2. The background

Early 1930's school architecture had been influenced by a trend
towards "open air" schools and by Board of Education suggestions
on design which had led to experimentation with new forms of
construction and layouts. New ideas about teaching were gaining
ground and there was a move towards designs which were more spacious
and intended to provide flexibility in use of space. The ethos of
the modern movement in architecture was leading to a less monumental
approach to civic design, and there was a belief in the possibility
of a harmonious marriage of functional architecture and industrial
technology which would also have a beneficent social impact.

The end of the Second World War left the United Kingdom with
problems of supply of schools for a quickly increasing young
population. A large number of schools had suffered damage, school
building had been neglected for some years, and the new philosophy
of the 1944 Education Act required each LEA to prepare a development
plan for educational building in its area. New regulations in
respect of school building were introduced in 1945, requiring
LEA's to comply with new minimum standards of building which meant
that many older buildings were in need of upgrading or replacing.
The raising of the school leaving age in 1947 also introduced a
need for more educational buildings.

There were several Working Parties and Committees appointed to
discuss Post-War building, some of which addressed themselves
specifically to school design and construction. Examples are the
committees chaired by Sir R. Wood[1] and W. C. Cleary[2]. It is
enough to point out here that, in the conditions of the Post-War
building boom new educational and architectural thought coalesced,
resulting in increased average areas per pupil, more adventurous
layouts, and a strong general belief that some measure of prefabrication was essential in order to improve the supply of schools.
Post-War school building therefore commenced with a tremendous
need for new building and an enthusiastic belief in the need for
innovation in the brave new world about to be constructed.

3. Areas and costs

Until 1949 there were no limits of cost per pupil, although cost
information was often expressed in this way. Proposals for school
buildings, accompanied by an estimated cost, were sent up from
Church Authorities and Local Authorities for approval. Clearly
there were regional variations in building costs, but apparently
comparable projects differed in cost by as much as 200 per cent[3].

The trend towards larger areas per pupil which had been accelerated by the "suggestions" contained in the 1936 Board of Education
pamphlet[4] received further impetus from the 1945 "School Building
Regulations"[5]. The County Architect of Surrey gave some figures
of average pre-war area per pupil[6] and post-war areas per place
have been obtained from the Pilkington report on Primary Schools[7].
To these data have been added some from Department of Education
and Science sources, converted to metric units and illustrated in
Graph number 1.

The Wood Committee had identified two concepts of school planning,
one involved layouts utilising a two-way planning grid and the other
involved a linear classroom solution to planning problems. Although
the former approach was developed by the schools building consortia,
after the example of Hertfordshire County Council schools construction[8], it was the latter approach which architects tended to adopt
in order to satisfy the regulations in the immediate Post-War period.
The linear classroom concept produced the "finger plan" schools
which, with their dispersed layouts demanded a large proportion of
circulation space, and high site areas.

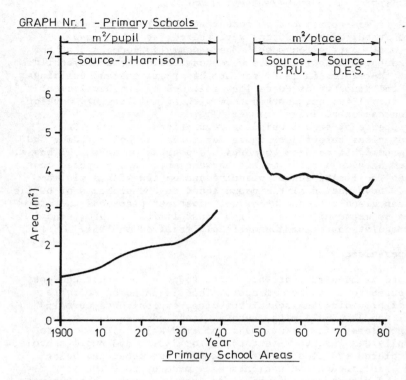

GRAPH Nr.1 – Primary Schools

Primary School Areas

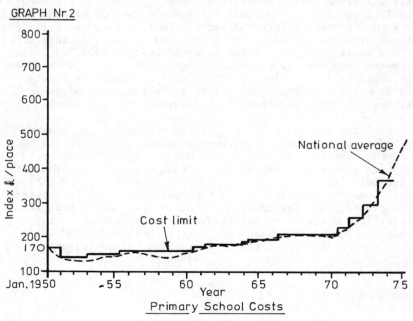

GRAPH Nr 2

Primary School Costs

233

By 1948 it was apparent that more restricted building sites must be used, that building resources were scarce, and that school buildings were costing too much. There was also concern that the large finger-plan school layouts were unsuitable for use by children. Government therefore, in 1948, set up the Architects and Buildings branch of the Ministry of Education, followed by the Development Group in early 1949, and charged them with stimulating new developments in education building.

The existence of school building regulations, and the LEA development plans referred to above meant that school building had to be programmed in advance to certain standards on an annual basis. This needed an agreed relationship between space and proposed cost in order to facilitate school planning and design with a minimum of delays. Devaluation of the pound added to the problems of public sector finance and the concept of net cost per place was therefore introduced by means of the setting of cost limits in 1950, school building regulations being appropriately modified in 1951.

4.   Cost per place

A brief note is necessary at this point to point out that the cost limit referred to is basic net cost. This is intended to represent the cost of providing the school building, exclusive of certain additional and abnormal costs, such as fees, adverse conditions and site problems. The relationship of allowable cost places to actual pupils designed for meant that, for 320 to 350 pupil schools, number of places equalled number of pupils, but above and below this range, incremental adjustments were made up to $\pm$ 40 for economies and diseconomies of scale in cost. Graph number 2 gives an overall picture of the National average cost per place, and the movements in the cost limit for Primary Schools. Graph Number 3 shows the average price per square metre of primary school buildings in England and Wales, and a published index of cost of building. The graphs cover the period of operation of the cost limit and the base for each index is January 1950 = 170. It is not suggested that too much detail can be read into the relative movements of these indexes, but the apparent trend is of interest. As an example, it was wondered if the cost per square metre of school building lagged behind the average cost of building generally, and whether this was due to the existence of the cost limit. Whatever the merits of this speculation are, it has been suggested[9] that average net profit on education building was in fact very low compared with much other building work.

Clearly, the dramatic fall in average cost per place in the early days of the cost limit is closely related to the reduction in the cost limit. The Ministry of Education certainly claimed[10] that great cost savings had been made when the projected cost of the school building programme without a cost limit was envisaged.

5. Success story?

The first cost limit, for 1950, at £170 per place was a significant reduction on the National average of £195 which had been monitored in the previous year, and a further reduction to £140 followed, before the long period of gradual increase to the year 1970 commenced. In fact, it took ten years for the cost limit to rise above the original £170. It might therefore be of interest to look briefly at the first few years of operation of the cost limit. Table 1 below shows the P.R.U.'s figures adjusted to decimal currency and square metres and also adjusted by the "Building" index[11]. Care has been taken in using the index due to the steep rise in prices in 1950/51 as well as in 1961, and it is believed that the figures given are reasonable.

PRIMARY SCHOOLS　　　　　　　　　　　AVERAGE　　　　　　　　　TABLE 1

Year	M of E limits £ cost/place	Average cost/place £	Area/ place m^2	Actual cost/m^2 £	Average cost £/m^2 at 1962 prices
1949		195.0	6.25	31.15	53.33
1950	170	158.0	4.86	32.50	54.01
1951	140	134.7	4.39	30.77	44.01
1952		131.8	4.09	32.17	41.31
1953	146	134.0	3.88	34.59	43.92
1954		136.9	3.92	34.86	43.77
1955	154	143.4	3.91	36.58	43.97
1956		148.8	3.83	38.95	44.46
1957		146.2	3.75	39.06	44.58
1958		145.0	3.79	38.31	43.73
1959		147.8	3.86	38.31	43.73
1960	164	155.0	3.91	39.70	43.98
1961	175	170.8	3.95	43.20	45.20

Caution has to be observed in the use of a general cost of building index for such adjustments, particularly as the size of schools, their dates of tender and local price variations in any group of data has been found to affect results. Having said this, it seems that, in the period during which the cost limit hardly rose over £170, actual costs per square metre rose by over 30 per cent. The initial reduction in area might have been thought to have accounted for the ability to keep within the cost limit, but in fact the Building Cost index rose by over 50 per cent, which means that other factors in the value for money equation were being affected.

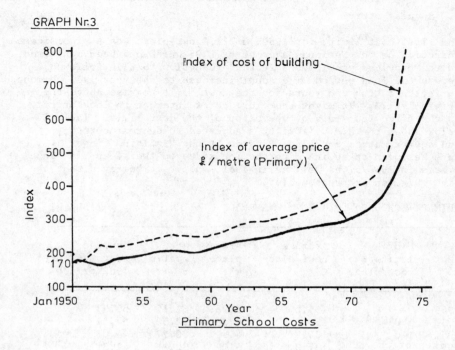

GRAPH Nr.3

Index of cost of building

Index of average price
£ / metre (Primary)

Index

Jan 1950   55   60   65   70   75
Year

Primary School Costs

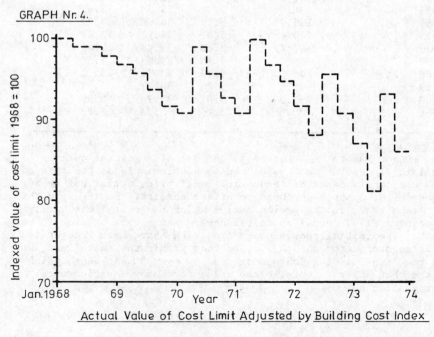

GRAPH Nr. 4.

Indexed value of cost limit 1968 = 100

Jan.1968   69   70   71   72   73   74
Year

Actual Value of Cost Limit Adjusted by Building Cost Index

236

6. Value for money

It is therefore clear that the existence of the cost limit concentrated design team's efforts on the achievement of value for money through more efficient design. As the cost limit provided a link between area and cost per place it was felt necessary to develop a system of apportioning forecast costs during design so that targets might be met. The system of elemental cost planning was therefore introduced by the Development Group, which included ex-Hertfordshire Quantity Surveyor, James Nisbet. The system was detailed in Building Bulletin No. 4 (1951) and was expanded by Nisbet in his book.[12] The elemental system was ideal for buildings such as Primary Schools with their more limited range of variables affecting basic cost, and suited the relatively slow and steady rise in cost of building which pertained from 1950 through to 1970.

Cost planning techniques helped to expose those design factors where economies could be easily made, floor area and perimeter of buildings being obvious targets.

The reduction of area achieved mainly by the reduction of circulation space caused a dramatic change towards compact layouts at an early stage in the cost limit's existence.

The scope for further reduction in cost was thus somewhat limited and this began to put pressure on service installations and quality factors. In particular, of course, the quality and prices of the schools built within the non-traditional constructional systems, including the school building consortia, caused considerable debate. It is apparent from the research that the constructional system of a proposed school frequently affected design factors, including layout, and therefore affected the way in which the design team could achieve value for money within the cost limit.

It should also be pointed out that there has always been a desire to design some schools for dual use, by pupils and by the community, perhaps also incorporating some measure of flexibility of layout. Whilst there were some success stories[13, 14] and DES tried to maintain overall quality of school building, it is probably the case that financial pressures reduced the opportunity to achieve the optimum value for money in individual Primary Schools. Of course, the cost limit was serving National objectives by sharing out the "education budget cake" and there was an urgent need to make school places available.

7. Pressure on the cost limit

Reference to Graph No. 2 shows that the cost limit, after a period of relative stability fell under serious pressure after 1969. Naturally at times there had been problems in meeting the cost limit, particularly in the early 1950's and early 1960's, but small adjustments had been sufficient to allow school building to continue,

GRAPH Nr. 5

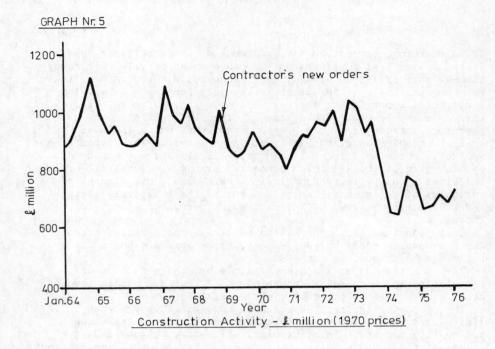

Contractor's new orders

£ million

1200
1000
800
600
400

Jan.64 65 66 67 68 69 70 71 72 73 74 75 76

Year

Construction Activity – £ million (1970 prices)

GRAPH Nr 6

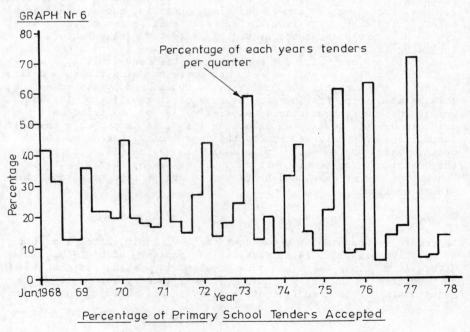

Percentage of each years tenders per quarter

Percentage

80
70
60
50
40
30
20
10
0

Jan.1968 69 70 71 72 73 74 75 76 77 78

Year

Percentage of Primary School Tenders Accepted

238

even if under cost pressures. Graph No. 4 takes the cost limit
at January 1968 as being 100 and, using the RICS quarterly class
D index of cost of building, traces its apparent fall in value of
nearly 10 per cent by early 1970, a period of two years. The
raising of the limit then saw an apparent and similar fall in its
value during the next year. Subsequent attempts to restore the
cost limit to at least its earlier value failed, the apparent value
falling by nearly 20 per cent in 1973. Clearly school building
could not be planned on such shifting sands of allowable cost and
the strains of trying to build schools were telling particularly
hard in the higher cost areas. Examples of comments on this topic
referred to "a heritage of sub-standard schools being built up"[15]
and "For reasons of economic policy Governments have chosen to
undervalue the effect of increased costs in order to regulate the
volume of public expenditure"[16]. The cost limit officially came
to an end in 1974 (in which year prices had just begun to enter a
period of relative stability and the system could have been continued
successfully). The January to June 1974 period was used to finish
off the 1973/74 programme year projects at which point the formal
cost limit system ended.

8. Demand

The use of the Construction Industry as an economic regulator has
helped to reinforce the very large swings in demand, which have
occured regularly[17]. Contractor's orders for new work shown in
Graph No. 5 give an indication of the fall in demand in 1973/74
which led to the more stable prices referred to above. School building, however, also has suffered from two other types of demand
fluctuation which actually caused some diseconomies in the
schools consortia programmes.

The first is "peaking" of demand usually in the first quarter of
each calendar year. Graph No. 6 is a histogram of a ten year period
showing the percentage of each year's tenders on a quarterly basis.
In 1977, for instance over 70 per cent of the calendar year's tenders
were dated in the first quarter of the year. This factor affected
the balance of any collection of annual average prices for schools,
particularly at times when prices were rising quickly.

The second demand factor was the cyclical swings in demand in
certain regions which seemed to indicate a lack of foresight in
overall planning. Five out of ten United Kingdom regions examined
by the author presented a pattern similar to the one regional
demand pattern shown in Graph No. 7.

9. Regional variations

It would be inappropriate to avoid the mention of regional variations
to the cost limit. The Ministry of Education, which became the
Department of Education and Science in 1964, had always said that
they did not wish to apply regional multipliers to the cost limit.
Schools Consortium architects had noticed regional variations from

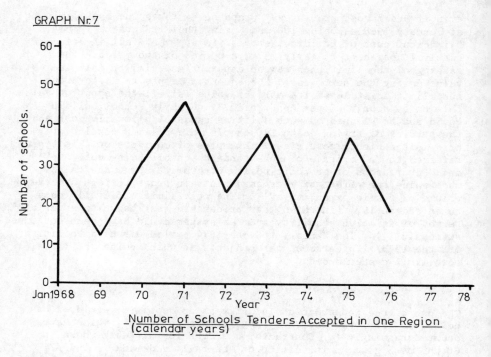

GRAPH Nr. 7

Number of Schools Tenders Accepted in One Region
(calendar years)

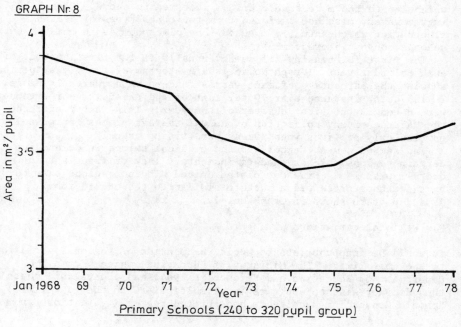

GRAPH Nr. 8

Primary Schools (240 to 320 pupil group)

very early days but it was felt that much variation arose from the
organisational differences in LEAs and from the ability or otherwise
of Architects to design economically in a system. Related factors
included demand, and perceived standards of performance for
particular localities. On the question of designer/cost relationships a Canadian study said that the designers name was the most
significant factor in schools costs in one region.[18]   Quite often
it was suspected that local variations within a region were greater
than many regional differences,[19] and the author's research supports
this view. Of course DES staff were quite aware of regional trends
in so far as they could be established and they tried to allow for
these in a less formal manner. Evidence of regional variations
was felt to be contradictory[20] and DES felt that "local variations
in cost limits would be beyond the wit of man to decide"[21]. The
author's study of regional variations of Primary School building
prices also concluded that there were some very broad trends, but
that other factors introduced considerable relative variability
between regions from year to year. The implication is therefore
that a published and detailed regional index might have become
self-fulfilling. That is, prices would tend to adjust to the cost
limit currently in force. What happens to quality of building
under conditions of pressure on cost limits is another story.

10. Quality

The measurement of quality is difficult due to the number of
subjective factors which might be involved. It is likely that the
development of the open plan school with its compact layouts and
optional uses of communal space was economic in capital cost terms
but it also was a physical manifestation of the progressive trend
in teaching. It was thought that flexibility in teaching needed
flexibility in layouts to allow the appropriate level of informality
for small group working.
    There are, however, mixed views about the real achievement of
"flexibility" in many of the open-plan or semi open-plan layouts,
particularly relating to noise and distractions caused to pupils
by people moving about the building. The absence, for instance of
internal circulation space, which saved money might, or might not,
be considered good quality dependent upon one's view of the
education process. It is, however, clear that falls in the value
of the cost limit particularly in the early 1970's caused many schools
to be built to lower than desirable space and other quality standards.
Graph No. 8 as an example shows the fall in area per pupil in a large
sample of primary schools up to 1974, and illustrates the upward
trend following the end of the cost limit. Possibly the only long
term quantitative verdict on the quality which was achieved in design
and construction would be the incidence of maintenance and running
costs on individual schools, and this is to be followed up in a
subsequent research project.

One of the factors which is often discussed is economy of scale.
It is usually the case that small schools are expensive and that
schools for older pupils consume more area per pupil. The author
attempted to ascertain if size of Primary School related to number
of pupils had a dramatic effect. Although yearly data cause some
variations, Graph No. 9 is typical and shows that cost per square
metre of schools tends to fall fairly quickly until the medium
sized schools (i.e. 240 to 320 pupils) are examined. After that,
size tends not to produce any significant economies of scale.
Graph No. 10 indicates, as might be expected, that area per pupil
falls until the larger schools (over 480 pupils) are examined.
The rise in area shown for the larger 1973 non-system schools was
observed in other years, but may have been due to the policy of
the type of LEA which tend to build the fewer, larger schools. It
will be appreciated that a rise in area per pupil will cause cost
per pupil to rise unless balanced by a significant fall in cost
per square metre.

11. Post cost limit

The position immediately after June 1974 was that each LEA received
an overall block grant financial allocation divided into sectors
which was to be related to a likely starts programme and DES
adjudicated on the proposed cost of projects without overtly applying
a specific cost limit. Analysis of data in the post 1974 quin-
quennium suggests that, on new schools, cost per unit of area was
vetted in relation to space per pupil on the basis of experience.
The intention of putting all major building projects into a three
year rolling programme unfortunately was not a success due to
financial cutbacks. It is important to note that the commitment
of Local Authorities budgets to schools building changed in
emphasis due to falling rolls in the mid 1970's. In many areas
much school building has become a problem of extending or
converting existing premises rather than new greenfield expansion
and value for money is often a question of looking at a proposal
from the point of view of the total facilities which will exist
on completion.

Recent legislation has introduced control of expenditure on an
annual spending basis as distinct from the previous concept of
starts. Prescribed expenditure is grouped under sector headings
and each Local Authority bids annually for its allocation, normally
not identifying individual projects in its bid. Central Government
responds with an annual financial allocation, and it has to be
noted that the Local Authority financial year does not coincide
with the calendar year. Naturally in looking at the amount to be
spent on education building DES try to assess potential demand for
future periods so that the three year demand projections of LEAs
are important.

DES, of course, still take a keen interest in the cost of schools
and has issued cost guidelines related to minimum teaching area (MTA).
MTA data are published by DES and relate to the proposed pupil

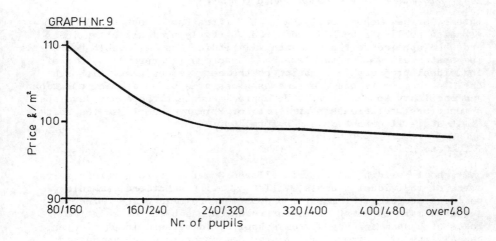

GRAPH Nr. 9

Price £/m²

110

100

90

80/160   160/240   240/320   320/400   400/480   over 480

Nr. of pupils

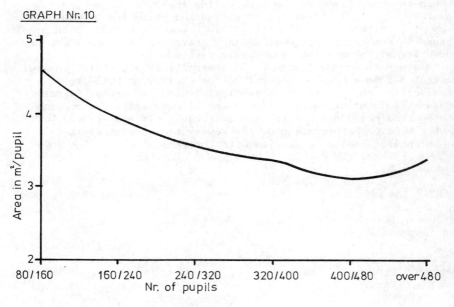

GRAPH Nr. 10

Area in m²/pupil

5

4

3

2

80/160   160/240   240/320   320/400   400/480   over 480

Nr. of pupils

1973 NON SYSTEM SCHOOLS
Banded by range of pupils

243

intake for new school building. ABCU (Area Based Cost Unit) data
are also published by DES, adjusted quarterly by means of an index
and when applied to the MTA data will indicate an allowable net cost.
The system allows for accelerated and automatic approval by DES of
individual projects in relation to the cost guidelines. Although
the MTA concept is not the same as gross area because of circulation
and ancillary space, it will be appreciated that there has developed
a functional relationship at least for Primary Schools so the cost
limit still lives but in a new form.

12. Conclusion

The school building cost limit allowed Government to curb excessive
costs of individual schools, whilst allowing a National schools
building programme of massive proportions to take place. The
author's main interest in recent research has been the relative
costs of system and non-system school building, and in particular,
the ability of design teams to provide appropriate accommodation
within the cost limit at any time during the 1968 period and
onwards. Regional variations in design and price factors have
been of interest during the early 1970's which, combined with a
change in demand projections, ended a quarter of a century of
operation of the original cost limit system.
    Possibly the major lesson to be learned is that, whilst capital
cost of buildings can be controlled, the quality of buildings
provided, or the provision of intrinsic value for money, is not
measurable in quantitative terms. Cost limits and index numbers
of "cost of building" are relatively crude tools and we need
better ones. The operation of the new system of cost control of
school building will be studied with interest.

References

1. Post War Building Studies No. 2: Standard Construction for Schools: HMSO: 1944.
2. Report of the Technical Working Party on School Construction: Ministry of Education: HMSO: 1948.
3. Britain's New Schools: Morrell and Pott : Longmans, Green: 1960.
4. Suggestions for the planning of elementary school buildings: Board of Education: Pamphlet No. 107: HMSO: 1936.
5. Standards for school premises regulations: Ministry of Education: HMSO: 1945.
6. Nursery, Infant and Junior Schools: J Harrison: RIBA Journal: November 1947.
7. The Primary School; an environment for education: Pilkington Research Unit, Liverpool University: 1967.
8. Prefabrication: R B White: HMSO: 1965
9. On getting paid, the right amount: M Barnes: Builders Conference: 1972.
10. The Story of Post-War School Building: Pamphlet No. 33: Ministry of Education: HMSO: 1957.
11. Cost Chart: The Builder: 19th November 1965.
12. Estimating and Cost Control: J Nisbet: Batsford: 1961
13. Department of Education and Science: The School and the Community: Design Note 5: HMSO.
14. Department of Education and Science: School and Community − 2: Design Note 14: HMSO.
15. Cost yardstick in a mess: President NFBTE: Building: 20th July 1973.
16. What Price Cost Planning: G Oak: Building: 16th May 1975.
17. J P Lewis: Building Cycles and Britain's Growth: MacMillan: 1965.
18. Schools in Nova Scotia: The Quantity Surveyor: May/June 1973.
19. Serial Contracting, Hants. Success Story: Building: 20th July 1973.
20. P A Stone: Building Design Evaluation: Spon: 1975.
21. Cost Limits for Educational Building: Building: 13th November 1970.

COST CONTROL OF BUILDING SERVICES - SOME LESSONS FROM PROCESS
ENGINEERING?

HELENE RYDING, Aston University
DAVID G. CHELMICK, MDA (Monk and Dunstone, Mahon and Scears).

This paper compares methods for cost control used in the building
industry with those in process engineering, and discusses their
suitability for application to a neglected area, building services
costs.

Cost Models

The exact cost, (or, more accurately, the price) of a building or
process plant is not known until its construction is complete.
During both the design and construction stages, various estimates
of the price are made, with the object of increasing accuracy of
prediction.   These estimates are made on the assumption that the
price of the finished project is related to the resources used in
the construction process: materials, labour, equipment etc.
Different models of this relationship are used at different stages
of the design and construction process, according to the availability
of suitable data, the accuracy of estimate required at that stage,
and the purpose for which the model is used.
 In the early stages, the models are very crude, often only rules
of thumb, and may be unvalidated in any scientific way.
 In the later stages of price estimation, the models employed are
often not actually viewed as models, but the 'true' price is
identified as that predicted by the model.
 It is therefore necessary to examine the models in use, for their
appropriateness to the particular stage of price estimation,
defined in terms of accuracy required at that stage, and data
available.

Cost control over time

Construction'costs' are identified by the date at which they
represented current market prices. They are then used as historical
cost data, updated by time series indices, usually published in the
relevant specialist construction or engineering journals. Thus
indices for building costs, and electrical and mechanical costs are
published by the Architects' Journal. In process engineering

application of these indices is reckoned to be accurate to $\pm$ 10%
over a 4-5 year period, and beyond this the accuracy falls off
rapidly.[1]

During the construction stage, cost models in both industries
make allowance for some increases in costs over time. In the next
part of the paper we shall assume, generally, that costs have been
adjusted for time differences.

First (rapid) estimate

In process engineering, new projects have to be evaluated at an
early stage for profitability, taking into account construction
costs, operating costs and future sales of the product. Since
designers usually operate the plants themselves, details of
operating costs are fairly easy to obtain.

The level of accuracy expected for this estimate, is $\pm$ 30% in
the process industry. The estimate is used to determine the
economic size of a process plant, in terms of output, or to give a
very rough budget figure for the construction cost of a building.
The models used in the two industries are quite different.

Process plants have quite clear economies of scale, and this
produces the 'six-tenths' rule used at this stage :

$$\frac{C_1}{C_2} = \left(\frac{Q_1}{Q_2}\right)^{0.6}$$

where C_1, C_2 are the costs of two plants with outputs Q_1, Q_2
$(Q_1 > Q_2)$ respectively. This rule can be used whenever the
historical cost of a similar plant is known, but the exponent varies
for different processes, and there are difficulties applying this
method to new processes.

In the building industry, this type of estimate is produced from
a model based on gross floor area :

$$\frac{C_1}{C_2} = \frac{A_1}{A_2}$$

where C_1, C_2 are the costs of two buildings with gross floor areas
of A_1, A_2 $(A_1 > A_2)$ respectively. Costs are compared for similar
types of buildings, e.g. schools, or offices. The model explicitly
assumes there are no economies of scale. Insufficient cost data for
similar buildings has been collected to establish this beyond doubt,
but it seems reasonably likely.

In both industries, allowance is made for different site
conditions, since the models are applied 'within battery limits'
or excluding 'external works', and suitable additions made to the
figure predicted by the model, to cover the costs of site
development.

Second (rapid) estimate

This estimate is needed to obtain authorisation to proceed with
detailed design at minimum cost, and the estimate gives a
prediction of the cost limit at an accuracy of about $\pm$ 20% in the
process industry.   At this stage, the design of the project is
available in outline.   In the process industry this is available
from the process flow chart, identifying the various stages of the
chemical process; in the building industry, by establishing the
various elements, (roof, structure, heating, electrical services, etc)
required by the building.

In the process industry this stage is receiving increasing
attention, partly because for new processes, it is in fact the only
way to obtain a first estimate.   The traditional way has been the
use of separate factors, e.g. principal equipment, piping
instruments etc, expressed as a percentage of the total.   This is
however dependent on historical data for similar processes.   Other
methods have now been developed, relating costs to 'functional units'
(e.g. distillation columns) or process steps.   Costs are compiled
from averages for similar units or steps, and adjusted for some
basic physical or chemical parameters of the process.   This can
involve a further application of the 'six-tenths' rule, often with
a more accurate exponent, e.g. Zevnik and Buchanan[2]:

$$C = 400 \ Q^{0.6} \ N \ \frac{ENR}{300} \ .10 \left[0.1 \ P_{max} + 1.80^{-4} (T_{max} - 300) + F_m \right]$$

where   C = capital cost
         Q = plant capacity
         N = number of functional units
       ENR = Engineering News Record Cost of Construction Index,
             base 100 = 1939
P_{max} = maximum process pressure
T_{max} = maximum process temperature
F_m = materials of construction factor (from 0 for mild steel
             or wood to 0.4 for precious metals)

This model applies to gas phase processes within battery limits,
plant capacity 10 million tons pa, temperature and pressure above
ambient.   Modifications of the equation are given for smaller
capacities and lower temperatures and pressures.

The advantages of this type of estimation model are that it can be
used before the design is well advanced, since only the main
parameters of the process are required; and that it can easily
be computerised.   There are disadvantages, however.   The models
have been prepared from a relatively small number of data sets, so
that their accuracy is sometimes in dispute.   Also the definition
of functional unit is not very clear.   Nevertheless, these models
have been tested and several writers (Bridgewater,[3] Allen and Page[4])
are optimistic about their development.

In the building industry, elemental analysis is carried out either on a percentage basis, based on historical data from similar buildings, or directly on the basis of cost per unit of gross floor area. The Architects' Journal produces elemental cost analyses based on the model:

$$C = \left(1 + \frac{K}{100}\right)\left[A \ \Sigma \ P_i + E\right]$$

where
C = cost of building
K = preliminaries %
A = gross floor area
P_i = cost/unit of gross floor area of ith element

The cost of most of the elements considered, (roofs, walls, finishes etc) is in fact largely determined by the area of each element, which is fairly obviously related to the gross floor area of the building. It is, however, less clear why the cost of services elements should be related to the gross floor area, except as a very crude measure of the size of building, and hence scale of services provision. The cost of air conditioning or water supply is only obscurely related to the gross floor area of the building, and is more likely to be determined by engineering parameters such as the outputs required from the services system, in a way very similar to process engineering. This problem and the small data sets available for prediction, for each building type make accuracy better than $\pm$ 50% difficult to achieve. The 'functional unit' type of model, requiring only basic technical information, would be of considerable help. Some very crude models of this type already exist e.g.

$$C = \Sigma \ N_i P_i$$

where
C = cost of electrical installation
P_i = cost per electrical point type i (e.g. light fitting)
N_i = number of points type i

Another more scientific attempt is the use of linear regression to relate the cost of services directly to building parameters other than gross floor area, or to some basic engineering parameters. McCaffer[5] reports good results for a heating model linking costs with heat and air flow, heat sources, distance heat has to travel and shape variables. An accuracy of 10-20% is claimed compared with a claim of 26% for traditional methods of estimating (no details of this were given). An electrical model gave an accuracy of $\pm$ 20% compared with 34% for traditional methods. Unfortunately none of the models used are published by McCaffer.

We believe that linear regression models based on approximate calculations of services loads (e.g. heat losses or gains) and some parameters covering the spread of distribution, are more appropriate at this stage. Models for different types of services systems, e.g. VAV or induction units, for air conditioning, would

enable cost comparisons for different technical solutions, as well
as prediction of costs at an early stage.   This has become much more
important now that operating costs must be taken into account.
Although the need to understand engineering calculations tends to
deter quantity surveyors, approximate methods are available for
calculation,[6] which are adequate for cost purposes at this stage,
e.g. Garrett[6], for air conditioning loads.   These approximate
calculations are easily computerised so that only building parameters
are needed as inputs.

   Research on cost models to be used by quantity surveyors on
micro-computers, at this stage of design, is underway at Aston
University under the joint supervision of the authors.

Third detailed estimate

In process engineering, an accuracy of 10% is claimed for this
stage, which is based on factor analysis.   Costs are predicted
from estimates of the costs of, or quotations for delivered
equipment, instruments and controls, piping, electrical
installations, buildings, utilities etc.   These costs are assumed
to be in a fixed ratio to the largest item, purchased equipment
(delivered cost), for a given type of process, (solid, solid/fluid,[1]
or fluid).   These ratios are available in published form e.g. Jelen[1]
to convert individual delivered equipment costs to installed
(battery limits) costs. Information is also available in this
source about economies of scale in equipment sizes.

   In the building industry, this stage of estimate can be
produced by the approximate quantities model :

$$C = \left(1 + \frac{K}{100}\right)\left[\sum_i L_i p_i + \sum_j A_j q_j + \sum_k V_k s_k + \sum_l T_l + E\right]$$

   Where C = cost of the building

  K = estimate of preliminaries %

  L_i, p_i = length and net installed price/unit length of
      elements priced by length

  A_j, q_j = area and net installed price/unit area of elements
      priced by area

  V_k, s_k = volume and net installed price/unit volume of
      elements priced by volume

  T_l = are lump sum items

  E = external works

At this stage of the building design, detail is available on most
of the elements of construction for the approximate quantities of
materials to be estimated and priced using historical data from
previous contracts, where information is readily available from
bills of quantities in a form immediately applicable to this cost
model.   (This is because the bill of quantities basically uses the
same model).

   However, this model is of little use for services at this stage of the design. The most cost-significant items are equipment, and the measurable items, (pipework, ductwork), are rarely designed in sufficient detail to measure at this stage. Even if they were, the current lack of bills of quantities for services make the model difficult to use due to lack of pricing data.

In practice, a model similar to the factor analysis of process engineering can be used for services. Equipment costs are estimated or quotations obtained, and ratios applied to give installed costs. Spons' M & E Handbook[7] suggests that for mechanical services, the materials: labour ratio is 70:30, and for electrical services 50:50.

Tender Stage

By this stage, a detailed design should be available. In the process process industry, although bills of quantities may be employed, the work in these is quite small, since the major items of cost are large items of equipment to be supplied on the basis of subcontracts already agreed. Chelmick's experience is that in process engineering, the design is much further advanced at this stage, compared with services designs. There is thus less likelihood of later changes, and hence less need for detailed measurement. Tender estimates are expected to vary by $\pm$ 5%.

In the building industry, the services design is frequently less advanced than the building design, at the stage when tenders for the latter are obtained. A prime cost sum for services is then inserted in the main bill of quantities. The estimate used for this sum may only be the second stage estimate discussed above, which has a very poor accuracy. It is this which may cause complaints that services engineers do not predict costs accurately. In fact,[8] the tools for anyone to do this are not readily available. Watson showed that between cost plan and tender stage, services costs could change by up to +114% and -45%, though these were exceptional cases in this sample. The method of cost planning was not stated.

Resistance to the use of bills of quantities for services has come largely from services subcontractors, and this is easily understood when the models for bills, and for traditional   estimation are compared. Bills require pricing in the form :

$$C = \sum_i L_i p_i + \sum_j A_j q_j + \sum_1 T_1 + P$$

where the sumbols are generally as for the approximate quantities model, and P is an addition for preliminaries and preambles. In fact services contractors appear to price on the following model, which reflects the way their costs arise:

$$C = X \left[(1 + \frac{MM}{100}) \sum_i M_i + (1 + \frac{ME}{100}) \sum_j E_j + (1 + \frac{ML}{100})R (\sum_i L_i + \sum_j L_j) \right]$$

where M_i, E_j are material and equipment costs for items i,j.

L_i, L_j are labour constants reflecting installation time in manhours, for items i and j

R is the cost of labour

MM, ME, ML are % mark-ups on materials, equipment, labour respectively.

X is a general % mark-up or down on the total cost, determined by market conditions, which gives the final tender price.

In order to convert the contractors' model into the bill model, a great deal of 'number-crunching' is required, which contractors, not unnaturally, try to resist.

A large proportion of the total services cost is in the equipment items. The current Standard Method of Measurement (SMM6), however provides for extremely detailed measurement of pipework and ductwork items which are not cost significant, and to which subcontractors often allocate identical prices. SMM6 could usefully be simplified. The labour constants used in the subcontractors' model are relatively constant throughout the industry, having come originally from the same source, and are known to quantity surveyors experienced in services. The main variables in tenders are thus the percentage markups and the market price factor. Considerable simplification would result if labour constants and quantities were agreed at the time of tender, and tendering was on the basis of the mark-ups.

Construction stage

In process engineering, this stage uses a refined tender estimate, with an accuracy of about $\pm$ 5%. Beyond this, the contractors profit may be in danger, although a 10% overrun for the purchaser's budget is considered not unreasonable.[9] For services contracts, assuming a reasonable design, there should only be small variations, caused by minor equipment changes, or slight rerouting of pipework or ductwork on site.

In process engineering, the start-up operation is estimated at up to 10% of the total cost, and can take 3 months to one year fulltime work[1]. Although services systems are of a smaller scale their increasing complexity has made the equivalent operation, commissioning, more important, more time consuming and more expensive for the subcontractor.

The SMM model relegates commissioning to an item of measured work, which subcontractors either do not price, or include at about 1% of the cost of that section of work. In fact, it more correctly belongs as an overhead item requiring substantial management and engineering skills, rather than craft skills.

The subcontractors' model has a tendency to mark up materials and equipment with an allowance for profit, whereas office overheads, supervision and design are covered by the labour mark up, without any real attempt to quantify overhead costs in terms of the contract's real needs. This item would normally appear under preliminaries, but these are often not priced, since subcontractors do not understand the concept. A change to operational estimating might be an answer, since the operations involved in services installations, (carcassing, installation of equipment, testing, and commissioning) are relatively independent of the type of system. Alternatively, a recognition of the cost involved, and a proper pricing of a preliminary item, as in process engineering, might ensure a more accurate reflection of the cost incurred.

Conclusion

In process engineering, a series of models have been developed, (and some tested), for several clearly identified stages of the construction process, with indication of the accuracy of prediction required at that stage. This has not been done for the building industry.

The methods used for cost control of services have remained largely undocumented. Quantity surveyors have tried to use cost models appropriate to the main part of the construction of buildings. These cost models are not always appropriate for services. Further work is needed to develop suitable models, with a comparable degree of accuracy to those used for general construction. In some cases, models based on principles used in process engineering may be more appropriate, both at cost planning stages and during construction.

The models developed in the building industry for the later stages have developed around specific contractual arrangement or practices, or professions and occupations and as such have become rigid. Difficulties in accurate prediction of services prices may be blamed on other parties in the design and construction process rather than problems with the modelling process.

In many cases the lack of detailed data on services costs hinders progress in model building and testing. This paper must therefore be seen as a plea for less 'commercial secrecy', and more co-operation in collecting and releasing data, particularly for services, so that progress can be made.

References

1. Jelen E C (1970), Cost and Optimization Engineering.

2. Zevnik F C and Buchanan R L (1963), Chemical Engineering Process, 59 (2) 70 February.

3. Bridgewater A V (1974), Rapid Cost Estimation in the Chemical Process Industries, 3rd International Cost Engineering Symposium.

4. Allen D H and Page R C (1974), The Predesign Estimation of Capital Investment for Chemical Plant, 3rd Int. Cost. Eng. Symp.

5. McCaffer R (1975), Some Examples of the Use of Regression Analysis as an Estimating Tool, The Quantity Surveyor, Dec. pp 81-86.

6. Garrett L A (1975), Estimating Air Conditioning and Cooling Loads, Building Services Engineer, 43 Aug. pp 89-93.

7. Spon's Mechanical and Electrical Services Price Book 1981.

8. Watson R B (1980), Counting the Cost, Building Services, February , pp 42-3.

9. A Guide to Capital Cost Estimation (1969), Institution of Chemical Engineers.

THE IDENTIFICATION AND USE OF SPEND UNITS IN THE FINANCIAL
MONITORING AND CONTROL OF CONSTRUCTION PROJECTS

TERRY PITT MSc FRICS,   Bristol Polytechnic
                         Wrightson Associates

1. Introduction

The ideas for this paper were stimulated by involvement in project
management for contracts valued between £100,000 and £5,000,000.
Although the ideas were developed for schemes which include a team
responsible for both design and build there is no reason why the
approach can not be applied successfully to cost control on a
traditionally organised project.
   In order to introduce project management techniques into contracts
smaller than those traditionally associated with project management
it was decided, very early on, to adopt a systems approach. This
approach could either be manual or computerised although the latter
offers obvious advantages especially when the cost and size of
modern micro/mini computers are considered. Therefore, in parallel
with the development of the ideas described in this paper, a computer
system for cost and resource control and management was developed.
   Experience has shown that the problem in applying the system
approach and hence computerising a task is not in the technicalities
of programming but in re-structuring the task or introducing new
concepts so that the approach can be applied. For example computerising the production of a Bill of Quantities without a standard
phraseology would be a very clumsy system. Similarly, attempting to
develop a project management system without a "common language" from
inception of design, through site production and finally for feedback
would present considerable difficulties. It was perceived that the
major problem to be tackled was either the identification or development of a "common language" for a project.

2. Project Information

Project information is presented in a variety of forms e.g. drawings,
specifications/bills of quantities and instructions. Each of these
forms tend to be structured in different ways e.g. drawings ranging
from layouts to details and generally giving information by location,
specifications generally structured by trade, bills of quantities
generally by SMM sections and instructions given when necessary.
There have been a number of attempts to co-ordinate

this project information such as by the use of CI/sfb elements for drawings, specification (NBS) and quantities. Most of these are not appropriate for cost and resource management as they are designed principally with the design team in mind and are not equally meaningful to the production team. A second defect is that they attempt to give a precise descriptive definition of an element. This was deemed undesirable as every construction is a unique solution to a unique problem and therefore elements should be flexible and adaptable.

Investigation highlighted two possible approaches which seemed to overcome the problems encountered so far. These were:-

> (i)   Activities
> (ii) Construction planning units

The idea of using Activities as described by Building Research Establishment was discounted as it appeared that they had generally lost credibility in the construction industry. This left construction planning units (CPUs). These were described in "Structuring Project Information" in 1972. The authors made two very important points:-

> (a) Construction planning units are based on construction method as distinct from the function performed by the component or part of the building: and

> (b) emphasis with construction planning units is on a particular project with which the team is concerned. It is considered more important to express the logic of that one project than define a set of rules for all circumstances; definition in detail and absolute consistency of application is not necessary, the constructional logic of the job in hand being paramount.

It was generally envisaged that there would be eight to twelve CPUs for a project.(A list of CPUs for a new build project suggested by the authors of Structuring Project Information is included in appendix A.) In order to manage a project successfully it was necessary for a further sub-division. These sub-divisions were referred to as spend units (SUs).

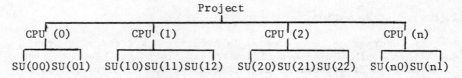

Diagram showing possible sub-division of a project

3. Spend Units

A spend unit similar to a CPU was a clearly definable "parcel" of work based on a construction method which could be an activity in a network. The problem now was that the imprecise nature of the definition which originally made the prospect of using spend units attractive was causing practical difficulties. A spend unit could be anything from install sanitary installation to fix backnut on W.C. cistern ballvalve in bathroom 3. Although the example is extreme it illustrates the necessity to investigate and develop a framework which would enable a project manager to identify spend units for a particular project. Bearing in mind that the principal task for the project manager is the efficient use of resources and one of the most difficult problems connected with the efficient forecasting and allocation of resources is variability. It was decided to use variability, or the reduction of it, as the criterion for determining the acceptability of the proposals.

As it was necessary to have a flexible and adaptable descriptive definition of a spend unit it was suggested that it would be more beneficial to identify the number and duration of spend units that would give an acceptable level of variability. Having no historical data to ascertain the consequences of various numbers of spend units of differing durations it was decided to use simulation.

A series of simulations were undertaken and the results are shown in graphs 1 and 2. The simulations were for a project with an estimated total cost of £100,000 over a 52 week period. Each spend unit estimate was calculated by dividing the total estimated cost by the number of spend units and multiplying this by a random number in the range of 0.5 to 1.5. Therefore on a project with a total estimated cost of £100,000 with 50 spend units each spend unit would be between £1000 and £3000. In addition it was assumed that estimates could be made with an overall variability of plus or minus 15% and an individual spend unit variability of plus or minus 40%. The formula for calculating the actual costs in the simulation were:-

$$AC(1) = EC(1) * R * \text{random number in the range .6 to 1.4}$$
$$AC(2) = EC(2) * R * \text{new random number in the range .6 to 1.4}$$
$$AC(n) = EC(n) * R * \text{new random number in the range .6 to 1.4}$$

Where

$AC(i)$ is the actual cost of spend unit (i)

$EC(i)$ is the most recent estimate for spend unit (i)

R is a random number in the range .85 to 1.15 and is constant for all spend units in each loop.

Finally it was assumed that the number of SUs commenced each month
were as follows:-

Month	Number Started (%)	Cumulative Number Started (%)
1	4.5	4.5
2	11.4	15.9
3	18.2	34.1
4	13.6	47.7
5	6.8	54.5
6	2.3	56.8
7	4.6	61.4
8	9.1	70.5
9	13.6	84.1
10	9.1	93.2
11	4.5	97.7
12	2.3	100.00

These figures were based on a programme for a hypothetical job.

The simulations also allowed the lag time between completion on
site of a spend unit and the ability to incorporate the information
from that into the remainder of the project. In order to ascertain
the effect of the number and duration of spend units this lag period
was held constant. For interest, however, for some simulations the
value for this lag time was varied so that the effect of this
"feedback lag" on the project could be looked at. The results from
these simulations are shown on graph 3.

As the simulations relied heavily on assumed data it was decided
not to undertake any further simulations to see the effect of varying the other variables such as total estimated project cost and
total project duration. It was decided to use the knowledge gained
from the simulations and apply it to a live trial project in an
attempt to improve the assumptions. It was hoped that these
"improved assumptions" could then be used to undertake further work
in this field.

A brief description of this trial project is given in appendix B.

4. Conclusions

Before drawing conclusions it must be reiterated . that the input
data was assumed and therefore it would have been dangerous to draw
detailed conclusions from the figures. What was relevant, however,
was the picture the results gave and from this a framework which
helped identify spend units for the trial project was highlighted.

Graph 1 showed that there were distinct benefits from increasing
the number of spend units especially upto a figure in the 50-60
range. Although benefits could be achieved beyond this range,
concern was expressed that if the system was to work effectively,
site staff must allocate resources and costs to the spend units and
if the number of spend units become too numerous this allocation

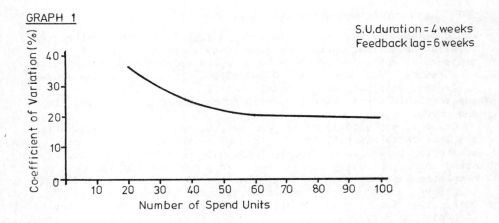

GRAPH 1

S.U.duration = 4 weeks
Feedback lag = 6 weeks

Co-efficient of Variation (%) vs Number of Spend Units

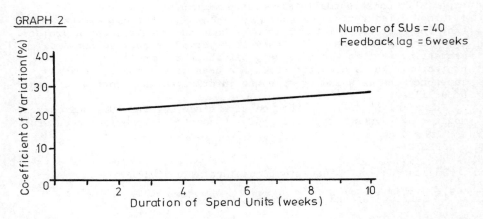

GRAPH 2

Number of S.Us = 40
Feedback lag = 6 weeks

Co-efficient of Variation (%) vs Duration of Spend Units (weeks)

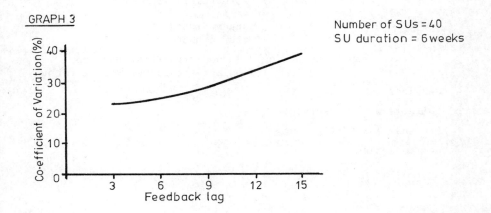

GRAPH 3

Number of SUs = 40
SU duration = 6 weeks

Co-efficient of Variation (%) vs Feedback lag

would become unreliable. It was also difficult to know whether the
results showed that there should be 50-60 spend units on all
projects irrespective of the size or spend units should always have
an estimated cost between £1000 and £3000. This will be the subject
of further work but for the trial project it is sufficient to
suggest 50-60 spend units each with an estimated cost between £1000
and £3000.

Graph 2, not surprising, suggested that the shorter the duration
the less variability there is. This of course assumes that no
monitoring takes place throughout the site work on an individual
spend unit. Obviously this is not always the case but the data on
completion must be more reliable than the estimates of estimates
that have to take place when monitoring part way through a spend
unit. Therefore one must conclude that the shorter the duration of
the unit the better although, this leads to problems with some items
especially those normally associated with preliminaries.

Graph 3 reinforces conventional wisdom that the quicker the feedback the more control there is over cost and resources on a project.
The shortness of these "feedback lag" periods supported the original
contention that it would be very difficult to manage the "smaller"
project without a computer which has been programmed to quickly
interpret the data and provide the information in a form which can
be utilized.

The trial project has been on site since October 1981 and current
monitoring suggests that the approach adopted is proving successful.

APPENDIX A

Suggested list of construction planning units:-

1 Setting out and excavation

2 Structure (load bearing) - foundations
                             - structural frame
                             - walls
                             - upper floors
                             - roof structure

3 Structure (non-load bearing) - non-structural envelope
                                 - non-structural sub-divisions

4 Services

5 Builders work to services

6 Finishes and fitting out

7 Commissioning

APPENDIX B

Project: Renovation and Conversion of existing mill building

Construction planning Units	Spend Units	
0 Preliminaries: General items	00	Site administration
	01	Insurances, rates and taxes
	02	Safty, fire health, and welfare
	03	Noise and pollution
	04	Drying the works
	05	Keep clean site and final clearance
	06	Testing materials
	07	Travelling/site expenses
	08	Photographs
1 Preliminaries: Temporary works	10	Plant,tools and vehicles
	11	Scaffolding
	12	Temporary accommodation
	13	Telephones
	14	Storage facilities
	15	Site services
	16	Temporary fencing and security
	17	Protection of the works
	18	Temporary and existing roads
2 Demolition	20	General demolition and removal
	21	Water tank
	22	Demolition for re-use
	23	External demolition
3 Fabric: External	30	Strip and repair roof
	31	Felt, batten, insulate and re-tile roof
	32	Timber treatment generally
	33	Walls/masonry
	34	Windows
	35	Doors
	36	Rooflights
	37	External decoration

4	Fabric: Internal	40	Internal structure
		41	Ground floor
		42	First floor
		43	Second floor
		44	Partitions
		45	Internal doors
		46	Joinery
		47	Internal finishes
		48	Floor finishes
		49	Staining/varnishing/painting
5	Services	50	Rainwater installation
		51	Internal pipework
		52	Sanitary ware
		53	Heating installation
		54	Gas installation
		55	Electrical installation
		56	Fire installation
		57	Burglar system
		58	Builders work
6	Fixtures and Fittings	60	Renovations
		61	Fittings
		62	Fixtures
		63	Ironmongery
		64	Signs
7	External works	70	Incoming services
		71	Drain runs
		72	Manholes/septic tank
		73	Access connection
		74	Internal roads, access routes and car parking
		75	Drain, ramp, paving, and tanking
		76	Timber ramp and footbridge
		77	Barn

SECTION VII

ESTIMATING AND BIDDING

ESTIMATING MARKET VARIATION

D T BEESTON, Department of the Environment (Property Services Agency)

1. *Introduction*

There is an obvious case for saying that, if quantity surveyors are aiming to estimate the contractor's tender, they should build it up in the same way as he does. This requires the estimation of the market addition separately from the contractor's cost.

Contractors would also benefit by formalising what they already do, so that it can be done more consistently and much of it by a lower grade of staff.

This paper outlines a bidding strategy which is a statistical representation of what contractors seem to be doing and which can be used by both contractors and quantity surveyors.

The strategy is a vehicle for disciplined judgement. It is not meant to replace judgement but merely to put it in its proper place - a supplement to, not a substitute for, calculation.

2. *The D Curve*

The strategy is anchored to a cost estimate which, in principle, excludes any allowance for market conditions. This cost estimate must be calculated by a consistent method. Whether or not it is actually measuring cost is less important than that it should be correlated with it as closely as possible. Its average error does not have to be zero, neither does it have to be known. Many users will use approximate quantities and a price book which may or may not be adjusted for inflation, location and other factors which they can allow for. In any case it must not be adjusted for subjectively assessed market factors. Whatever method is chosen it should not be changed unnecessarily. If it is, the basis of the strategy will have to be recalculated retrospectively.

The D curve is prepared from the results of a large number of bidding competitions. For each competition the lowest bid is compared with the estimated cost and the percentage difference is called D.

$$D = \frac{\text{Lowest Bid} - \text{Estimated Cost}}{\text{Estimated Cost}} \; \%$$

If the strategy is being operated by a contractor he must exclude his own bids.

This variable D is recorded for all contracts and its cumulative distribution obtained. Table 1 is an example. This distribution is plotted, as in figure 1, to produce the D curve. P is the proportion of competitions with greater than the stated values of D. In use it has other meanings as well.

In principle the strategy consists of deciding on an appropriate value of P to represent keenness and reading off the corresponding value of D. This percentage is added to the cost estimate to give the bid or estimate.

For a contractor, P represents his desired probability of winning the competition, bearing in mind that the higher the value he chooses the lower the profit he makes if he wins. This trade-off can easily be made the subject of a separate simple strategy. Otherwise he can repeat the process of using the bidding strategy with different values of P and calculate for each his expected profit. Expected profit is calculated by multiplying the probability of winning by the profit if the bid wins. The probability of winning must be expressed as a number between 0 and 1 by dividing the percentage by 100.

For a quantity surveyor, P represents his desire to avoid errors in one direction more than the other. If he has no preference he will use P = 50%. If underestimates are twice as objectionable as overestimates he will use P = 33$\frac{1}{3}$%.

In fact these values of P are not used as they stand. First they have to be adjusted for the tendering micro-climate. The macroclimate, which applies to all competitions at that time, is allowed for after the D curve has been used.

3. *Micro-climate*

Small adjustments to P are made to reflect what is believed to be the level of keenness of the bidders for the particular contract in comparison with keenness in the general run of contracts at that time. It will depend on the attractiveness of the contract. If it is thought that bidding will be especially keen P should be increased. If it is thought that it will be less keen than usual P should be **de**creased. The adjustment should be small and will seldom need to exceed 5%.

4. *Macro-climate*

The general or background level of market keenness is measured by the current average value of D, the variable upon which the strategy is based.

$$D = \frac{\text{Lowest Bid} - \text{Estimated Cost}}{\text{Estimated Cost}} \, \%$$

The value of this variable for every competition should be plotted against the date of the competition to provide an index. Movements of this index would be far too erratic to be clearly indicative so instead of plotting the individual values they must first be smoothed. Exponential smoothing is the appropriate method.

The principle of exponential smoothing is to obtain a new index as a weighted mean of the old index and the new value of D. It has some of the properties of a moving average but recent values have more influence than older values. The example at the end makes the method clear.

The choice of smoothing weights depends on the rate at which data arrive. Experiments with the base data trying various weights will quickly show which smooth the movements enough without suppressing the response. The larger weight should be on the old index, so 0.9 and 0.1 would be reasonable ones to try first. It may be necessary to put even more weight on the old index; perhaps as much as 0.95 in a field where results are unusually erratic.

If a sudden change in the market is expected the balance of the weights should be temporarily shifted for a few calculations to quicken the response to the changed level. As soon as the new level is reflected the weights must revert to their usual values.

No "age correction" is necessary because there should be no long-term trend in the D index, but there is no objection to using one if it is decided that a medium-term trend should be reflected fairly promptly. Books on forecasting give the method.

To adjust the value of D for macro-climate after reading it from the cumulative frequency curve it has to be increased by the amount by which the current value of the smoothed D index exceeds the median (ie 50%) value of the curve. In fig. 1 the median value is +8% so if the D index is running at +10.1% a value of D read off as -5% will adjusted to -2.9%.

5. *Number of Tenders*

Both in the production and use of the D curve the bid should be adjusted for the number of tenders.

A greater number of tenders tends to reduce the lowest bid. To calculate the amount of the reduction, use can be made of the shape of the distribution of bids for a contract. Apart from a negligible degree of upward skewness it is closely represented by the normal distribution. This means that if we know its standard deviation we can calculate the effect of the number of tenders as half of the effect on the range.

The range, expressed as a proportion of the standard deviation, is given in table 2.

To measure the effect on the lowest bid of changing the number of tenders, the ranges corresponding to each number of tenders can be read from table 2 and halved. The difference between the half-ranges for the two numbers of tenders gives the average difference between the lowest bids. Range is expressed in table 2 as a proportion of the estimated standard deviation of the population of bids. This is obtained by applying the estimated population coefficient of variation, which is a percentage, to the median for the particular competition. The median is chosen because there are occasional erratic bids and because the distribution is slightly skewed, but the skewness is so little that the arithmetic mean could be used if more convenient. This could be so if a computer is employed.

The population coefficient of variation can be estimated from all available competitions by expressing each bid as a percentage of the mean, or median, for that competition and calculating their overall coefficient of variation. This method has the advantage that the shape of the distribution can be examined at the same time. Otherwise just as satisfactory is calculating the individual coefficients of variation and averaging them.

By this method, before calculating the D values for the purpose of drawing the D curve, the lowest bids can be adjusted to their equivalents for a standard number of bids. The standard can be any convenient number, but it saves effort if it is the commonest number encountered.

The method of adjusting D for the number of tenders should also be used, in reverse, when the strategy is in use. It should be applied when a value of D has been read from the cumulative frequency curve and has been adjusted for macro-climate. The value of D relates to the standard number of tenders and can be adjusted for the number in the current competition by the following procedure. Unfortunately the explanation is necessarily elaborate and some may prefer to skip straight to the formula which follows it.

The difficulty is that, because the competition has not yet taken place, the median bid is not known. It is required to convert the estimated population coefficient of variation to the standard deviation when only the estimated cost and the value of D for the standard number of tenders are known. These must be used to estimate first the lowest of the other competitors' bids and from that the median bid.

The best estimate of the expected value of the lowest bid is obtained from the D curve by reading off the value of D corresponding to a P value of 50% adjusted for the micro-climate. This value of D is used to adjust the estimated cost to give the expected value of the lowest bid for the standard number of tenders.

To estimate the median bid from the lowest bid apply the current estimated population coefficient of variation to half the range for the standard number of bids in table 2, subtract this product from 1 and divide the result into the expected value of the lowest bid.

The estimated population coefficient of variation can be applied to this estimate of the median bid to give the estimated population

standard deviation. With this the effect of changing the number of
tenders can be found from table 2 by multiplying it by half the
difference between the ranges for the two numbers of tenders.
A symbolic representation of the calculation is as follows

V%	=	estimated population coefficient of variation of bids for the same contract
E	=	estimated cost using a consistent method
D(50+m)%	=	the value of D from the D curve corresponding to a value of P of 50% adjusted for the assessed micro-climate
R(S)	=	the value read from table 2 for the standard number of tenders
R(N)	=	the value read from table 2 for the actual number of tenders

The addition to the lowest bid for changing the number of bids
from S to N is:

$$\frac{\dfrac{R(S) - R(N)}{2} \; E \; (1 + \dfrac{D(50 + m)}{100}) \; \dfrac{V}{100}}{1 - \dfrac{R(S)}{2} \dfrac{V}{100}}$$

This formula has to be used frequently so it would be worth
writing a program for a calculator or a computer.

6. *Updating The Population Coefficient Of Variation*

The population coefficient of variation changes much less than D and
can be assumed constant if little refinement of the strategy is
required. But if effort can be spared it ought to be monitored.

The best way is to keep a record of the estimated population
coefficient of variation for each competition. Exponential smoothing
with heavy loading of old values will provide a figure for adjustments for number of bids. It will also indicate when it is necessary
to redraw the cumulative frequency curve.

The best smoothing weights would probably be the same as those used
for monitoring D. Although the variability of the coefficient of
variation is less, heavily back-loaded smoothing is indicated because
quick response to change is not required.

7. *Summary Of The Bidding Strategy*

7.1 Preparation

Estimate the population coefficient of variation of bids using all
available competitions, at least 30. The easiest way is to estimate
it from each competition individually and average them.

From each competion exclude own bids and use table 2 to adjust the lowest bid for the number of tenders to give the equivalent for a 4 bid competition, if 4 has been chosen to be the standard number. This requires that the standard deviation be obtained by applying the estimated population coefficient of variation to the median, or arithmetic mean, of the bids in the competition.

For each competition calculate

$$D = \frac{\text{Lowest Bid} - \text{Estimated Cost}}{\text{Estimated Cost}} \ \%$$

excluding own bids.

Form the cumulative frequency distribution of the adjusted values of **D**.

Smooth the cumulative frequency curve to obtain the D curve. Its horizontal axis is D and the vertical axis is renamed to become the desired probability of winning (P).

Record the values of D in date order and exponentially smooth them to give a D index which is kept up to date.

Similarly exponentially smooth the estimated population coefficients of variation and keep this average up to date.

7.2 Using The Strategy

Calculate an estimated cost using the same objective method as usual when calculating D.

If a contractor, decide on as low a probability of winning as is acceptable. Two or three probabilities may be chosen and parallel calculations made. The one finally selected would be the one which maximised profit when the products of profit times probability were compared. If a quantity surveyor, choose a value of P which represents the desired probability of under-estimating.

Adjust the desired probability by an amount representing the assessment of the micro-climate.

Look up the value of D corresponding to the chosen, adjusted probability of winning (P).

Adjust D for the macro-climate by adding to it the amount by which the current smoothed D index exceeds the value of D when P is 50%.

Calculate the bid suitable for the standard number of bidders from this adjusted value of D as follows.

$$\text{Bid} \ = \ \text{Estimated Cost} \ (1 + \frac{D}{100})$$

For the purpose of calculating the bid the estimated cost must be the one calculated objectively by the standard method. A more refined or appropriate estimate of cost could be used for other purposes, such as calculating profit.

Finally adjust the bid for the number of bidders in the competition using the current exponentially smoothed average estimate of the population coefficient of variation in the formula at the end of Section 5. It is important to do this adjustment for the number of tenders last, to minimise the recalculation necessary if the number has to be changed.

8. *Testing And Tuning*

Any new strategy should be tested on past data and then run in
parallel with existing methods. It is not enough to judge the performance of a strategy using the same data as that upon which it was
based. Tests on such data would show the strategy in too good a
light.

The first requirement is to see whether its performance is as good
as existing methods. If not it must be considered whether this is
because its performance is more erratic or whether its average is
wrong and it merely requires a simple addition or subtraction from
all its results. The appropriate analogy is with the rifle. If it
is sufficiently consistent it can easily be zeroed, but if it is
erratic it is probably useless.

Tuning, in the case of this strategy, consists of getting used
to choosing values of P which truly reflect the bidder's wishes.

When assessing the success of this or any other bidding strategy
(or any estimating strategy) it is important not to pay attention
to individual results. There are bound to be chance occasions for
a contractor's estimator when the tender was exactly right (just
below the second lowest) or, for a client's estimator, when he
was very close to the lowest tender. Although these may be good
moments to ask for a pay increase, they do not alone indicate much
about the value of the strategy. Only long run results based on
carefully kept records can form a basis for accepting, rejecting
or tuning a strategy.

The best method of assessment is to run the two methods, new and
existing, side by side for at least 40 competitions. The criteria
for the comparison must be decided in advance and, for contractor's
estimators, it is not obvious what they should be. The best is
probably total profitability, but contractors have other objectives
as well and each would have to decide how best to measure the extent
of his achievement.

9. *Example Of Preparation And Use Of The Bidding Strategy*

9.1 Preparation

The strategy will be based on the 76 competitions for a particular
type of work, which are the subject of table 1.

For each competition the population coefficient of variation of
the bids in the competition had been estimated by using the
$\sigma(n-1)$ button on a calculator, dividing by $\bar{x}$ (the arithmetic mean
calculated at the same time) and multiplying by 100. The arithmetic
mean of these 76 estimates of the population coefficient of variation
was 6.8% and their median was 8.2%.

The most recent 20 figures were:

11.4% 5.9% 5.8% 6.9% 7.3% 5.4% 6.7% 6.8% 7.9% 10.2%

8.1% 6.0% 4.9% 13.8% 10.1% 3.9% 5.8% 6.5% 8.3% 7.0%

Own bids were included.

For each competition the value of

$$D = \frac{\text{Lowest Bid} - \text{Estimated Cost}}{\text{Estimated Cost}} \%$$

had been calculated using a consistent estimating method. Own bids were excluded. Adjustments of lowest bids to a standard number of tenders of 4 has already been made.

The D curve has been drawn in figure 1. Although the frequency distribution is reasonably symmetrical the frequencies are more concentrated at the centre than in a normal distribution.

The reason for this shape may be a change in the dispersion of the D values so that the distribution is a mixture of distributions. However, this is only a guess and until it becomes clear that the D curve should be based on only some of the data, shown to be a better guide to the future, it is best to use all of them as they stand.

The D values for the latest 20 competitions were:

-8% 6% 4% 13% 3% 0% -2% 3% 40% 8% 2% -10% 31%

5% 12% 10% 1% 22% 6% 18%

Exponential smoothing of the values of D need only be done on the latest, say 20, values provided that a very high or low starting "old" value is not used. In this case an old value of 4% will be used, it being the average of the first 5 values in the above 20. Weights of 0.9 and 0.1 will be used. The smoothed values are:

```
 4 x 0.9 + (-8) x 0.1  =  2.8
 2.8 x 0.9 + 6 x 0.1    =  3.1    (in fact 3.12 is
                                   held in the
 3.1 x 0.9 + 4 x 0.1    =  3.2    calculator and
                                   used in the next
 3.2 x 0.9 + 13 x 0.1   =  4.2    line)
 4.2 x 0.9 + 3 x 0.1    =  4.1
 4.1 x 0.9 + 0 x 0.1    =  3.7
 3.7 x 0.9 + (-2) x 0.1 =  3.1
 3.1 x 0.9 + 3 x 0.1    =  3.1
 3.1 x 0.9 + 40 x 0.1   =  6.8
 6.8 x 0.9 + 8  x 0.1   =  6.9
 6.9 x 0.9 + 2  x 0.1   =  6.4
 6.4 x 0.9 + (-10) x 0.1 = 4.8
 4.8 x 0.9 + 31 x 0.1   =  7.4
 7.4 x 0.9 + 5  x 0.1   =  7.2
 7.2 x 0.9 + 12 x 0.1   =  7.6
 7.6 x 0.9 + 10 x 0.1   =  7.9
 7.9 x 0.9 + 1  x 0.1   =  7.2
 7.2 x 0.9 + 22 x 0.1   =  8.7
 8.7 x 0.9 + 6  x 0.1   =  8.4
 8.4 x 0.9 + 18 x 0.1   =  9.4
```

If it is decided that weights of 0.95 and 0.05 should be tried it would be necessary to cover the latest 40 values. (A reasonable number is twice the ratio of old weight to new i.e. 2 x $\frac{0.95}{0.05}$)

It seems that the D values had been in a low period but may have returned to values close to the median of all 76 which was 8.2%

Exponential smoothing of the latest 20 estimates of the population coefficient of variation would begin as follows. The first old value is, again, the average of the first 5 values.

```
7.5 x 0.9 + 11.4 x 0.1  =  7.9
7.9 x 0.9 + 5.9 x 0.1   =  7.7
            etc.
```
and ending with 7.4

9.2 Use

The competition for which a bid is required is one which the contractor would like to stand slightly more than the usual chance of winning. He has a comfortably full order book so to squeeze this job in he would have to slow down some others. On the other hand the work is of a type that he likes, the client is a reasonable one and the architect easy to work with. These influences almost cancel each other out but leave a slight balance in favour of more than average keenness.

He usually hopes to win about a fifth of the competitions for which he enters so for average keenness he would use a desired probability of winning of 20%. In this case he decides to use 30%. He guesses that the contract would be fairly popular with other bidders so he needs to enter the D curve with a P value slightly higher than 30%. He decides that the micro-climate requires it to be raised to 33%.

The value of D corresponding to a P value of 33% is 14%.

The current value of his D index (the exponentially smoothed average of the values of D which he calculates from the results of all competitions for which he estimates) was calculated above and stands at +9.4%. This is 1.2% above the 50% point on his D curve so this must be added to the 14 so far calculated.

The total of 15.2% is the amount by which his bid should exceed his estimated cost, so his bid would be E + 0.152E = 1.152E if the number of other tenders were 4. He believes there are 6 other contractors intending to bid. To preserve his desired probability of winning he will have to reduce his bid. The amount is calculated as follows. He may wish to repeat it for other possible numbers of tenders.

The latest exponentially smoothed average of the estimates of the coefficient of variation of the population of bids is 7.4%

To use the formula given at the end of section 5 the following are required.

```
V  =  7.4%
D(50 + m)%  =  D(53)%(including the allowance for microclimate)
```

$$= 5\% \text{ (reading the value of D}$$
$$\text{corresponding to P=53\%)}$$

$$R(S) = R(4) = 2.06 \text{)}$$
$$R(N) = R(6) = 2.54 \text{)} \text{ from table R}$$

The amount by which he should change his bid to allow for there being 6 other bidders instead of the standard 4 is:

$$\frac{\dfrac{R(S) - R(N)}{2} \ E \ (1 + \dfrac{D(50 + m)}{100}) \dfrac{V}{100}}{1 - \dfrac{R(S)}{2} \ \dfrac{V}{100}}$$

$$= \frac{-0.24 \times E \times 1.05 \times 0.074}{1 - 1.03 \times 0.074} = -0.020 \times E$$

Therefore his bid should be reduced from 1.152E to 1.132E.

Further calculations of his expected profit and other financial considerations may make him less keen on winning than he originally was. In that case he should rework the calculation with a lower level of keenness, say 25%, instead of 30%. This would produce a bid of 1.172E. This may give him sufficient profit to justify a level of keenness of 25%. If not, perhaps a further reduction to 20% producing a bid of 1.212E would be acceptable.

Experience of using the D curve would allow an acceptable harmony between profit and keenness to be achieved quickly.

It may often be the case that a contractor would be less keen on winning a contract if he realised how much profit he could make with an acceptably lower probability of winning. He may be willing to enter a larger number of competitions and rely on chance to give him the necessary proportion of wins.

10. *Further Development*

There is more that could be done to extend the strategy to maximise profit and to take into account costs incurred by estimating. The method described here could form the core of a comprehensive strategy suited to the firm's policy.

A computer program could be writen which required as input only the amounts tendered by all bidders in previous competitions, the estimated costs and smoothing weight changes to provide the basis of the strategy. In use the inputs would be the desired probability of winning, the assessment of the micro-climate and the number of other bidders.

The program could be extended to cover other factors, such as financial considerations, but it is important that it should output intermediate stages. It must not become a black box whose working is not understood by the user, because he must be able to see when an assumption upon which the method is based no longer holds.

Table 1. Data for the D curve

$D = \dfrac{\text{Lowest bid-estimated cost}}{\text{estimated cost}} \%$	Number of Competitions	Cumulative Frequency	Cumulative % P
-40% -	1	76	100
-30% -	4	75	99
-20% -	6	71	93
-10% -	9	65	86
0% -	23	56	74
10% -	14	33	43
20% -	7	19	25
30% -	6	12	16
40% -	3	6	8
50% -	3	3	4
60% and over	0	0	0
	76		

Table 2. Range as a proportion of population standard deviation in samples from a normal distribution

Number of Tenders	Average range $\div$ Standard Deviation	Number of Tenders	Average range $\div$ Standard Deviation
2	1.13	10	3.08
3	1.69	11	3.17
4	2.06	12	3.26
5	2.33	13	3.34
6	2.53	14	3.41
7	2.70	15	3.47
8	2.85	16	3.53
9	2.97	17	3.59

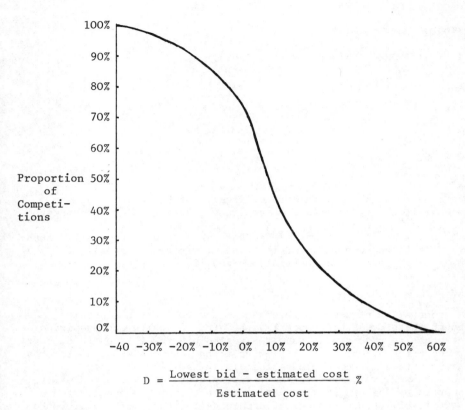

$$D = \frac{\text{Lowest bid} - \text{estimated cost}}{\text{Estimated cost}} \%$$

Fig. 1   Cumulative frequency curve (The D Curve)

A BIDDING MODEL

MARTIN SKITMORE*, University of Salford

'If (a contractor) knew the tender figures of his competitors he
would be in a very advantageous position indeed.' Oxley and Poskitt[1]

'I love chaos, It is a mysterious, unknown road with unexpected turn-
ings. It is the way out. It is freedom, man's best hope.'
Ben Shahn - quoted in Industrial Design, 13, 16 (1966).

1. *Introduction*

The procurement of new construction in the UK is predominately
through the process of competitive bidding, in which each aspiring
contractor is required to submit a sealed bid for the contract, that
is, the price at which he is willing to perform the work. After the
closing date all the envelopes containing the bids are opened and the
contract usually awarded to the lowest bidder. The level of price
submitted by the bidder thus determines not only his anticipated
profit level but his chances of obtaining the contract and it is to
this end that the science of bidding strategy is directed.

The construction industry bidding procedure is characterized by
three major factors limiting strategic considerations. Firstly, as
mentioned above, is the sealed bid, typical of closed bidding situations. Closed bidding implies that competitors' bid simultaneously
effectively preventing a game theoretic approach, an approach usually
identified with manoeuvres based on changing information about
competitors' behaviour.

Secondly, construction is largely a "one-off" process carried out
by what is sometimes aptly referred to as a custom industry. Econometric type price optimizing models aimed at balancing volume, and
profits through economies of scale have therefore invariably been
found inappropriate.

Finally we find our industry is burdened with a factor of considerable importance in quantitative analyses - uncertainty. Uncertainty
at all levels. Uncertainty in the prediction and recording of output

---

* The mathematical formulae are presented with the assistance of
Dr Ernest Wilde, University of Salford.

levels of site activity; uncertainty caused by the knowledge that
post contract design changes may disrupt work programmes; to
uncertainty of the identity, number and actions of competitors in the
next bidding competition.

Recognition of these three factors has prompted this writer to
consider afresh the most useful constituents of a model of the bidding situation for construction projects. The approach has been to
analyse the practical issues involved and, in consideration of the
type, volume and availability of data, to propose a synthetic model
capable of reasonably representing what we generally agree to be
'reality'.

2. *Major Variables*

The major variables in any specific construction bidding competition
are seen as:-

(a) the number of bidders;

(b) the distribution of each bidder's possible bids.

Each bidder's frequency distribution is usually considered to
really be the joint distribution of two component variables - the
cost estimate and the mark-up - both of which are strictly unique for
each competitor. The variability of the first component variable,
the cost estimate, is attributed to three factors:

(a) Inherent unpredictability - due to unaccountable or unpredictable fluctuations in the performance of site activity such as those
fluctuations caused by varying weather conditions, site management,
individual performance levels etc.

(b) Uncertainty of (i) the exact nature of the task, due to
inadequate design specification or lack of experience of the particular construction methods required, and (ii) the prediction of future
cost levels, such as experienced in fixed price contracts or, where
contractual escalation provisions exists, in allowing for possible
shortfalls.

(c) Errors - such as arithmetical errors, omission or duplication
of costs etc. (The use of the independently produced Bill of Quantities in the UK excludes the possibility of variations in quantity
counts.)

Ideally, the variability in the above items should be covered by
means of a contingency allowance in the bid. The value of such allowance would of course depend upon the bidder's assessment of his cost
estimate variability and his risk attitude.

The influence of the mark-up component on the overall bid distribution is dependent on the bidder's pricing policy. The application
of a constant mark-up percentage necessary to achieve a reasonable
profit level irrespective of any strategic considerations will have no
effect on the bid variability. A variable mark-up policy, perhaps the
most common policy, is a function of job desirability and possibly
strategic manoeuvring to reduce money "left on the table" (the difference between the lowest and second lowest bid). Job desirability is
influenced directly by many factors - favoured project types within

the bidder's expertise area, prestigeous projects etc. The bidder's work load has an indirect influence as it rather depends on the bidder's expectation of securing work elsewhere, the urgency of the situation and the degree of optimism which in turn is largely dependent upon the actual and perceived state of the market.

Strategic adjustments because of the relative threat of competitors, if any, probably depend on the bidder's attitude to the use of the various mathematical models proposed by researchers. More likely any such manoeuvring will be based on a subjective assessment of the opposition.

Bid variability is also affected by a special feature of job desirability where interest is too low to justify the effort and expense of preparing a detailed cost estimate. In this case it is possible that a cover price be taken or, where a variable mark-up policy is adopted, a rough cost estimate made with a relatively high mark-up (Fig. 1).

mark-up policy	tender action			
variable mark-up	no bid	cover price	rough estimate + high mark-up	detailed estimate + low mark-up
constant mark-up	no bid	cover price		detailed estimate

low   high

job desirability

Fig. 1. Job desirability continuum

3. *Relative Distributions*

The difference between each bidder's distribution of potential bids for a contract is in the shape and relative central tendencies (referred to here as proximity). (Fig. 2)

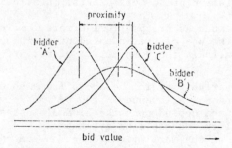

Fig. 2. Shape and proximity of competitors' bid distribution

Difference in shapes is attributed to differential levels of all the factors discussed above with the exception of (a) inherent unpredictability. Uncertainty, errors, job durability, strategic considerations will all be contributory factors in shaping each firm's bid distribution.

The proximity of the distributions is determined by the bidders' relative efficiency (sometimes called competitive advantage) in the absence of collusion.

Relative efficiency differences are attributed to differences in managerial and construction skill and techniques, the facility to obtain cheaper materials, more efficient labour etc and, in specific situations, because of particular building functional and constructional types, employers, designers, geographical locations, contract types etc.

4. *Simplifying Assumptions*

It is clear that a model of the construction bidding situation must contain certain simplifying assumptions to reduce the system to a manageable size in terms of demands on data and mathematical techniques.

(a) Collusion and Cover Prices

The two first and most obvious assumptions must be that collusion and/ or the use of cover prices is a sufficiently infrequent event to prevent the models invalidation. Collusion, whilst not strictly an illegal practice, is nevertheless not readily admitted for obvious reasons. The little evidence available suggests that it is only likely to exist in highly specialized areas of construction where the bidders virtually monopolize the field – a rare occurrence. Cover prices are, however thought to be a more frequent event, and researchers have adopted a variety of methods to seek their exclusion. All the methods are however rather arbitary and as, by their very nature, cover prices are supposed to look realistic, it is reasonable to suppose that their presence is not likely to significantly distort the model.

(b) Randomness and Independence

The lack of any meaningful evidence demonstrating the existence of systematic relationships between the dispersion of bids and any possible causal variables strongly encourages the assumption that bids can be theoretically treated as being drawn at random from a distribution of possible bids. Statistical analyses of bids for construction projects performed by Whittaker[2], McCaffer[3], Skitmore[4], Flanagan and Norman[5], Johnston[6] and others have so far indicated a slight influence of contract size, number of bidders and market conditions but not of sufficient impact for predictive purposes.

Job desirability, particularly when prompted by a falling work load is often said to influence a contractor's bidding behaviour, but as Fuerst[7] has shown, the overall variability of bids is likely to be so great as to be largely unaffected by any systematic mark-up manipulation.

(c) Distribution Shape
The shape of the bid distributions are generally accepted to be of the error-type, such as is usually represented by the "normal" or "Gaussian" distribution as shown in Figure 2. Data analysed by Gates[8] and others, of the distribution of estimated and actual construction costs of individual contractors support this view and, as mentioned above, the cost estimate distribution is likely to be the dominant factor.

(d) Consistency and Stability
The final, and perhaps most controversial assumption in this and the majority of other bidding models, is that the model is static and stable. That is to say that the distribution parameters are fixed over a period of time and bidding competitions. It is argued that the only indication of possible future events lies in the extrapolation of past data. If this is unacceptable then any static bidding model will be useless. But then the usual methods of cost estimating and tender price forecasting which also depend upon past data must be criticized similarly.

5. *The Model*

(a) Parameter Estimation
Before the model shown in Figure 2 can be tested, it is necessary for its parameters to be estimated. As the simplifying assumption has been made that each bidder draws his bid from a "normal" distribution, the sole parameters required are the mean (μ) and variance (σ^2) for each distribution A, B, C,... thus fully defining variables A (μ_A σ_A^2), B (μ_B σ_B^2), C (μ_C σ_C^2),.... Unfortunately, as each contract is of a different size, the parameters cannot be obtained directly but must be estimated from the data available by some scaling technique which will eliminate the size variable. Of the various possible approaches to this problem, a method deriving the parameter estimates from the joint ratio distribution of pairs of competitors bids is being developed and is described in the appendix.

(b) Predictions
Having fully described the model in Figure 2 it is now possible to ascertain the probability of an individual bidder winning the next competition. Given that the identity of the bidders is known then the probability P_W of a bidder A winning is given by the general formula:-

$$P_w \equiv \int_{-\infty}^{\infty} P_A(x) \quad \int_{-\infty}^{x} P_B(y)dy \int_{-\infty}^{x} P_C(z)dz... \quad dx \qquad (1)$$

Where the p.d.f. of variables A, B and C is $P_A(x)$, $P_B(y)$, $P_C(z)$ respectively.
   Which, for normally distributed variables, becomes

$$P_w \equiv \int_{-\infty}^{\infty} \frac{1}{\sqrt{2\pi}} e^{-\frac{1}{2}x^2} \left\{ \int_{-\infty}^{\frac{\sigma_1 x + \mu_1 - \mu_2}{\sigma_2}} \frac{1}{\sqrt{2\pi}} e^{-\frac{1}{2}y^2} dy . \right.$$

$$\left. \int_{-\infty}^{\frac{\sigma_1 x + \mu_1 - \mu_3}{\sigma_3}} \frac{1}{\sqrt{2\pi}} e^{-\frac{1}{2}z^2} dz \ldots \right\} dx \qquad (2)$$

It can be seen that the probability of winning, contrary to widely held belief, is dependent on the standard deviations and relative means of the individual variables.

If bidder A's distribution consists of his *cost estimates* then it is an easy step to calculate the probabilities of winning for all possible mark-up additions or multiplications, and thus the expected monetary values. Maximizing the expected monetary values will then indicate what is frequently referred to as the optimum mark-up level.

Whether this optimum mark-up level is necessarily the best mark-up for the bidder to apply depends largely upon the relationship between the bidder's cost estimate distribution and the costs he will actually incur should he win the contract. Several researchers including Fine[9], Weverbergh[10] and Morrison and Stevens[11] have pointed out that winning bidders are likely to be underestimated. In terms of this model this suggests a bias of winning bids from the left hand tail of the distribution (Fig. 3a). The effect of this bias on future profits is dependent on the parameters of the distribution of possible actual costs.

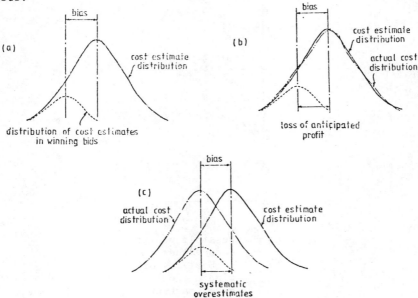

Fig. 3.   Countering bias by systematic overestimates

If the mean of the actual cost distribution coincides with the mean of the cost estimate distribution then the bias in the winning bids will result in a decreased level of profit to that contained in the mark-up. This disparity in mark-up and actual profit, coined by Fine as Break Even Mark-up is in this case effectively the bias in the winning bids (Fig. 3b). Systematic overestimation (Fig. 3c) can compensate to some extent and is shown by Weverbergh to exist in practice. Further "fine tuning" must be accommodated in the mark-up level before a true picture of likely profits can be seen at the various mark-up levels. It is apparent therefore that the selection of an appropriate sub-optimal strategy depends upon the bidder's risk attitude as, for a given mark-up, a *range* of possible profit outcomes is associated with the probability of winning the contract. Figure 4 illustrates a simplified version of the situation of a contractor bidding from a uniform cost estimate distribution of range ±10%. The upper and lower limits of all possible anticipated profits are therefore ±10% a given mark-up. The most likely profit/loss will be biased towards bids containing underestimated costs in proportion to the level of mark-up as indicated, being more biased with high mark-ups (c.f. Weverbergh[10]). From this it is a simple matter to calculate expected profit/loss for a range of mark-ups for any risk preference of the bidder and hence the "optimum" mark-up.

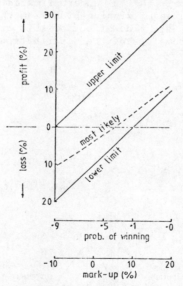

Fig. 4.   Possible profit/loss values and probability of winning for a given mark-up

(c) Sensitivity Tests and Approximations

An essential stage is the development of predictive models is to investigate the effects of changes in the model's specification. Experience has shown that a large amount of data is needed before any stability can be expected. Difficulties are also anticipated in obtaining sufficient a priori data for practical applications. It is manifestly apparent that some degree of approximation will be essential.

The difficulties in estimating the distribution parameters from sample moments of the joint distribution is the major mathematical consideration. The assumption that a normal or lognormal distribution exists, though convenient from a purely mathematical viewpoint, may not hold in reality. The model provides a basis for evaluating the effect of loosening this restriction. Similarly, Friedman's[12] assumption of independence, another convenient mathematical assumption, can be critically examined. The situations where bidders have a similar view of job desirability, social cost or even operate similar bidding strategies cannot be discounted (though the latter complication raises issues normally dealt with in game theory).

Finally exists the problem of the identity and number of competitors. It is not anticipated that collection of historical data is likely to be a problem. Most bidders for construction projects know the value of bids attributed to certain competitors either through the client, designer, some intermediary organization such as Builders Conference or simply "through the grape-vine". The major difficulty comes in assessing the identity of the bidders in the next competition. Where some competitors are known, but not all, the remaining competitors distributions could be assessed subjectively (the subjective probability of beating the individual provides enough data for a reasonable estimation of the parameters). Alternatively the unknown bidders can be assigned a collective distribution based on a method of pooling data on infrequent competitors. One major advantage of this model is that data can be collected on all bidders in the market, irrespective of the collector's participation in the competition. Pooling, however has its limitations since a particularly dangerous competitor may be bidder in the pool, resulting in an over-generalization of the models predictions, for it is apparent that the competitors with the highest probability of winning have the biggest influence on the overall results.

Where nothing at all is known about the identity of the competitors they could be guessed or, again, pooled data used resulting in the attendant disadvantage described.

It has been suggested that, in the event of not even knowing the identity or the number of competitors, that the number of competitors could be estimated by regressing other variables such as job size, market conditions, client etc and applying pooling techniques again. It is not likely however that aggregation to this degree could be anything other than misleading. A further method, proposed by Weverberg[10], involves establishing the distribution of the *lowest* bids on past projects, irrespective of identity of bidders. The difficulties in estimating the individual distribution parameters has inhibited the full development of the model, but it is possible that, in situations where information regarding the identity of the competitors is lacking, such a technique could be adopted.

6. *Static v. Dynamic Models*

Dynamic models have generally been discussed in relation to the bidder's corporate strategy, more specifically the effects of losing the contract in a dynamic environment. The approach has been to assign an opportunity cost to this event, the opportunity cost being generated by an overall corporate model, of which the bidding model is a subsystem. The model outlined, in producing ranges of anticipated profit levels, needs no modification for input into any corporate strategy. In this respect, the model does not, nor is intended to, make any decisions involving utilitarian aspects of the company's activities.

The other aspect of dynamicism in bidding, which is fundamental to the model's ability to describe the 'real' processes involved, lies in the supposed stability or even existence of the frequency distributions postulated. The apparent lack of use of such models developed since Friedman's first attempt in 1956[12] suggests that industry is either too ignorant to apply the techniques or perceptive enough to see their inapplicability in reality. Bidding strategy has unfortunately been beset by controversy and the sometimes rather complex mathematics involved even in simple static models has not helped. It is perhaps only natural therefore that the science should be treated with great caution until the major issues have been resolved.

7. *Conclusions*

A model has been described wherein the major elements of bidding for construction contracts have been incorporated. Recent investigations into cost estimating suggest that random processes largely dictate the eventual outcome of bidding competitions. The model however does rely on a certain stability to exist - that each bidder draws independently from a unique and identifiable distribution with fixed parameters and, strangely, that the values are random. "Strangely" because in randomness there exists a great deal of predictability. As Peirce[13] comments when imagining what a completely chance world would be like "... certainly nothing could be imagined more systematic". It is not proposed here to enter into any lengthy discourse on the role of chance in construction except to note that, repeated statistical analyses on construction cost/price data have failed to eliminate the chance element. It seems appropriate therefore to accept that chance is the main factor and use the information profitably.

The model consists of a series of theoretical unique frequency distributions from which each bidder draws his bid. The shape of each distribution represents the variability of the bids caused by uncertainty, errors, job desirability etc. The difference between the means (or proximity) represents the relative efficiency of the competitors organization. A method is described for estimating the parameters of each distribution from the joint frequency distribution of each pair of bidders and from this the optimum mark-up can be calcu-

lated. From the established relationship between estimated and actual
cost, a range of likely achieved profit levels can be calculated for a
given mark-up for the bidders selection.

The model can be tested under a variety of conditions involving
modification of assumptions and restrictions on data acquisition.

The main objective to the model's industrial application is that
directed at all such static models, but it is possible it will provide
a useful aid in managerial decision taking, particularly with the
increasing use of micro-computers facilities for storage and manipulation of data.

8. *Further Applications*

In the course of preparing this paper it has occurred to the writer
that further possible applications may be available.

It is apparent, for instance, that if a price forecaster (eg.
Quantity Surveyor or designer) were to record his forecast and the
bids subsequently obtained, together with bids on other contracts
where data is available, it would be possible to use the model to
predict the probability of each of the potential bidders winning the
next contract, and thereby enable the selection of a suitable short
list of bidders, and also an assessment in probabilistic terms of the
value of the lowest bid relative to the forecaster's own estimate.

Furthermore the model can also be used in any competitive situation
where values are available of previous competitions between competitors and where random processes are expected to operate within each
competitor's inherent differences. Such applications may perhaps
include sports such as athletics, shooting, rowing, golf etc to assess
overall probability levels. Additional activities such as games and
gambling where combinations of skill and chance are involved may also
be applicable.

References

1. Oxley, R. and Poskitt, J. (1971), *Management Techniques applied to the Construction Industry*, 2nd ed., Crosby Lockwood Staples, ISBN 0 258 968478.
2. Whittaker, J. (1970), *A Study of Competitive Bidding with particular reference to the Construction Industry*. Thesis presented for the degree of Doctor of Philosophy, The City University, London.
3. McCaffer, R. (1976), *Contractor's Bidding Behaviour and Tender Price Prediction*. A Doctoral Thesis submitted in partial ful-filment of the requirements for the award of Doctor of Philosophy, Sept., Loughborough University of Technology.
4. Skitmore, R.M. (1981), *Bidding Dispersion - an investigation into a method of measuring the accuracy of building cost estimates*. Thesis submitted for the degree of Master of Science, University of Salford.
5. Flanagan, R. and Norman, G. (1982), *Chartered Quantity Surveyor*, Mar., p.226-227, "Making good use of low bids".

6. Johnston, R.H. (1978), *Optimization of the selective competitive tendering system by the construction client*. Transport and Road Research Laboratory Report 855, D.O.E.
7. Fuerst, M. (1977), *Journal of the Construction Division*, ASCE, Vol. 103, No. CO1, Mar., p.139-152, "Theory for Competitive Bidding".
8. Gates, M. (1967), *Journal of the Construction Division*, ASCE, Vol. 93, No. CO1, Mar., p.75-107, "Bidding Strategies and Probabilities".
9. Fine, B. (1974), *Building*, 25/10/74, p.115-121, "Tendering Strategy".
10. Weverbergh, M. (1977), *Competitive Bidding - Games, Decision Making and Cost Uncertainty*, Doctoral Thesis, Universitaire Faculteiten Sint-Ignatius te Antwerpen, UFSIA/78*03141.
11. Morrison, N. and Stevens, S. (1980), *Construction Cost Data Base*, 2nd annual report of research project by University of Reading, Dept. of Construction Management for Property Services Agency, Directorate of Quantity Surveying Services, Department of the Environment.
12. Friedman, L. (1956), *Operations Research*, Vol. 4, p.104-112, "A Competitive Bidding Strategy".
13. Peirce, C.S. (1980), *Collected Papers*, ed. C. Hartshorn and P. Weiss, cited in Book A, "Randomness and the Twentieth Century", In *Risk and Chance*, ed. Dowie, J. Lefrere, P. OUP.

Appendix

Let us assume that three random variates A, B, C, are independently log normally distributed.

Given distributions of the ratios A/B and A/C, we can find distributions of log A/B and log A/C.

Now

$$E(\log A/B) = E(\log A - \log B) = E(\log A) - E(\log B) \tag{3}$$

and

$$E(\log A/C) = E(\log A - \log C) = E(\log A) - E(\log C) \tag{4}$$

If we assume a value for $E(\log A)$ then $E(\log B)$ and $E(\log C)$ can be determined from (3) and (4).

Then as

$$Var(\log A/B) = Var(\log A - \log B) = Var(\log A) + Var(\log B) \tag{5}$$

$$Var(\log A/C) = Var(\log A - \log C) = Var(\log A) + Var(\log C) \tag{6}$$

$$Var(\log B/C) = Var(\log B - \log C) = Var(\log B) + Var(\log C) \tag{7}$$

And thus, considering (5), (6) and (7) simultaneously

$$2\,Var(\log A) = Var(\log A/B) + Var(\log A/C) - Var(\log B/C)$$

$$2\,Var(\log B) = Var(\log A/B) + Var(\log B/C) - Var(\log A/C)$$

$$2\,Var(\log C) = Var(\log A/C) + Var(\log B/C) - Var(\log A/B)$$

But if log x is normally distributed with mean μ and variance σ^2 then

$$E(x) = \exp \mu + \tfrac{1}{2}\sigma^2$$

$$Var(x) = E^2(x) \; e^{\sigma^2} - 1$$

So we can calculate E(A), Var(A), E(B), Var(B), E(C), Var(C).
This technique can clearly be extended to cases containing more than three variables.

ANALYSIS OF THE PRELIMINARY ELEMENT OF BUILDING PRODUCTION COSTS

COLIN GRAY M Phil MCIOB University of Reading

Introduction

Design teams need to estimate real cost. This is more difficult
in the construction industry than in manufacturing industry generally,
which has an enormous wealth of research and literature devoted to
the analysis of its costs. Why can not the same techniques be
applied? The answer is that there are major differences between
manufacturing industry and construction, the main one being the
separation of the design process from the construction process and
it is from this that the problem of real cost determination arises.
(Hardcastle, 1980).

In manufacturing industry, the inter-related processes of product
design, component manufacture and final assembly are normally
carried out within the same organisation. Problems of cost can be
resolved within the common design and manufacturing management
structure by modifying either the design, production or the material
and component specification, recognising the inherent inter-
relationships and objectives of the whole process. Construction on
the other hand is not managed in such an integrated way. The design
process is extremely complex, involving many separate bodies,
specialisms and sub-contractors. With the increasing size and complexity of building projects, the professions within the construction
industry who constitute design teams are in danger of becoming
divorced from both the building contractor and the user. (Building
Design, 1980).

It has long been held by the design team professionals that the
constructor must take responsibility for selecting the method of
construction to be used and thus is the only one who knows its real
cost. However, the information upon which the contractor bases his
estimate is usually less than that which the design team has and
this raises the question of why the design team cannot perform
similar calculation at an earlier stage in the design process.
Preliminaries is the element most directly influenced by the choice
of construction method and if its real costs can be estimated during
design it would provide a basis for estimating the real cost of
design decisions. Therefore research was undertaken (Gray, 1981)
to analyse the methods used by constructors to estimate the pre-

liminaries element of their construction costs to see if there is an
approach which the members of the deisgn team could use.

Definition

The following definition of a preliminary will be used throughout:

A preliminary is the application of a resource (men, machines,
materials or money) which will not be incorporated within the
permanent building work but which is necessary for the construction
of the permanent work (excluding temporary support works).

The exclusion of temporary support works is necessary because
most estimators consider it with the work to be supported and do
not include it in the preliminary section of a bill of quantities.

The Existing Methods of Predicting Preliminaries

The research started by considering the current methods used by
quantity surveyors working in design teams to predict the price of
preliminaries.

For a variety of reasons it has become general practice for
design teams to predict the cost of the preliminaries of a project
as a percentage addition, to the elemental cost, related to the
value of the project (Ferry D.J. and Brandon P. 1980). This is
because historically preliminaries have been a small cost in
relation to overheads and profit which are normally expressed in
percentage terms. Since the development of estimating techniques
based on the code of Estimating Practice (ClOB, 1979) from 1968
onwards the trend has been to price plant, staff and all of the
other direct site costs within the preliminaries. The effect has
been to expand the content of the preliminaries to be within a
range of 12.6% to 64.4% of project cost (Flanagan, 1980) or 8.6%
to 21.6% (BCIS, 1980, these figures are the range of the mean of
each years figures from 1973 to 1979, from a sample which showed
wide variability). Both of these studies showed wide variability
within the ranges which did not significantly alter when particular
groups or types of building were analysed separately.

The major data source for prediction, that of the reported costs
in tenders therefore, suffers from wide variability with little or
no correlation between preliminaries and project value (Flanagan
1980). Although bills of quantity detail the items to be included
in the preliminaries (SMM 6 lists 67 items) the trend has been for
contractors to include a few all embracing sums which may have an
element of strategic pricing for commercial reasons.

Whether the values, produced by the contractor before any
adjustments are made, could be used as the basis of a prediction
via the percentage method was examined, because if it was possible
then this would allow a simple accurate method of prediction, providing the data could be made available.

Analysis of Contractors Preliminary Pricing

The method used for this snalysis has been to test, statistically, sets of data provided by two leading contractors, who have adopted the methods of estimating recommended in the ClOB code of estimating practice. They each made available a set of unadjusted preliminary calculations. A sample of 24 contracts were supplied by contractor A and 20 contracts supplied by contractor B. These were analysed for content within each of the sub-categories of the total preliminaries and adjusted where necessary to produce two similar sets for analysis.

Tables 1 and 2 contain the descriptive statistics for the two sets of data, placed in order of magnitude of the mean value of each preliminary item. It is noticeable from the two data sets that the first five items of staff, plant, scaffold, accommodation and temporary electrics from the majority of the total element cost, 87.9% and 82.9% for contractors A and B respectively. If cleaning is included the figures are 89.3% and 89.0% respectively. One can conclude therefore that nearly 90% of the cost of preliminaries is contained within six items. The subsequent analysis and research has therefore concentrated on these six items.

As the relationship normally used by the professional quantity surveyor is preliminaries as a percentage of project value, the tests have been to attempt to discern a relationship between the six preliminary items and the project value, Table 3.

The first test, using Pearson 2nd product moment correlation, was between the total preliminaries and project value. A correlation co-efficient for contractor A's data set of .0779 and contractor B's data set of .1484 indicates there is no correlation between preliminaries and total project value. Further tests were made to test the same relationship but for particular types of building use, offices and schools for instance, and for particular types of construction, for example concrete frames and structural steel frames.

The correlation only became more consistant when the number in the sample reduced to 4 or 6 buildings which then became unreliable becauße the sample was too small. In the larger sample sizes the low R^2 indicated the wide variability of the relationship and the unreliability of using them for prediction. Contractor A's set in most cases was diametrically opposed to Contractor B's set, i.e. the slopes of the regression are reversed. However, from a visual inspection it was noticeable that there may be a curvilinear relationship between the plots. This was investigated for the series of plots in which this relationship was more pronounced. By considering a variety of equations it was attempted to produce a regression equation which improved the fit. This was unsuccessful, in that each case required a different equation and that generally the sample size was too small and the plots too disparate.

The analysis has shown that preliminaries when expressed as a percentage are very variable with a wide range of value. The variability is not significant affected by the size of project, its type of construction or complexity. There is also no significant correlation between the preliminary percentage and the value of the project or the type of construction. Data from bills of

quantity is therefore an unsuitable basis for the professional
quantity surveyor to use for his prediction of preliminary costs.

The Contractor's Method of Pricing Preliminaries

The contractor has recognised the problem of using lump sum
additions and adopts an individual assessment of each project to
calculate very carefully the work items and thus cost, which each
buildings design will require for its construction.

The methods of analysis adopted by the contractor must therefore
establish the full cost implications of each design which is done
using a variety of skills as shown in the procedure, Figure 1.

Even a cursory examination of the 45 sets of preliminary calculations sheets used by Contractors A and B to support the figures
in the above analysis reveals a consistency of method even though
the figures give a wide range of results. Nearly every item is
generated from a calculation, the use of lump sums is very limited
and rarely exceeds £100 in any instance. That there is an underlying method is apparent and it is therefore important to determine
it. Each of the six largest preliminaries were examined in detail
with the objective of determining the scope of the preliminary, the
quantitive assessment and information source, the price information
any other information necessary and the method in which it is used
for the calculation of the price of the preliminary.

The Method of Analysis

The composition of the sets of preliminaries were discussed with
the estimators from the two contractors, A and B who showed that
the system of calculation followed the pattern of; a) determine
the scope of work or the extent of the problem, b) determine the
practical solution, c) quantify the resource, and d) price the
resource. From the 45 sets of analysis and with the help of
specialist sub-contractors and suppliers each of the above four
stages have been examined to produce a method of calculating each
preliminary item. This produces the system typified by the analysis
of external scaffold, Figure 2, which will be followed through each
stage as an example.

The Scope of the Work
The construction of the building presents the problem to be overcome, in this case in three separate sections. Within each section
is a series of headings which by following them though as a check
list develops into a systematic pattern of analysis of the design
to enable a full description of the problem to be developed. For
each set of changing circumstances a description, in terms of the
required resources and a measurement of the building for which they
are required can be determined. In this way the analysis is very
detailed and consequently quite precise.

The Practical Solution

Within the access industry there are at least six alternative
methods of providing access, therefore there is a need to be aware
of the capabilities of each system. Even within the systems there
are options and Figure 3 describes a pattern of questioning necessary to obtain the most satisfactory technical solution to the provision of a particular form of scaffold solution from which a
specification can be established. This is typical of the patterns
of questioning used in all the preliminaries and suggests the level
of detail to which the development of the description of the scope
of the work must be taken.

The Measurement of Work

With items such as scaffold it is relatively straightforward to
determine the quantity of work by measuring the area of the building
for which the access is required. However, items such as staff and
plant have a less positive relationship with the scale of the
building but nonetheless in every preliminary the scope can be
measured from information from the building or from another preliminary.

The Price of the Work

Once the specification and quantity are available the pricing takes
the form of unit rates for each item. This is a composite item of
individual resource costs for providing the item for the period it
is required on site. It is only at this point that a selection is
made of the specific equipment which will satisfy the solution.
It is a feature of the UK construction industry that via the plant
hire industry practically any item of equipment is available. This
puts the emphasis on selecting the correct solution rather than
fitting the solution to the available equipment. The larger contractors will hold a stock of commonly used equipment, but for the
remainder quotations from suppliers and sub-contractors are used.

By analysing each of the preliminary items it has been possible
to produce a flow chart for each, Figure 4, shows the schematic
system developed for the pricing of scaffold. Similar systems have
been developed for each of the 6 major preliminary items. The
features of each flow chart is that the scope of the work is a combination of the requirements of the building, the data for which is
available from the design drawings, and practical process requirements, gantries, fans, etc. derived from the requirements of other
preliminaries. At the estimating stage the quantities are priced
from freely available commercial price information together with
periods of hire derived from the construction programme.

Analysis of Pricing System

A study of the six schematic pricing systems reveals the following
significant features.

The Inter-relationship of Preliminaries
Figure 5, demonstrates that the information required to determine
the scope of each preliminary item is derived from a combination of
sources, the most important being the type of construction and the
size of building. However, each preliminary is composed of a
different combination of requirements including information from
other preliminaries thereby giving it a unique nature. Thus site
accommodation depends on decisions about staff.

Commonality of Design Information
Figure 6, describes briefly the extent of the information that is
necessary to determine the scope of work of each of the preliminary
items. It can be seen that a set of 1:100 drawings normally produced by stage D, of the RIBA plan of work (RIBA Management handbook,
1967) would contain the required information. The implication is
that via the flow chart models (Figure 4) there is a direct link
between the design of a building and the cost of the preliminaries.
The extent of any preliminary cannot however be read directly from
the drawings as each is a compound of differing aspects of the
project. However by using a systematic method of analysis the
scope and specification of the required resources can be developed
from the early design drawings.

Figure 6 Design data information requirements.

The Site
     Total Site Area   Height, Size and Construction
     Building Footprint Area   of Adjacent Structures
     Temporary Road Area

Number of Buildings per Project (consider
each building separately).

For Each Individual Building

     Type of Primary Structure   Elevation Components:
     Profile of Each Elevation   Number
     Number and level of Each Floor   Type
     Dimensions of Each Floor Level   Fixing Location
     Maximum Building Height   Type of Internal Fixing
     Expansion Joint Location   Components

The Importance of Construction Programme Information
The costs of each of the individual preliminaries are a composite
of: capital equipment cost, hire costs, staff costs, operator and
labour costs and one-off unit costs. However, as 88% of the costs
(Contractors B sample estimates) are derived from weekly or daily
charges, programme time information is essential.
    However, a knowledge of the overall duration of the project or
of individual buildings within a project is insufficiently precise
for estimating preliminary costs.
    The overall duration identified by the project start and com-
pletion dates are required only for particular items of staff,

temporary accommodation and temporary electrics, but only to
determine part of their costs. The majority of the costs are
related to the duration of specific tasks in each building. The
information required is shown in Figure 7.

Each building must be studied individually to ascertain the combination of components and therefore the sequence of tasks as the
start and finish times depend upon the form of the construction.
Therefore the time period for inclusion within the calculation of
preliminaries costs will vary from building to building.

Figure 7 Programme period deliminators for inclusion of time
periods into preliminary calculations.

Project Start and Finish Date
for Each Building:
Start and Finish Date
Commencement of:
        Each Section
        Superstructure above ground
        Floor Level
        Reinforced Concrete Structure
        Cladding
        First Fixing
        Floor Cycle Duration
Completion of:
        Roof Level Plant Rooms
        Cladding
        Floor Screeds
        Finishes
        Section
        External Works

Conclusion and Future Work

This paper has been concerned with an examination of the current
system of valuing the preliminaries element of a construction
estimate, at the pre-bid stage.

From the analysis of the method of predicting preliminaries
adopted by the professional quantity surveyor, it is clear that
there is a consistency of approach throughout the pre-bid stages,
of a percentage addition. The research has shown, however, that
preliminaries, when expressed as a percentage, are very variable
with a wide range of value. This variability is not significantly
affected by the size of project, its type of construction or complexity. There is also no significant correlation between the
preliminary percentage or its sub items and the value of the project
for all or any particular type of construction.

A careful analysis of the way preliminary items are priced, by
contractors, suggests that a small amount of data derived using the
normal measurement techniques used by the professional quantity
surveyor from stage D design drawings and a small amount of data
devised from a strategic construction programme provides all of the

project information required for the satisfactory calculation of preliminary prices. All of the other data that is required is readily available to the contractor and quantity surveyor alike. Therefore if the quantity surveyor can develop methods of producing strategic construction programmes, they have the capability of estimating preliminaries at or before stage D of the design development.

The conclusion has not been tested but the relationships identified during this research programme, are being developed into a simplified set of procedures for use by the professional quantity surveyor, to enable him to more accurately assess the real costs of the relationship between a building's design and the construction method required to build it.

ACKNOWLEDGEMENTS

Thanks must be given to the Chartered Institute of Building, Queen Elizabeth II Silver Jubilee fund for assisting with this project which is part of a larger study to investigate the techniques for determining the relationship between design and construction.

PRELIMINARY	MINIMUM	MAXIMUM	MEAN	STD DEV	KURTOSIS	SKEWNESS
STAFF	18.300	36.750	26.057	5.291	-0.909	0.164
PLANT	10.090	38.200	21.974	5.149	-0.691	0.450
SCAFFOLD	6.690	30.100	14.829	5.848	0.953	0.898
TEMPORARY ACCOMMODATION	7.100	20.800	11.931	3.164	1.982	2.574
TEMPORARY ELECTRICS	5.700	14.000	8.302	2.350	0.029	1.088
CLEANING	2.400	10.300	6.020	2.152	-0.473	0.376
HOARDING	0.380	6.400	2.358	1.677	0.437	1.102
WATCHING AND SECURITY	0.000	4.620	1.757	1.252	-0.007	0.289
TELEPHONE	0.800	1.830	1.290	0.271	-0.297	0.430
TEMPORARY ROADS	0.000	4.900	0.635	1.156	10.208	3.030
WATER FOR THE WORKS	0.200	1.000	0.546	0.228	-0.794	0.357
DRYING OUT	0.000	2.000	0.320	0.557	4.041	2.101
PROTECTION	0.000	2.500	0.260	0.733	6.829	2.827
TESTING	0.000	0.720	0.196	0.234	-0.048	1.023
INSURANCE	1.090	6.170	3.142	1.450	-0.611	0.549
FROST PRECAUTIONS	0.000	0.900	0.191	0.282	0.830	1.403

Table 1 Contractor A Table of Descriptive Statistics

PRELIMINARY	MINIMUM	MAXIMUM	MEAN	STD DEV	KURTOSIS	SKEWNESS
STAFF	18.700	42.200	31.162	6.791	-0.769	- 0.374
PLANT	11.300	29.100	19.634	5.019	-0.809	- 0.167
SCAFFOLD	8.500	38.500	18.833	8.860	-0.400	0.736
TEMPORARY ACCOMMODATION	8.600	15.700	11.587	1.827	-0.380	0.293
TEMPORARY ELECTRICS	2.600	14.200	6.725	3.092	0.162	0.745
TEMPORARY ROADS	0.100	5.800	2.475	1.932	-1.354	0.517
HOARDINGS	0.300	4.500	1.733	1.059	1.669	1.365
WATCHING AND SECURITY	0.000	3.200	1.546	0.785	-0.399	0.337
CLEANING	0.000	2.800	1.387	0.755	-0.548	0.390
TELEPHONE	0.500	3.100	1.308	0.617	1.550	1.182
FINAL CLEAN	0.000	2.300	1.167	0.494	1.188	0.197
PROTECTION	0.000	1.300	0.479	0.341	0.710	1.196
TEMPORARY PLUMBING	0.100	1.000	0.378	0.212	2.263	1.528
PUMPING	0.000	1.600	0.362	0.421	2.552	1.549
SAMPLES	0.000	1.400	0.342	0.272	9.902	2.681
DRYING OUT	0.000	0.700	0.171	0.227	-0.212	1.047

Table 2   Contractor B   Table of Descriptive Statistics

PRELIMINARY	CONTRACTOR A				CONTRACTOR B			
	CORRELATION COEFFICIENT	R^2	CONSTANT	SLOPE	CORRELATION COEFFICIENT	R^2	CONSTANT	SLOPE
TOTAL	.0778	.0061	9.5503	.0865	.1424	.0203	9.4010	.3026
STAFF	.0955	.0091	29.7499	.3268	.0282	.0008	25.8623	-.9364
PLANT	.0155	.0002	18.9636	.2821	.3848	.1481	17.7788	1.9698
SCAFFOLD	.0434	.0019	20.1482	.0987	-.2787	.0777	16.9567	-1.0236
TEMP ACCOMDTN	-.4599	.2115	12.2309	-.3538	-.4320	.1866	13.7160	-.8586
TEMP ELECTRICS	.3726	.1389	5.3714	.5253	-.4250	.1806	9.6059	-.6272
CLEANING	-.4140	.1714	1.8729	.2221	.4393	.1929	4.7853	.5939

Table 3 Relationship between project value and preliminary percentage

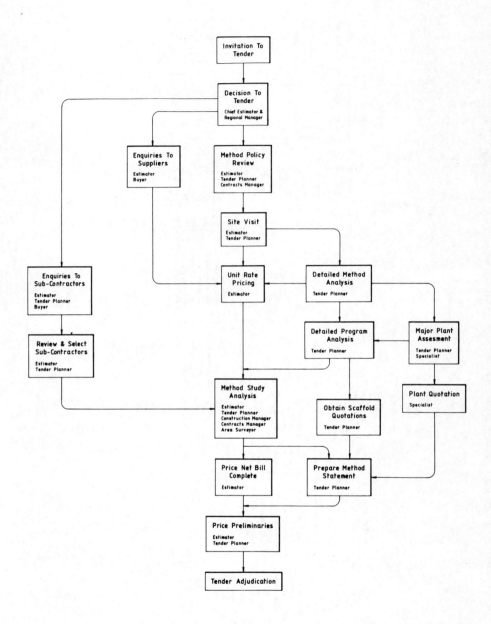

Figure 1   Contractors' Estimating Procedure and the Personnel
involved

301

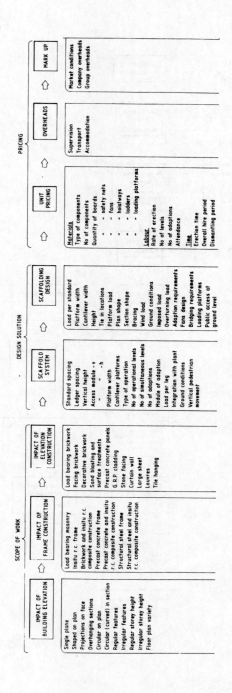

Figure 2   The Stages in Pricing External Access Scaffold

302

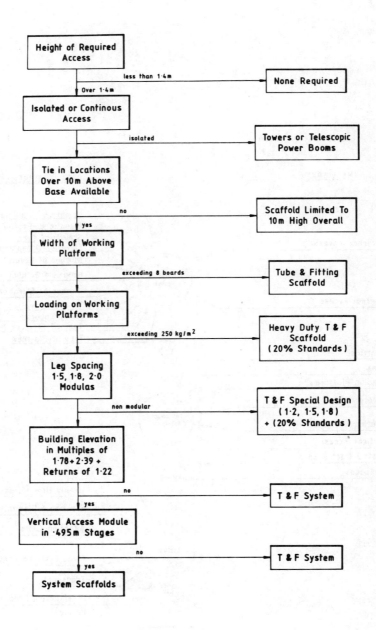

Figure 3   Scaffold System Selection Decision Tree

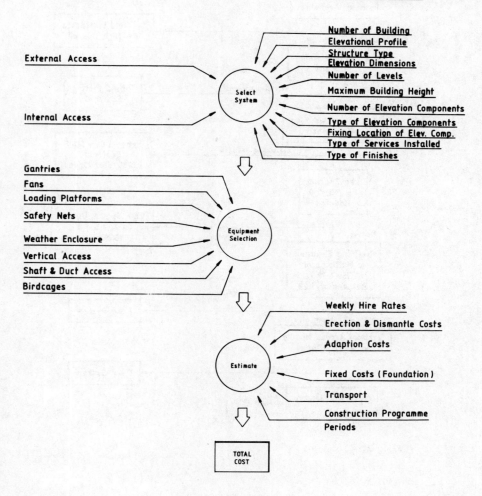

THE WORK

External Access

Internal Access

Gantries
Fans
Loading Platforms
Safety Nets
Weather Enclosure
Vertical Access
Shaft & Duct Access
Birdcages

DATA REQUIREMENTS

Number of Building
Elevational Profile
Structure Type
Elevation Dimensions
Number of Levels
Maximum Building Height
Number of Elevation Components
Type of Elevation Components
Fixing Location of Elev. Comp.
Type of Services Installed
Type of Finishes

Select System

Equipment Selection

Weekly Hire Rates
Erection & Dismantle Costs
Adaption Costs
Fixed Costs (Foundation)
Transport
Construction Programme
Periods

Estimate

TOTAL COST

Figure 4   Schematic System for Pricing Scaffold Preliminary

304

PRELIMINARY \ INFORMATION SOURCE	CONSTRUCTION TYPE	BUILDING SIZE	MECHANICAL PLANT	SCAFFOLDING	HUTTING	STAFF	TELEPHONE	TEMP ELECTRICS	ACCESS ROADS	HOARDINGS	WATCHING	DRYING OUT	CLEANING	CASING & PROTECTING	WATER	INSURANCE
MECHANICAL PLANT	●	●														
SCAFFOLDING	●	●	●		●					●						
HUTTING						●										
STAFF	●	●	●													
TELEPHONE						●										
TEMP ELECTRICS	●	●			●				●		●	●		●		
ACCESS ROADS	●															
HOARDINGS			●									●				
WATCHING	●	●														
DRYING OUT	●															
CLEANING	●	●			●											
CASING & PROTECTING	●															
WATER	●															
INSURANCE	●	●														

Figure 5 Plot Showing the information sources for determining the scope of work of preliminaries

References

B.C.I.S. (1980) STUDY OF PRELIMINARY PERCENTAGES,
    Study 17A R.I.C.S.

CODE OF ESTIMATING PRACTICE 2ND EDITION (1979)
    The Chartered Institute of Building Ascot.

Ferry D.J.and Brandon P. (1980) COST PLANNING OF BUILDING
    4TH EDITION, Crosby Lockwood, London.

Flanagan R. (1980) TENDER PRICE AND TIME PREDICTION FOR
    CONSTRUCTION WORK, Phd Thesis submitted to University
    of Aston in Birmingham.

Gray C. (1981) ANALYSIS OF THE PRELIMINARY ELEMENT OF
    BUILDING PRODUCTION COSTS, M Phil Thesis submitted to
    the University of Reading.

Hardcastle C. (1980) COST PLANNING IN THE 1980's?
    The Quantity Surveyor Journal, pp 261-266.

RIBA Management Handbook (1967) RIBA Publications, London.

SPONS ARCHITECTS AND BUILDERS PRICE BOOK (1980) edited by
    Davis Belfield and Everest, Spons Publishing, London.

SPONS MECHANICAL AND ELECTRICAL PRICE BOOK (1980) edited by
    Davis Belfield and Everest, Spons Publishing, London.

STANDARD METHOD OF MEASUREMENT OF BUIDLING WORKS 6TH
    EDITION (1979) RICS.

THE MEASUREMENT OF CONSTRUCTION SITE PERFORMANCE AND OUTPUT VALUES FOR USE IN OPERATIONAL ESTIMATING

ALFRED H WOOTTON BSc AIQS
Department of Construction and Environmental Health
University of Aston in Birmingham

Production orientated evaluation methods have, over recent years, been the subject of considerable debate and argument. The construction industry has hitherto relied upon a fairly rigid and precedental approach in the manner in which information for tendering purposes is collated and presented. This, in turn, has conditioned the procedures and methods by which competing firms prepare and evaluate their tender bids.

Such formal and somewhat inflexible procedures have persisted for many years, however, the past decade has witnessed a ground swell of opinion which encourages and indeed, in some quarters, advocates a radical change in the manner in which the construction industry evaluates its work proposals. Most of the changes that have taken place to date tend to be contractor based rather than from the 'professional' side, although valuable work was carried out at the Building Research Establishment during the 1960's.

From a personal involvement in the 'design and build' sector of the industry during the late 1950's and early 1960's the author developed and applied a variety of measurement, documentation and evaluation methods that closely related to the operational requirements and erection sequences of the firm's construction activities. Such an approach generally proved advantageous to the company since data inputs and outputs owed allegiance to construction method rather than to standard methods of measurement or accountancy biased cost-control systems. A few clients required tenders to be detailed in Bill of Quantity form so a compromise format was produced which utilised the current Standard Method of Measurement rules but collected and presented work in erection sequence order plus grouped elements. A rapid approximate estimating service developed as a by-product and operated to a + or -5% accuracy limit. The control of costs was appreciably improved, although mis-allocation and other site management "error's" occurred.

Studies relating to production performance on construction sites have been carried out at Aston since 1962. The intention of this paper is to describe the progress made to date in attempts to identify the problems and difficulties that arise from the adoption of a production orientated bias in the design and development of operational estimating systems.

Data retrieved from construction sites or performance outputs have generally reflected the element of 'control' that has been experienced on site, both in physical and financial terms. Control systems usually compare 'input' against 'output'. Although, in practice, these values are more often than not measured in different ways using differing terms. This renders such comparisons invalid and, in addition, it has been found(1) that the measurement terms are at complete variance with the working method in use on site.

Many currently operated cost control systems fail through incompatibility of the 'input' and 'output' measurement terms and even when the systems are subjected to the discipline of computerisation the problems do not disappear.

Because cost control systems are, generally speaking, accountancy based and orientated, they tend to collect too much information which is both irrelevant and ineffective from a site manager's point of view. He is assailed and oppressed by financial control demands and requirements, many of which he can do little to satisfy. Cost information related to progress arrives too late for any action other than maybe remedial but, in any case, is usually in a form that defies resolution since the terms are irrelevant to the prevailing circumstances. We have found many cases where the scope available to the site manager to exercise financial control is considerably less in overall terms than the system envisages or demands. Examples quoted later illustrate this observation.

A close correlation exists between the use of the terms 'control' and 'exploitation' on construction sites. The latter term can be used to express the degree of flexibility that exists within a contract and to which the site manager can address himself. The greater the exploitation factor the greater the need for site management to effectively control. Such exploitation factors can be identified, measured and built into the contractor's evaluation method at the bid stage. The element of exploitation is common to both the estimating and contract management functions. That its measurement should be in mutually relevant terms for both functions is obvious but in addition to its financial dimension a further common denominator is that of time.

Site managers operate concomitantly under or within the constraints of time and cost, owing allegiance to both, but being devoid of directives on which to desert when failure conditions appear. The fault of many. planning techniques is that basically they are optimistic; they assume success, but few, if any, are capable of measuring or identifying alternative action courses when possible failure appears to be in the offing. Parameters for control purposes are therefore a priority and preferably these should be construed using a simple transfer process of extracted data from the initial evaluation system.

Many contractors prepare method statements and pre-tender programmes at the bid preparation stage and although this is a valuable exercise, many opportunities are lost since few firms actually continue by setting up an effective and realistic time/cost control system based upon the pre-tender consortium reinforced with data on previous performance output values, etc.

Perhaps we are measuring the wrong things. What do we need to measure; what input data is required; and what output information should we expect?

The starting point from which to create answers to these questions lies in an estimating approach that embodies the operational order of the work carried out on site and takes into account the nature, timing and circumstances under which the work is contemplated. Such an approach should differentiate between those items where the control element is geared against loss (in its many forms) and those items which afford scope for exploitation.

Comparative studies indicate that the proportion of total value of a contract, capable of being subjected to on-site 'direct' control, ranges from 6% to 20% and on average 1% or less of total contract value is affected when a 10% improvement in 'direct' employees' output is achieved from whatever cause. The following example serves to illustrate the point.

	£
Contract Sum	1,185,187.50
Profit	56,437.50
	1,128,750.00
Overheads, etc	53,750.00
	1,075,000.00
Preliminaries, etc	75,000.00
Cost of Work	1,000,000.00
Sub-Contractors)	
(Nominated and domestic)	600,000.00
Own work	400,000.00
Materials, etc	240,000.00
Labour costs	160,000.00
Directly employed	

Analysis of controllable elements in the labour cost produced a factor of 35% as being the proportion 'uncontrollable'. These items emerged as a constant factor related to the attendance of the personnel involved irrespective of whether they actually worked or not. The remaining 65% related directly to the work items and were capable of being controlled so that improved output could be measured in work item productivity gains with consequential time benefit accrual.

The contract included allowances for material wastage and, although often insufficient, a target saving of 3% - 4% on the overall wastage allowance was possible. The ascertained controllable costs from the site manager's point of view appeared thus:

	£
Materials wastage (3%)	7,200.00
Labour cost (65%)	104,000.00
Controllable cost	111.200.00

Representing 9.95% of the pre-profit cost of £1,128,750.00 and 9.38% of the contract sum. In either case a 10% improvement in productivity would produce a change in contract cost of less than 1%.

Certain preliminary items together with attendance items and adept plant manipulation and the like could also produce further improvement but in 'controllable' terms it would be minimal.

Fundamental problems exist in changing a traditional 'unit-rate' system of estimating to one geared to operational and production sequences. The difficulties are transient only insofar as system definition is concerned. The major obstacle must be in overcoming the industry's reliance upon the existing precedental measurement systems which have had a restrictive and even coercive effect upon its method of approach. The industry has been conditioned to produce evaluation and control systems which match these standard formats. Some reference books on pricing construction work, now in their 107th and 154th years of publication, bear witness to these facts.

The operational bill of quantities format was innovatory but appeared over-complex to many and again required a major shift in the thinking and approach of most contractors' estimators.

Research so far indicates that, when tenders are prepared using both the conventional bill of quantities format and the operational estimating approach, final tender amounts do not vary greatly; in fact, most results show a + or -7% range of variability. The variance in some cases has been as low as + or -2%.

What is striking is that within these ranges of low variability there are wide divergencies between comparable sections of the tenders. Groundworks, external services, concreting items and, to a lesser extent, brickwork seem to be the major work areas where differences occur.

Examples follow (Figures 1 to 5) which serve to illustrate these differences. In the main, they constitute the 'controllable' areas of site construction work and go unnoticed unless specific control arrangements are introduced.

FIG. 1. (REF.2)

Activity Description	Man Hours		Overall Duration (Weeks)	
	Contractors Estimate	Actual Observed	Contractors Estimate	Actual Observed
Pile and Pile caps	1900	300	5	28
Ground Floor Slab	960	330	2	29
First fixings, first floor	1800	920	6	35
Suspended ceiling, first floor	600	280	3	22

FIG. 2. (REF. 3)

MATERIAL	SECTION OF WORK	% WASTE	ESTIMATE
CONCRETE	TRENCH FILL (A)	14·3%	10%
— " —	——·—— (B)	12·4%	10%
——·"——	FLOOR SLABS	4·6%	10%
HARDCORE	——·"——(A)	66%	10%
——·"——	——·"——(B)	45%	10%
COLOURED MORTAR	OVER A LARGE CONTRACT	67%	10%
FACING BRICKS	——·"——	19·8%	10%

FIG. 3. CONCRETE (REF. 3.)

DESCRIPTION	MEASURED	USED	DIFFERENCE
1. 22.5 N Concrete to man-holes incl. bed and surround to precast units	10 M³	27 M³	– 17 M³ (170%)
2. 10:1 Mix - Concrete Cavity fill (some 21 N from slabs used)	15 M³	12 M³	+ 3 M³
3. 15 N Concrete footings	29 M³	50 M³	–21 M³ (72%)
4. 21 N Concrete slabs.	51 M³	58 M³	– 7 M³ (13%)
	105 M³	147 M³	– 42 M³

FIG. 4. HARDCORE (REF. 3)

DESCRIPTION	MEASURED	USED	DIFFERENCE
1½" DOWN - Road adjacent	NIL	72 M³	-72 M³
3" DOWN - Car Park	66 M³	66 M³	—
2" DOWN - Bungalows and driveways.	520 M³	670 M³	-150 M³ (30%)
M.O.T. TYPE 1 OR 2 - Roads and Crossovers	230 M³	410 M³	-180 M³ (72%)
TOTAL	816 M³	1218 M³	-402 M³ (50%)

FIG. 5. MATERIAL LOSSES AT VALUATION STAGE

DESCRIPTION	COST £	VALUE £		PROFIT	LOSS
		MATERIALS ON SITE	MEASURED		
Preliminaries	1320	—	460		860
Tipping fees	1453	—	1400		53
Hardcore (2075 tonnes) (1300 M³ in B/Q)	5777	—	3487 (716 M³)		2290
Concrete (147 m³)	2222	—	1596 (105 M³)		626
Concrete sundries	257	86	51		120
Common bricks	3253	1056	2204	7	—
Engineering bricks (400)	38	—	—	—	38 *
Brickwork Sundries eg. Sand, cement etc	3448	2742	800	94	—
Blocks	2059	1568	491	—	—
Facing bricks (30.000)	2061	1540 (22000)	709 (9140)	188	—
Window & Doorframes	2488	1680	977	169	—
Trusses	1152	1152	—	—	—
Roof Timbers	368	196	—	—	172 *
External doors	466	480	—	14	—
Drainage — Metalwork	478	270 }	617	37	—
PCC Goods	707	335			
Strutting	190	—	167	—	33
Pipes and Channels }	2690	1224	1316	—	150
Granular fill (287 tonnes)	1086	80	400		606
Sundries & Plant loss	527	—	—	—	527
TOTAL	32000	12409	14675	599	5565

(REF. 3.)   * Material on site measurement error.

Site managers are aware that two sets of performance criteria require their conformity, ie time and cost. In the interests of one or the other, but rarely together, they take action on decisions based on their current position and its problems set against forward projections. Any guidance information on likely decision outcomes will be helpful, especially if such information can be gleaned through initial evaluation sources and can be seen to re-appear in the control loop of their management system.

An estimator, not faced with this dichotomy, can consider for instance 600 No column bases of identical or variable shape and size and express the outcome in a variety of forms. The site manager has to view the same bases having in mind a variety of conditions and circumstances that are likely to alter at any time from his programme and resource point of view, and he could well generate many alternative approaches for executing the work. A graphical method for developing such alternatives was published by Gomez and Wootton(4) in 1978 and has been used both in the UK and the USA for this purpose.

In this connection research has shown that the estimator's initial progno stication holds good for probably 50% - 70% of the particular activity involved, especially when dealing with groundwork situations. The remaining percentage of the work gets completed as and when convenient or when the labour and/or plant is free from more critically significant activities. In these cases observations show that both time and cost escalate but, because the work is now being carried out at a time when the resources would otherwise be slack, the exploitation factor originally envisaged, has vanished and although the time factor must have receded in order of importance, the cost factor will show up against the profitability figures for the work. Sometimes, because of the distortions and discrepancies in current target and bonus systems, such losses are further increased for bonus will be paid on such work even though no time gain or advantage is possible or can be made use of.

If an estimator includes an overall average output value to cover such disruptive factors then the chances are that his tender bid will be uncompetitive and unsuccessful. Yet it is vitally important that such disruptive factors take place and should be taken into account. As stated previously, the use of an operational estimating method does not appear to increase or decrease overall tender bid values disproportionately so there is scope for a programme related understanding to be incorporated in the estimators assessment. Apart from representing the situation envisaged with greater accuracy there is the added benefit that the site manager can be made aware of the varying production rates included and could take action accordingly.

Experiments with site management decision aids have been carried out; one has been to use a time-keyed daily activity chart which contains a chronological list of major or primary events that must take place in order for a controlled performance to follow. Its use has meant that activity ranking has to take place and daily programmes are devised to determine primary, secondary and subsidiary order activities. Some activities are dependant upon plant output, ie a concrete batching plant supplying five or six work stations, labour activity change at a work station avoiding non-productive time delays and subsequent disruption, material supply demanding the use of cranage at a specific time. It can be shown that if the primary elements are controlled then most work fits into place. Secondary items are also subject to surveillance but of a lesser degree while subsidiary items rarely need a daily check.

When a large volume or quantity of work has to be executed over a period a 'speed' diagram can be used to indicate the performance outputs that have been included in the estimate and these diagrams take into account factors of design that do not find expression in the normal Bill of Quantities approach.

Take the following example of plasterboard where fixing area is measured in m^2, the material is supplied in standard sheets, a proportion being required to be cut and fitted to a profile.

Figure 6 shows the type of 'S' curve that one normally sees for most activities. Figure 7 is probably nearer the truth in that a large (85%) percentage of the materials can be fixed rapidly, the remaining (15%) percentage will take almost as long. These charts could be set out by the estimator as a basis for feedback data on performance and equally the site manager would be aware of the costs necessary to keep in reserve to enable him to complete the work within the budget or target.

FIG. 6.   (REF. 5.)

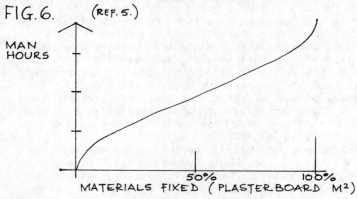

FIG. 7.   (REF. 5.)

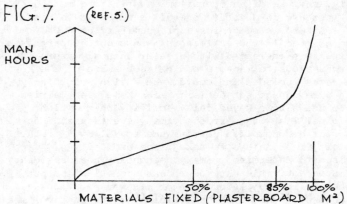

Finally, Figures 8 - 15 have been extracted from actual studies carried out which compared traditional estimating methods (ie unit rate build-ups based on Bill of Quantity formats) and operational estimating methods (based primarily on programme evaluation).

Figures 8 and 9 are for a small Local Authority contract executed in 1977/78 and clearly shows that although the overall contract costs were similar, considerable variation occurred between comparable operations.

Figures 10 - 15 relate to a speculative light industrial development contract and Figure 13 shows the comparison for this development between the traditional form and the resource form of estimating. Figures 14 and 15 give a breakdown of costs for each method.

In all cases studied so far, the resource or operational method of estimating has facilitated the collection and expression of production data more easily than the traditional method. Cash flow measurements have been extracted and profiled without difficulty. The time value of money as a resource becomes apparent and under actual contract circumstances the effects of delays and disruptions from whatever source can be rapidly assessed and evaluated at a rate analogous to the timing and circumstances of the occurrence and its possible disruptive effect.

The feedback of information would be enhanced, the incidence of mis-allocation would diminish and control efforts could be related, without unnecessary dilution, to the items that were most significant in achieving a profitable outcome and client satisfaction.

References

1   Ceislik R C P, "Variations in Productivity with Contract Duration", Undergraduate project study report, Department of Construction & Environmental Health, University of Aston, 1979.

2   Roderick I F, "Examination of the Use of Critical Path Methods in Building", Building Research Establishment, Current Paper 12/77, March 1977.

3   Willis N J, "A Move Towards Accurate Estimating", Undergraduate project study report, Department of Construction & Environmental Health, University of Aston, 1980.

4   Gomez-Chadwick, Patricio and Wootton, Alfred H, "Journal of the Construction Division - Proceedings of the American Society of Civil Engineers", Vol 104, No C O 1, March 1978.

5   Butcher S J, "Control of Construction Work", Undergraduate project study report, Department of Construction & Environmental Health, University of Aston, 1982.

6   Ovenden, David R, "Operational Estimating", Undergraduate project study report, Department of Construction & Environmental Health, University of Aston, 1982.

OPERATION	TOTAL COST A	B	GROUNDWORKS + UNLOAD A	B	BRICKWORK A	B	JOINERY A	B	PLANT LEASE A	B	LORRY A	B	MATERIAL A	B	WASTE A	B	SUBCONTRACT	PC SUPPLIES	PC CONTRACT
Reduce dig / clear site	495	317	226	75					200	225	65	117							
Excavate / concrete founds.	916	1952	170	275					270	250	175	334							
Floor slabs	1491	1939	529	465					500				788	783		183			
Brickwork to foundations	1706	1935	84	75	506	606							1180	190		190			
Superstructure brickwork	7536	6732	84	300	4310	4300							2788	424		424			
Structure timbers	1899	1895	12	150			512	270								77		250	
Fibreglass	286	326	30	75									230	29	77	29	236		
First fix joinery	2145	2114	110	150			345	270					1633	41		41			
Second fix joinery	2281	2226	18	150			727	540					1156	420		120		260	
Plumbing + heating	4152	4152															3652		508
Plastering	2105	2105															2105		
Glaze tiling	236	236															236		
Floor finishes	742	742															742		
Floor screed	456	456															456		
Glazing	255	255															255		
Painting	903	983															983		
Electrical	1245	1245															1245		475
Felt roof	276	276															276		
Roof tiles	1723	1723															1723		
Services and drainage	4234	4030	800	695					500	900	50	117	1254	200		200	2761		900
Retaining + blockwork to ...	2653	2522	176	75	1177				76	350		17	1097	110		110	276		754
Landscaping	1464	1666		990													716		
Groundworks	1766	2648	527	815					130	750		32	749	104		104		150	
Tarmac	200	1200															200		
Preliminaries	5613	5695		455															
TOTAL	44534	44060	3754	3956	400	1250	1680	1000	470	2375	340	640	3904	400		400	2473	460	2635

FIGURE 8

A = NEW METHOD B = BASIC METHOD

* = work which cannot be directly compared

EVALUATION OF THE NET PRICE (REF 5)

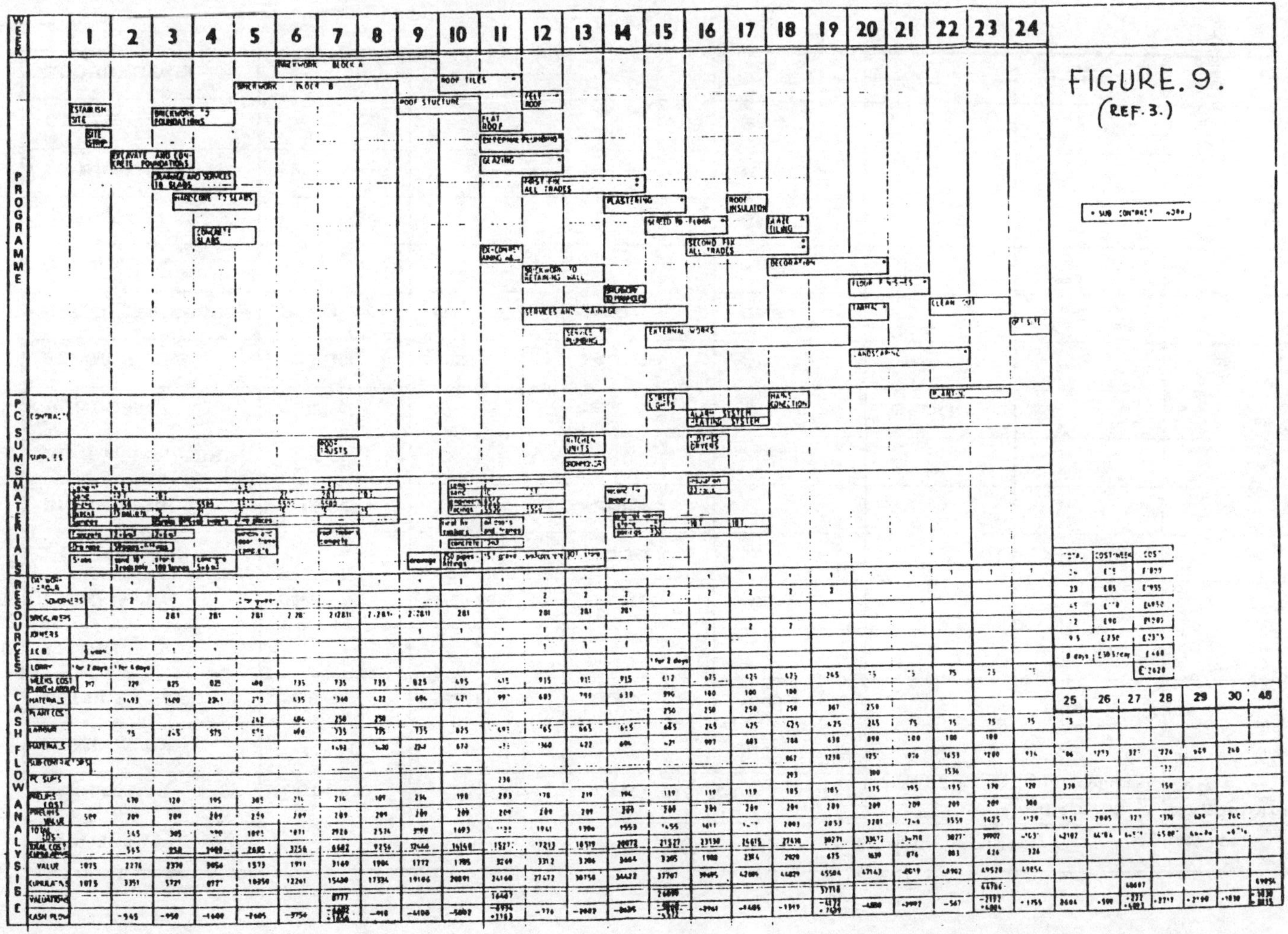

FIGURE. 9.
(REF. 3.)

TABLE 1

Operation	Labour	Plant	Materials	Sub Contract	Prime Cost Sums		Provisional Sums
					supply	sub contract	
DEMOLITION				922			
REDUCE LEVEL	1855	1440	2230	7979			
PILING						51000	
PILE CAPS	13967	2302	17928	4852			
P.C.C. FRAMES		1000	1000			145000	
BRICKWORK	61260		49097				
ROOF and CLADDING				75923			
WINDOWS	1115		53		8500		
FLOOR SLABS	15568	2130	48518	2027			
INSITU STAIRCASES	2124	61	1461	286			
STEELWORK				4696			
BALUSTRADES				666			
MARLEY RAIL				164			
GLAZING				284			
DOORS STEEL						8500	

FIGURE. 10. (REF. 6.)

FIGURE. II. (REF. 6.)   sheet 2 of 3

TABLE 1

Operation	Labour	Plant	Materials	Sub Contract	Prime Cost Sums		Provisional Sums
					supply	sub contract	
WOODWORK	17113		20737				
PLASTER				9244			
ELECTRICS						14500	
PLUMBING				14560			
SUSPENDED CEILINGS				8391			
PAINTING				7976			
TILING				6320			
DRAINAGE	22773	4516	44776	1321			
B.W.I.C.S.	5381	936	8225	69			500
ROADS	3494	2187	3474	3358			
LANDSCAPING				402			4500
BOUNDARY WALLS							1000
INSTALLATIONS							2700
WATER CONNECTIONS							7000
GAS CONNECTIONS							45000

TABLE 1   FIGURE . 12 . (REF. 6.)   sheet 3 of 3

Operation	Labour	Plant	Materials	Sub Contract	Prime Cost Sums		Provisional Sums
					supply	sub contract	
ELECTRIC CONNECTIONS						8500	
STREET LIGHTING						6500	
EXTERNAL WORKS	9631	1239	29541	5677			
TARMACADAM				6913			
GENERAL PRELIMINARIES							41700
CONTRACTORS PRELIMINARIES	31159	24630	15155				
CONTRACTORS STAFF	34760						
DAYWORKS							3250
DAYWORKS % ADDITIONS							4500

TABLE 2	FIGURE: 13. (REF.6.)			
	TRADITIONAL		RESOURCE	
OPERATION	LABOUR	PLANT	LABOUR	PLANT
REDUCE LEVEL	1855	1440	1366	1461
PILE CAPS	13967	2302	11661	2168
BRICKWORK	61260	—	59796	—
WINDOWS	1115	—	1169	—
FLOOR SLABS	15568	2130	20311	3603
INSITU STAIRS	2124	61	2120	—
WOODWORK	17113	—	18294	—
DRAINAGE	22773	4516	13743	3210
B.W.I.C.S.	5381	936	2540	192
ROADS	3494	2187	3919	2104
EXTERNAL WORK	9631	1239	11794	906
SUB TOTAL	154281	14811	146713	13644
CONC. PLANT ABSTRACTED				1463
TOTAL 1	154281	14811	146713	15107
TOTAL 2	169092		161820	
DIFFERENCE	7272 or 4.30%			

Chart ref.	Resource Requirements	Value of Plant	Value of Labour
D	Excavate for pile caps etc. 2 No. Mustang 120 for 89.27 Hrs	£1399.75	
	2 No. Plant operators		£696.31
	2 No. Labourers		£610.61
E	Backfill to excavations 1 No. Mustang 120 for 75.78 Hrs	£594.12	
	1 No. Plant operator		£295.54
	1 No. Labourer		£259.17
	1 No. Whacker DVPN75	£58.35	
	1 No. Labourer		£259.17
F	Trench sheeting 8 No. Labourers for 34.48Hr		£943.37
G	Break off pile caps to level 4 No. Breakers for 58.00 Hrs	£116.00	
	8 No. Labourers		£1586.88
H	Formwork 8 Joiners for 85.50 Hrs		£2831.76
	2 Labourers		£584.82
I	Make a start on fixing bolt boxes 1 No. Joiner for 40.00 Hrs		£165.60
J	Complete bolt boxes 8 No. Joiners for 47.30 Hrs		£1566.58
	2 No. Labourers		£323.53
K	Concrete to pile caps 3 No. Labourers for 58.00 Hr		£709.92

WEEKLY RESOURCE REQUIREMENTS.

FIGURE. 15. (REF. 6.)

WEEK 7	COST	WEEK 8	COST	WEEK 9	COST
2 No. Mustang 120	£627.20	8 No. Joiners	£1324.80	3 No. Conc Labourers	£489.60
2 No. Plant operators	£312.00	2 No. Labourers	£273.60	2 No. Joiners	£331.20
2 No. Labourers	£273.60	3 No. Conc Labourers	£489.60	1 No. CAT 951	£288.00
8 No. Joiners	£1324.80	1 No. CAT 951	£288.00	1 No. Plant operator	£156.00
2 No. Labourers	£273.60	1 NO. Plant operator	£156.00	2 No. Labourers	£273.60
1 No. Joiner	£165.60	2 No. Labourers	£273.60		
1 No. CAT 951	£238.00	1 No. Breaker	£20.00		
1 No. Plant operator	£156.00	3 No. Labourers	£410.40		
2 No. Labourers	£273.60				
1 No. Breaker	£20.00				
3 No. Labourers	£410.40				
WEEKLY COST	£4124.80		£3236.00		£1538.40
CUMULATIVE COST	£31144.12		£34380.42		£35918.82

323

THE PREDICTION OF EXPENDITURE PROFILES
FOR BUILDING PROJECTS

BY DR A C SIDWELL BSc(Hons) ARICS, MCIOB, MBIM
   MR M A RUMBALL BSc(Hons) ARICS
Lecturers in Construction Management and Building Economics in the
Department of Construction and Environmental Health
The University of Aston in Birmingham.*

Introduction

This paper reports research into the accuracy of formula to predict
'S' curve expenditure profiles for construction projects. There is a
need for clients to know the pattern of financial committment for the
future, and for those clients with a number of building projects, the
timing and distribution of the expenditure is of prime importance.
    The production of 'S' curves may be approached in two ways. The
first is from the basis of the contract programme (e.g. the Gantt bar
chart) where the sequence of activities is translated into money terms
Lott(1). The second approach is to base the 'S' curve on the valuations of work in progress, the assumption being that these reflect the
building activity. This may be calculated as the work proceeds, but
work by Hudson(2) and Drake(3) suggests a mathematical formula which
will plot an 'S' curve given only the contract sum and the contract
duration. This may therefore be applied in advance of start on site
to predict expenditure.
    The use of one standard formula is bound to present problems due to
variables in construction projects. These might include the type of
building, the height, shape and design characteristics of the building,
the value of the contract, land costs, external works, individual
contractors pricing characteristics and external environmental
influences.
    This research examines the predictive ability of the formula and
parameters suggested by Hudson(2):-

$$Y = S \left(x + Cx^2 - Cx - \frac{1}{k} (6x^3 - 9x^2 + 3x) \right)$$

Where Y = cumulative monthly value of work executed before
          deduction of retention monies or addition of fluctuations

---

Footnote *

The authors wish to acknowledge work done by M J Baldwin and S A Frost
who conducted the case studies during their undergraduate work and
Mr A L Traill, Senior Lecturer in the Department in Economics and
Statistics, for assistance with the statistical analysis.

$$x = \frac{\text{month (m) in which expenditure Y occurs}}{\text{contract period (P)}}$$

s = contract sum

c and k = standard parameters

Preliminary Study - 10 projects

A preliminary study of 10 projects was undertaken to investigate the application of the formula. Data were obtained from contacts with industry. No bias was sought in the sample, which encompassed a wide range of building types varying in cost between £60,000 and £9 Million, Baldwin(4). Information was collected on:-

(1)   original contract duration
(2)   original contract sum
(3)   interim valuations
(4)   date of practical completion
(5)   final contract sum (less fluctuations)

Graphs were drawn for each of the 10 case studies of the actual cumulative expenditure (interim valuations) against the actual contract time. On each was superimposed the expenditure 'S' curve which would have been predicted by the formula, using the original contract sum and original contract duration. Fig 1(a) illustrates a typical curve. Table 1 gives details of the 10 case studies and a summary of the results of the comparison of actual expenditure profile with predicted profile.

In no case was the predicted curve correct in the first instance. Five projects followed the curve fairly closely for the first few months before deviating, but the remaining five deviated from the start of the project.

Revised prediction curves were plotted in all cases and in a further five cases a second revised curve was needed before any degree of congruence was achieved between the actual expenditure curve and that forecast by the formula.

Clearly in all 10 cases the formula would have been unsuccessful at predicting the expenditure profile before the construction commenced. However, when the forecast is revised in the light of actual expenditure some use could be made of it. These preliminary case studies do not examine the operation of the formula in depth, but they do suggest some reasons for the discrepancies:-

(1)   inclusion of large contingency sums in the original
contract sum
(2)   extensions of time
(3)   delay caused by contractors
(4)   variations
(5)   a technical problem of the choice of points on the
actual expenditure curve used as a basis for the revised
curve.

NO	PROJECT TYPE	COST	Duration (Months)	Revisions necessary to forecast expenditure		Cause of divergence
				1st Revision at month	2nd Revision at month	
1	7 Detached Houses	£ 66,000	13.5	6	–	–
2	Public House	£ 68,000	10	1	–	Variation of omission
3	22 Semi-detached Houses	£ 162,000	25	1	–	Contingency sum
4	2 Eight storey blocks of flats	£ 204,000	18	5	11	–
5	2 Blocks of flats	£ 420,000	18/22	1	8	Contractor in liquidation
6	61 Mixed dwelling units	£ 560,000	25	5	–	Variation of omission
7	16 Storey office building	£3,303,000	32	17	22	Design problems, extension of time
8	Office block	£4,000,000	33	9	12	–
9	9 Storey housing development	£6,704,000	32	14	–	Extension of time
10	Hospital	£8,935,000	37	15	20	–

TABLE 1 - Preliminary Study

COST CATEGORY	INDUSTRIAL	COMMERCIAL	HOUSING	REFURB- ISHMENT	EDUCATION	HOSPITAL	TOTALS
75,000 - 120,000	1						1
120,000 - 300,000	1		1		1		3
300,000 - 1,200,000	5	9	1	2			17
1.2m - 2.0m	2		1		1		4
2.0m - 3.0m	4	1					5
3.0m - 4.0m		1				1	2
5.0m - 6.0m	2						2
6.0m - 6.5m	1		1				2
7.0m - 7.5m		1					1
8.0m - 8.5m	1						1
TOTALS	17	12	4	2	2	1	38

TABLE 2 - Second Study Sample Characteristics

Second Study - 38 Projects

Experience gained in the preliminary study showed that the ability of
the formula to forecast expenditure profiles could be influenced by a
number of variables, some of which may not reasonably be forseen
before start on site.

A major problem is the change that frequently occurs in the
contract sum and contract period. When this happens, new forecasts
may have to be produced and the results from the preliminary study
suggest that the revised forecast may be a good predictor of the
actual expenditure.

In order to study the characteristics of the 'S' curve in greater
detail, the second study sought to eliminate the variables of contract
sum and duration by examining the formula prediction as applied to the
data obtained after completion on the actual final contract sum and
actual duration. This concentrates attention on the shape and dispen-
sation of the curve. The actual gross monthly valuation figures were
compared directly with the monthly figures predicted by the formula,
Frost(5).

Details of the sample of 38 projects which included various building types and costs are given in Table 2. Again no bias was intended
in the sample which was obtained from contacts within the industry.
Data were collected for:-

(1)   actual contract duration
(2)   final contract sum (less fluctuations)
(3)   interim valuation figures (less fluctuations and retention)

Analysis

The data were analysed using the University's ICL 1905 computer
program VA13 which provided scattergrams of the data. The Interactive
Data Analysis (IDA) program developed by the University of Chicago was
used on the HP2000 to derive the regression lines of Y on X with
values for m and c and the 't' test of tm and tc.

Graphs were prepared for each case study showing the actual expenditure curve and that forecast by the formula. This permitted macro
examination of the relative shapes of the curves. However, these
graphs were of limited use for analysis because they are based on
cumulative values which may include cumulative errors. A more useful
approach is to plot the individual values for each monthly period. If
the actual valuation (observed) is plotted against the forecast
(expected) then a scattergram of points is obtained which in fact,
bears no resemblance to an 'S' curve.

Figures 1 - 6 illustrate the type of scattergrams obtained by
plotting the individual observed valuations against expected values.
A regression line has been drawn through the points for purposes of
illustration. The normal, cumulative, 'S' curve is drawn for
comparison.

Figure 1(a) illustrates the case where forecast expenditure follows
the actual curve exactly. Under these circumstances the points on
the scattergram, figure 1(b) will all lie on the one straight line at
45° passing through the origin.

The equation for the regression line is given by:

Y = m x + c

where Y   =   the expected values (E)
       X   =   the observed values (O)
       m&c =   constants

In this particular case:

    (Expected) E = O (Observed)
therefore   C = Ø
then   m = 1

Figure 2(a) illustrates the case where observed expenditure is below
the expected curve but follows the shape of the curve and is more or
less parallel to it (except at the ends). In this case the observed
valuations are lower than forecast. The scattergram in figure 2(b)
has a line at 45° but intercepts the 'y' axis at c. Therefore, in this
case the constant c is a measure of the gap between the curves, the
difference between the expected and observed values.

Figures 3(a) and (b) illustrate the reverse condition, but for the
case where the observed valuations are above the forecast line.

Figures 4(a) and (b) illustrate the expenditure profiles and regression line for case study 31. They indicate what happens to the
regression line if the 'S' curves cross or in particular, because the
'S' curve is cumulative, where the slope of the observed curve changes
direction in relation to the expected curve. The regression line is
flatter and where it is to the left of the 45° line then the expected
values are greater than the observed values. Where the regression
line is on the right of the 45° line the observed values are greater
than the expected values.

Figures 5(a) and (b) illustrate the expenditure profiles and regression line for case study No. 5. They indicate that the observed
valuations were always less than expected, that is, the   contract
was behind schedule all the time. Again the value of C gives an
indication of how far apart the two curves are, but this is confused
by the slope of the line which, not being parallel to the 45° line,
indicates that the gap between the two curves varies.

Figures 6(a) and (b) illustrate the expenditure profiles and regression line for case study 20 where the valuations were always greater
than expected. One might speculate that the bills of quantities had
been front loaded.

The Slope of the regression line

If the observed valuation's 'S' curve follows the forecast curve
exactly then, the regression line slopes at 45° through the origin.
The constant C is equal to zero and the coefficient m must be equal
to 1. When the slope of the regression line is other than 45° this
will indicate a discrepancy between the expected and observed values.
It is the slope of this line which indicates this change. Since the
slope of the line is governed by the coefficient m, then the discrepancy between the expected and observed expenditure profiles may be
measured by the amount by which m deviates from the value 1.

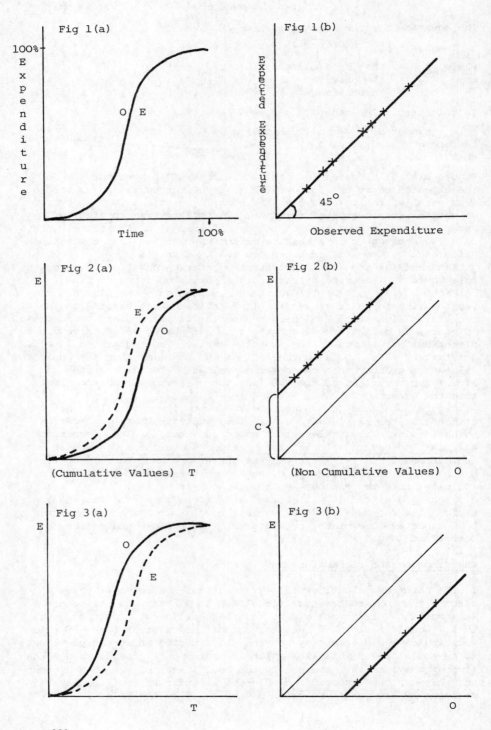

Fig 1(a)

100% Expenditure

Time   100%

Fig 1(b)

Expected Expenditure

45°

Observed Expenditure

Fig 2(a)

E

E

O

(Cumulative Values)   T

Fig 2(b)

E

C

(Non Cumulative Values)   O

Fig 3(a)

E

O

E

T

Fig 3(b)

E

O

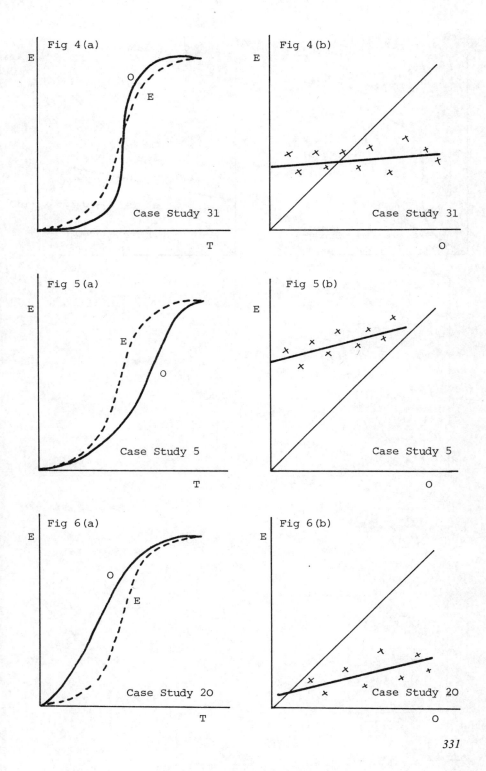

Fig 4(a)   E   O   E   Case Study 31   T

Fig 4(b)   E   Case Study 31   O

Fig 5(a)   E   E   O   Case Study 5   T

Fig 5(b)   E   Case Study 5   O

Fig 6(a)   E   O   E   Case Study 20   T

Fig 6(b)   E   Case Study 20   O

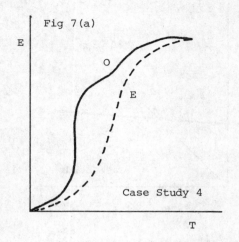

Fig 7(a)

E

O

E

Case Study 4

T

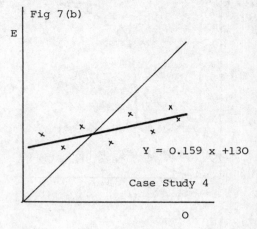

Fig 7(b)

E

Y = 0.159 x +130

Case Study 4

O

This may be estimated by scrutiny of the coefficient m in the table of results, Table 3. Clearly where the coefficient is low the forecast expenditure curve may be expected to be a poor predictor of the actual expenditure. Figures 7(a) and (b) illustrate case study No. 4 where the observed expenditure profile is quite different from the expected.

The slope of the regression line is 0.159x and is significantly different from the 45° line.

A level of significance for m

The success of the formula as a forecast of actual expenditure may be measured by the discrepancy between the regression line of observed against expected and the 45° line. This may best be measured by the extent to which the value of m in the regression equation deviates from 1.

We therefore wish to test the null hypothesis that m does not diverge significantly from 1. The alternate hypothesis being that m does not equal 1.

The IDA regression program gives tm and tc where t is the ratio of difference between 1 and the estimated slope to the standard error of that difference. tc is analogously defined:

$$t_{m-1} = \frac{m-1}{\sigma m-1}$$

$$t_c = \frac{c}{\sigma c}$$

Table 3 gives the results, for a two tailed test at a level of significance of 95%. The critical values of tm-1 must be obtained from 't' distribution tables, the number of valuations less one being the degrees of freedom.

Where tm-1 is greater than the critical value we must reject the hypothesis that m = 1 and by implication accept the alternate hypothesis. Only where m is equal to 1 can we conclude that the formula is a valid predictor of the actual expenditure profile.

Results

The statistical test indicates that in only 8 cases out of the 38 studies was the forecast expenditure a valid predictor of the actual expenditure. Tables 4, 5 and 6 are contingency tables which examine the results in relation to contract value, building type and contract duration. They suggest that the forecast may be more accurate with low cost projects in the cost categories £120,000 to £1.2M with shorter durations of less that 30 months. However, these results are not statistically significant. No firm conclusions may be drawn concerning building type.

Despite the 't' test, visual observation of the actual cumulative distribution curves suggests that for all practical purposes the curve reasonably models a further 7 cases. It may be felt that the 't' test is unduly sensitive.

Contract No.	Duration 'n'	m	c	tm	tc	tm-1
1	14	0.23176	99.531	2.064	5.523	6.8418
2	15	0.4982	70.847	3.787	3.420	3.8144
3	14	0.51789	86.082	4.284	3.555	3.9880
4	17	0.15918	130.490	1.465	6.206	7.7384
5	14	0.16483	370.810	0.969	4.112	4.9098
6	20	0.25556	190.500	1.254	3.337	3.6529
7	30	0.70360	55.028	5.305	2.070	2.2348*
8	45	0.43246	78.423	5.163	5.699	6.7757
9	13	0.61868	4.254	2.913	1.668	1.7954*
10	9	0.32140	75.876	0.831	1.641	1.7546*
11	14	−0.01105	26.150	−0.071	5.141	6.4963
12	9	0.14971	9.2021	1.189	5.241	6.7536
13	14	0.27956	35.401	1.826	3.865	4.7057
14	14	0.58169	6.191	3.959	2.598	2.8470
15	29	0.54280	126.62	3.385	2.648	2.8512
16	60	0.43599	66.387	4.458	5.207	5.7670
17	41	0.26522	57.803	3.026	7.182	8.3834
18	45	0.32913	49.142	3.902	6.914	7.9535
19	34	0.30016	42.111	5.318	9.771	12.399
20	9	0.27418	36.452	0.912	2.172	2.4143
21	10	0.078267	33.090	0.386	3.703	4.5458
22	14	0.42481	34.717	2.296	2.734	3.1088
23	14	0.31031	62.222	1.299	2.705	2.8871
24	8	0.67754	111.65	2.280	1.030	1.0851*
25	10	0.16670	37.315	0.785	3.265	3.9007
26	20	0.045794	33.733	0.249	4.587	5.1884
27	18	0.60247	13.881	2.634	1.630	1.7380*
28	16	0.30056	21.333	2.305	4.336	5.3640
29	6	0.68899	36.544	4.118	1.655	1.8589*
30	22	0.097004	49.259	0.703	5.428	6.5441
31	16	0.057755	17.899	0.270	3.892	4.4049
32	14	0.21978	22.428	1.314	3.849	4.6647
33	26	0.25128	19.390	1.455	3.906	4.3354
34	33	0.39446	17.561	3.467	4.632	5.3222
35	24	0.16137	53.624	1.644	7.017	8.5438
36	12	0.54092	9.4289	1.856	1.460	1.5752*
37	22	0.67898	14.650	3.548	1.566	1.6775*
38	21	0.38122	35.477	2.286	3.439	3.7105

TABLE 3 - Results

n   = Duration in months, the number of valuations & data points
m   = coefficient of the slope in regression equation
c   = constant c in regression equation
tm   = t test of m against 0
tc   = t test of c against 0
tm-1 = t test of m against 1
*   = values of tm-1 significant at 95% level

Cost Category (£)	Number of Cases		Totals
	Significant	Not Significant	
75,000- 120,000	-	1	1
120,000- 300,000	2	1	3
300,000-1,200,000	4	13	17
1.2m - 2.0m	-	4	4
2.0m - 3.0m	1	4	5
3.0m - 4.0m	-	2	2
5.0m - 6.0m	1	1	2
6.0m - 6.5m	-	2	2
7.0m - 7.5m	-	1	1
8.0m - 8.5m	-	1	1
TOTALS	8	30	38

TABLE 4 - Relationship between contract value
and accuracy of expenditure forecast

335

Building Type	Number of Cases		Totals
	Significant	Not Significant	
Industrial	3	14	17
Commercial	2	10	12
Housing	2	2	4
Refurbishment	-	2	2
Education	1	1	2
Hospital	-	1	1
TOTALS	8	30	38

TABLE 5 - Relationship between building type
and accuracy of expenditure forecast

Contract Duration (Months)	Number of Cases		Totals
	Significant	Not Significant	
0- 5	-	-	-
6-10	3	4	7
11-15	2	10	12
16-20	1	5	6
21-25	1	3	4
26-30	1	2	3
31-35	-	2	2
36-40	-	-	-
41-45	-	3	3
46-50	-	-	-
51-55	-	-	-
56-60	-	1	1
TOTALS	8	30	38

TABLE 6 - Relationship between duration of contract and accuracy of expenditure forecast

Conclusions

This research has examined the validity of using a mathematical formula to forecast expenditure profiles on construction projects. A method of analysis and statistical test is suggested which permits objective measurement of the success of the formula.

The results indicate that there are cases where the formula cannot be shown to be a poor predictor. This occured in 8 out of 38 case studies. However, there is no statistical evidence of consistent success for any particular project characteristics.

There appear to be two main problem areas:

(1) **Project Characteristics**

The application of a standard formula presents difficulties when projects are so diverse in nature, for example in design, quality, duration, external works, and environmental conditions.

The poor success of the formula in this problem area may be mitigated by examination of the parameters in the equation. However, examination of these case study results suggest that it is the variability of the expenditure profiles which cause the inaccuracies because in the minority of cases is the actual expenditure curve merely a different shaped 'S' curve. In the main, the disruption is due to hiccups in the expenditure curve.

(2) **Problems during the construction phase**

Changes frequently occur as the project progresses, for example variations, extensions of time, weather and other external influences. These necessitate revised forecasts and it would be true to say that if the forecast is revised often enough it is bound to approximate to the reality. The preliminary study indicated a level of success for the formula. However, this may be because a number of revisions were made.

References

1. Lott, B.D. 1981 'A systematic approach to monitoring progress' The Chartered Quantity Surveyor, November 1981 pp 88-89
2. Hudson, K.W. 1978 'DHSS expenditure forecasting method' The Chartered Quantity Surveying Quarterly volume 5 No 3 Spring 1978 pp 42-45
3. Drake, B.E. 1978 'A mathematical model for expenditure forecasting post contract' paper presented to the CIB W-65 second International Symposium Organisation and Management of Construction' Haifa, Israel, November 1978, pp II-163 to II-183
4. Baldwin, M.J. 1979 'An investigation into the formula method of forecasting capital expenditure on construction works' Final year dissertation for the degree of Building Economics and Measurement (unpublished)
5. Frost, S.A.W. 1981 'An investigation into the application of the 'S' curve and in particular the DHSS formula method, to different market sectors of the Construction Industry' Final year project for the degree of Building Economics and Measurement.

SECTION VIII

COMMUNICATION & COMPUTERS

THE DEVELOPMENT OF A STANDARD USER INTERFACE FOR APPLICATION SOFTWARE
FOR USE IN THE BUILDING INDUSTRY OF THE U.K.

R B MIDDLETON, B.Sc., Ph.D., Newcastle upon Tyne Polytechnic
C HARDCASTLE, B.Sc., M.Sc., Newcastle upon Tyne Polytechnic
A P CUNNINGHAM, B.Sc., Newcastle upon Tyne Polytechnic

1. *Introduction*

As a graduate in an area outside the sphere of the building industry
much time has been spent researching into the whole spectrum of
activities and processes within it. This, I feel, has resulted in an
unbiased approach, which has been an asset in the formulation of a
software tool that will benefit the industry as a whole rather than
an isolated area of it. The research commenced in December 1980 and
is expected to finish November/December 1983.

2. *Background Perspective to Research*

The development and specialisation of the Quantity Surveyor, from the
simple role of "MEASURER" in the past to Economist of the building
industry of the present, is indicative of an industry typically
moving towards an increase in functional specialisation. Over the
last thirty years especially, the participants and their associated
information systems have become more selective in the dissemination
of information, supplying to each area only that necessary to support
it. Participants have lost sight and significance of their roles in
the overall structure and communication of the industry and, to some
extent, fail to appreciate the effect they have on it. For example,
according to the U.S.A. General Accounts Office July 1978 in their
report "Computer Aided Building Design" [1] some 75-90% of the costs
of a building are determined by the time the working drawings have
been prepared. Assuming similar results in the U.K. to what extent
does the Architect know about the true implications of the design upon
cost allocation or construction methods?
    Inherent in the U.K. building industrys fragmentation, special-
isations and seemingly increasing professional divisions are the
associated problems of the lack of information standards. Resultant
difficulties arise from having to pass information through a number
of professional boundaries, or interfaces, that have differing data
and information requirements, with respect to content and format.
Also, it becomes difficult to ensure that the information that is
disseminated is of the correct type and in the most suitable form for
the purpose or function of the receiver of the information.

For example, the translation of drawings and specifications into
Bill descriptions involves the complete analysis of a construction
project into a mass of descriptive and quantified raw material. From
this, information could be extracted not only for estimating but for
cost planning, cost control, valuations, ordering, construction
programming and project programming. However this mass of information
is swallowed up in the processing to satisfy the primary purpose of
the Bill, identified by previous researchers as being that of a
contract document providing a uniform basis for the submission of
tenders. The resultant Bill, whilst constraining all, reveals
nothing.

For many years there have been numerous attempts at solving the
communication problems resulting from the lack of standards, such as
the Universal Decimal Classification, Cl/sfB, Co-ordinated Building
Communication, Construction Planning Units, B.S. 3589 Glossary of
Building Terms, the Construction Industry Thesaurus, the National
Building Specification and the changing emphasis of the standard
Method of Measurement all testify to an awareness of communication
problems and the need to overcome them.

The introduction and application of computers to the building
industry was seen as the solution that everyone had been waiting for.
The first international symposium on information flow in the building
process was held by the CIB in Oslo in 1968. The title was: "The
information flow in the building process – classification and coding
for computer use" [2]. However the real moves forward came later in
1970 from a symposium in Rotterdam entitled: "Some problems of
information flow in the building process" [3]. Numerous working
parties and proposals evolved as a consequence of this symposium.
The results of those have helped the industry by providing theoretical solutions and solving problems in isolated areas of interest.

The results after twelve years is however that computers have had
relatively little impact on the industry. Where they have been
introduced they only serve to mechanise existing manual systems,
primarily in the larger firm.

In general where the application of computers has grown and
developed, it has been in an unco-ordinated fashion. For example,
there are a number of standard phraseologies and specification clauses,
as well as Bill of Quantities systems. Also there are numerous
computer programs that set out to solve the same problems in a variety
of ways, the N.E.D.O. program, for example. As a result, the computer
solution has in many cases not revolutionised but perpetuated the
existing lack of standards as well as introducing a multiplicity of
its own.

In many computer applications there is the use of arbitrary and
non-standard rules and the use of coded data formats for inputs and
outputs to programs which depend to a large extent on the computer
system used. This reflects the characteristics and limitations of
the various hardware and software that was, and is available. The
computer users of the industry require systems that will accept input
and generate output in the terms and layout which they, and not the
systems designers, find most natural and convenient.

In the light of this slow response to the use of the computer in
the building industry the hapazard development of systems and the
arrival of micro computer technology, a European study project was
instigated in 1978. [4]. A consortium of international firms and
computer experts made up the team. It attempted to find those aspects
of the building industry and process to which the application of
computers was relevant. It identified areas in which computers could
have significant effect, as well as the problems inhibiting wider
application in the industry. It also tried to assess the implications
and benefits of overcoming the problems.

The study made a number of recommendations for action to be taken
by the European Community aimed of achieving a basis for the effective application of computers within the building industry of the
community.

This research addresses three of those recommendations. The
recommendations were that:

"the E.C. should –

Support the specification of a standard input/output language
for application software in the building industry – Recommendation
3" (p. 49)

"the E.C. should –

Support the specification of a standard for the description and
structuring of information exchanged between participants,
computers and software in the building process – Recommendation
5" (p. 50)

"the E.C. should –

Support the development of an input/output software system based
on the standard input/output language – Recommendation 6" (p. 50).

3. *Formulation of Objectives*

With these recommendations as a foundation to the research. Various
solutions were feasible. A language, if developed, could be used to
define the operations which are performed on data supplied by the
various users. The steps could be formulated in the terms of this
language in order to define the operations to be performed on the
given data structures.

Alternatively, it would be possible to define the questions on
prompts which would have to be output by an application program in
order to get the correct input from a user to perform any operation
he requires.

The first approach is applicable to batch processing whereas the
second is suited to the interactive mode of computing. This interactive mode is preferred by most users. In the future this trend will
increase and therefore the research has emphasised this workstyle.
An important consideration is that the untrained person cannot easily
access machine stored information as this requires a specialised
knowledge of the methodology and semi-formal language used to store
or retrieve the information. Hence, access to the computer, or machine

stored information, is limited by a communication barrier. A compounding difficulty is that the ultimate users of the computer stored information may not reasonably be expected to become proficient in a computer language in addition to their other duties. This may have been a dominant factor in the industry not accepting computers in the past. After thorough consideration of the problems it was decided that the development of a language was not a reasonable solution, but the alternative approach of the development of an interface was reasonable. This would be an interface between a persons normal written communication medium (a natural language for his discipline) and the more formal language normally used when accessing information from a computer.

4. *Aims of Research*

The objective of the research project is therefore:

"The development of a standard user interface for application software for use in the building industry of the U.K."

The term interface denotes those parts of the User and Computer Communication medium that not Hardware and that can usually be varied by program control.

The aspects covered by the research are as follows:
  i) Format, layout etc of displayed material
 ii) Sequence of user/computer messages
iii) Logical progression through the users
     application area
 iv) Structure and retrieval of data.

These aspects are all dependent upon the particular USER, his FUNCTION and the OBJECTIVE of that function.

The emphasis here is not on the PROGRAMMING language is use, but the Syntax, Semantics, Dialogue between the COMPUTER and the USER.

Due to the constraints of time and the complexity of the research one particular information flow was isolated. The flow chosen was that considered to be of major importance in the industry, i.e. that from Designer through to Contractor with emphasis upon production resources. This flow has often been considered as containing an information gap or as being an area of frequent breakdown in communication. Other data flows must necessarily be considered in order to allow for subsequent expansion of the project into these areas.

The plan of work for the research is to design a database for the information used in this data flow and following on from this an interface will be designed to access this data base. This interface can also be driven by an application program when information is required from the user.

5. *Methodology*

The efficiency of the construction process depends upon the ability of its participants to communicate correct and adequate information to each other throughout its complex systems and procedures.

This communication is at present done via classifications,
conventions and information tools – the majority of which are manual
systems. Computerisation of the information flow and associated
processes make these systems an important consideration in the
specification of standards for the representation of information for
each user group.

It was decided to base the research upon the users information
requirements and the ways they access this information, rather than
analyse existing computer orientated systems. Many of these have
been developed within the limitations of the hardware and software
of the time. Many of the original criteria for computer system
development have changed with the introduction of micro computer
technology and the availability and portability of hardware and
software.

With this approach the results of the research should be able to
be implemented upon any computer system.

The general approach to the research has followed the guidelines
as in FIG. (1).

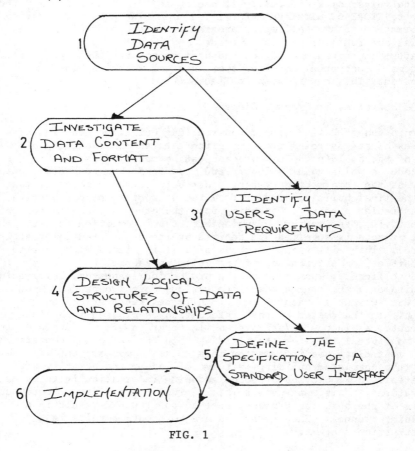

FIG. 1

6. *Analysis of major Classification Systems*

Information which emanates from design proposals through to resource requirements in production, has been examined by inspection. of its structure of classification.

As a classification indicates the relationships between items of knowledge, a study of the classification structures has helped to define those items of knowledge which I feel the industry considers important, and the relationships which are possible between them. These have identified:

  i) Particular user groups in the building industry and their
     information requirements in terms of content and format
 ii) The preferred information structure, or patterns, within
     each user group area.

From this, sets of rules will be produced for the manipulation and retrieval of information within a particular group and between defined user groups.

The rules, or standards, will become INTERFACE ROUTINES for the user to STORE or RETRIEVE information from a data base or provide information to an application program in ways compatible to his normal and familiar working methods.

After an initial survey of the user groups and their associated systems a conceptual model of their systems relationships was prepared. This can be seen in Figure (2).

7. *Conclusions at Present Stage*

The process of designing and constructing buildings consists of a series of stages going from inception through to completion and use of the building. This process can be modelled as sets of procedures which require inputs and transform them to provide outputs. Many of the information systems contribute to the conceptual models of the final building product from the viewpoint, or perspective, of a particular process. Depending upon the process, such as designing, quantifying, estimating, planning etc, the information required for that process is supplied and defined by its associated information system. This will specify groups of building items with particular properties, or attributes, of the building fabric.

Specifically, when considering production resources, information requirements are broken down into various units, dependent upon context, which can then be individually described or costed. For example in the design context information is classed in building elements, as in the Cl/sfB system, in order to aid the designer in his conceptual image of the building. In the surveying context the Q.S. must class his information in a suitable manner that will help cost the completed building in terms of labour and quantity of material. E.R. Skoyles proposed grouping information in terms of operations in his many papers [5] in order to facilitate conceptual views of the building process, that in total would result in a building product. The operations are information units in a construction context.

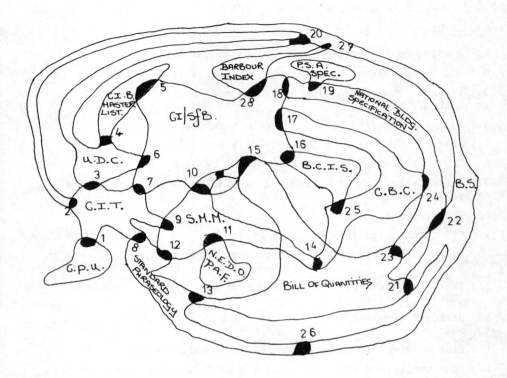

FIG. 2   Information Systems of Major Interest in the Building
Industry

Key to Intersections

1. Similarity in Facet for Operations and Activity.

2. CIT terms tend to follow the B.S.I. Glossary B.S. 3589.

3. Present attempts to link up coding and terminology of U.D.C. with CIT.

4. List adapted from U.D.C., used to structure documents.

5. Master list adapted for use in Cl/sfB Table 4.

6. U.D.C. building tapes used in Cl/sfB Table 0.

7. Attempts to link up coding and terminology. Some comparable facets.

8. CIT terminology links are expected.

9. CIT terminology links are expected.

10. Bills of Q. using Cl/sfB can be restructured in accordance with S.M.M.

11. Cross reference boundaries have some in common with S.M.M.5. (However many do not correspond).

12. Standard term for every feature which S.M.M. requires to be measured.

13. List of standard phrases and descriptions of work.

14. Can compile and code traditional B of Q from CBC in accordance with S.M.M.

15. Bills of Q. can be structured using facet structure of Cl/sfB.

16. Element facets very similar. Can be re-organised to have close correlation.

17. Considered a particular application of Cl/sfB w.r.t. organisation of Project Information.

18. Specif. Clauses grouped under Cl/sfB classific. so as to provide co-ordinated cross-reference.

19. P.S.A. specification developed from NBS 1973. Also, separates workmanship and product clauses.

20. U.D.C. classified as B.S. 1000 for use in the Building Industry.

21. Specifications in preambles reference B.S.

22. Product/material clauses reference B.S.

23. Bill of Q. can be structured using NBS.

24. C.B.C. references NBS specifications. Labour/product/materials.

25. For labour costing: component facet similar in B.C.I.S. to first facet in CBC.

26. Standard phraseology cites B.S. standards in specification level.

27. Barbour Index - classified under Cl/sfB system and references B.S.
28. standards and codes.

The basis of the information in all the systems is the same. The difference may be put down to context, structure and level of information detail required.

After inspection of the systems and consultations with the participants it was concluded that though inter-relation exists the systems were not compatible and it was not possible to make one universal system, or create one standard interface to a database. However, aspects of many of the existing systems will be utilised in the sets of related information structures used in the interface. This will give the interface ease of use and familiarity to the user as well as an added flexibility to incorporate any new structure or relationship that may evolve.

The intention is for a data base to hold the source information, such as that produced by the Quantity Surveyor and Architect. Participants in the building process who require access to this information will do so via the USER INTERFACE. The interface will produce screen formats particular to each user that are called in response to prompts given to the user. The interface will have the relationships as in the data base, but each access to the data by a user will be in terminology and logic familiar to that particular user. Investigation of the information systems will help to define the various access methods.

For example in Figure (3) if the SOURCE user has input on item of data, access paths related to other users will require a different view, perspective or facet of that item.

An extension of the research would be to provide the capability for the interface to be driven by an application program. If the required data is stored in the data base the application program would access the data base directly, where as if the user was required to provide information the application program would use the interface. See Figure (4).

In conclusion, the interface is a general purpose routine which is important in the context of promoting computer acceptability within the building industry. It will do this by fulfilling the need for a software tool in the development of applications within the industry. It should benefit the programmer and program user having input and output routines specifically designed for the building industry.

The report "Relevance of information syntax to computer data structures" [6] set out to find if information science could be related to computer science in a way that would benefit the industry. The intention was to create a metasystem that would aid data coordination. Though the concepts involved were correct I feel the design of an interface is the optimum solution to the problem of unifying information science and computer science rather than a metasystem.

This research has independently arrived at the same conclusions, at a conceptial level, as an E.E.C. predevelopment study on input/out conventions [7], carried out by R.I.B. in Stuttgart, W. Germany as well as those of the Danish researcher Olaf Kayser |8| from the Danish Engineering Academy. After corresponding with these institutions about this work and subsequently visiting them in Europe

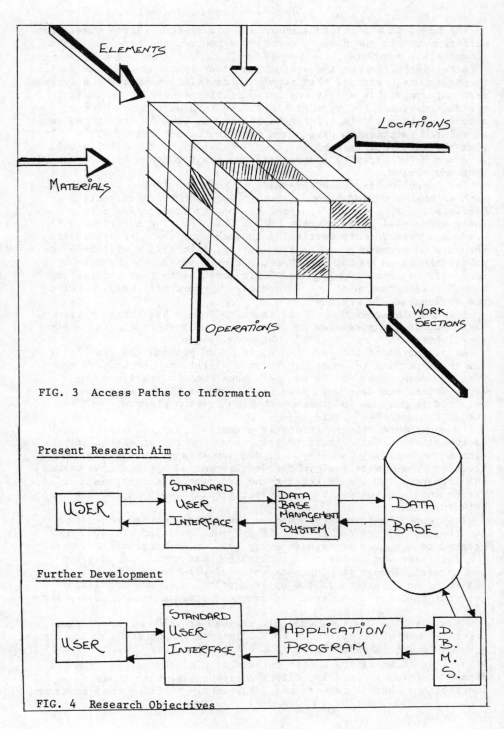

FIG. 3 Access Paths to Information

Present Research Aim

FIG. 4 Research Objectives

to explain it to a greater extent, negotiations are at present underway with both these institutions to collaborate on a larger scale
project. This should lead to the implementation of a more general
input/output interface which would be used for most processes in the
building industry both in the U.K. and Europe.

8. *References*

[1] U.S.A. General Accounts Office.
"Computer Aided Building Design"
July 1978.

[2] International Symposium Norway
"Information Flow in the Building Process - Classification
and Coding for Computer Use"
CIB Report BB-NBRI 1968.

[3] International Symposium Rotterdam
"Some Problems of Information Flow in the Building Process"
CIB Report 14A BC 4013 1970.

[4] EEC Study Project
"The Effective Use of Computers Within the Building Industries
of the European Community"
Ref T/1/77 1979.

[5] BRS
"Introduction to Operational Bills" 1964
"Practical Application of Operational Bills" 1966
"Preparation of Operational Bills" Paper 10 1967
"Introducing Bill of Quantities Operational Format" 1968
E.R. Skoyles.

[6] Directorate of Architectural Services Data Co-ordination Branch
"Relevance of Information Syntax to Computer Data Structures"
P.S.A. 1978.

[7] EEC Predevelopment Study: Final Report
"Feasibility Study of Common I/O Conventions for the Building
Industry"
RIB December 1981.

[8] Danish Engineering Academy
"A Scheme for Separating I/O from Programs", submitted to
C.A.D.
Olaf Kayser 1982.

PRACTICAL APPLICATION OF SOFTWARE PACKAGES FOR
THE QUANTITY SURVEYOR

COLIN McCARTHY, McCARTHY, LILBURN & PARTNERS

Introduction

The objective of this paper, is to provide the professional quantity
surveyor with a framework, within which he can introduce the
Computer Technology, into the practical day-to-day running of an
established office and also within the following parameters. Firstly,
the Software to be used shall be standard off-the-shelf programmes.
Secondly, a limited capital expenditure, thirdly, a specified period
of time and fourthly, with the minimum amount of disruption to
office practice.
    The paper has been written from personal experience, in which the
above parameters have been more or less kept to and it is hoped that
the following paper will give those, considering intergrating a
Computer into their practice, guidance with which to approach this
aim.
    In preparing the paper, a number of assumptions have had to be
made.

A. The type and size of professional practice, it is envisaged that
the system to be discussed would be best suited to small and medium
practices, with between 3/6 technical staff and 2/3 administrative
general office staff.
B. The work load of the practice consists of, work from government
agencies and private clients, in the ratio of 80/20%.
C. The decision to introduce a Computer has been taken and also the
necessary hardware has already been purchased. The hardware purchased is an Apple Plus II micro-computer with a 48K, 12" green
screen monitor, Epsom MX-100 printer. In order to use a wider
variety of software on the Apple II, it was decided to purchase a
Z-80 Card, which enables CP/M to be run, and also 80-column Videx
card, the above system has sufficient capacity for the majority of
Quantity Surveying applications.
D. Possibly many may consider the most important, the budget for
Software, which has been established at £1,500.00, this is a
realistic amount with which to start and will be sufficient to
purchase all the standard off-the-shelf software,necessary to
implement the system.

E.   A defined period of nine months has been set for completion of implementation of the system, within the practice.

F.   The number of staff involved has to be limited at first, so that once they understand the system, other staff members can be trained, the assumption is, the nucleus of the Computer Section, will be composed of a Partner with limited computer knowledge and a member of the general office staff, who is principally the operator, with no computer knowledge.

The parameters together with the various assumptions, having been established, the paper has been divided into four main sections, which analyses the steps required to be taken, in order to design and implement the computer system successfully.

1.   Identification of Procedures to be Computerized
2.   Software Selection
3.   Application of Software
4.   Execution of System

1.   Identification of Procedures to be Computerized

The first and most important task to be completed before any Software is purchased, is to establish exactly what office practices and procedures are readily adaptable to computerization. One of the greatest pitfalls, is to buy software packages without fully under-standing one's requirements. In order to determine these, it is best to divide up the office's operations into two main divisions.

A.   General Office Administration, which broadly is composed of Accounts, Job Costing, Contract Records, etc.

B.   Technical or quantity surveying operations, such as preparation of Preambles and Preliminaries, Cost Data, etc., if one asks the follow-ing questions, do any existing office operations, contain or require any of the following steps;

1.   Preparation of Standard Documents
2.   Repetitive Records
3.   Up-dating of Information
4.   Storage of Information
5.   Reference Sources
6.   Mathematical Calculations
7.   Periodic Reports

Those operations or tasks which are easily suited to immediate computerization are readily identifiable.

With the above, we now have an outlined framework from which we are able to prepare our list.

A.   General Office Administration

N.B. Within this Category, there are certain programmes, for example, word-processing which shall be also used in Category B, applications. The aim of preparing these lists is to identify the operations and the common programmes which can be applied to as many operations as possible.

The following are those areas, which my own practice has identified and has subsequently computerized.

1. Standard letters — A. Fee Reminders
                          B. Valuation letters
                          C. Circulars
                          D. Offers of Employment
                          E. Replies to job applications
                          F. Cover letters
2. Practice Brochure
3. Practice Information Forms
4. Address and Telephone Directory
5. Contract Records
6. Store-Room Inventory
7. Practice Accounts
8. Staff Payroll
9. Fee Account Records
10. Job Costing
11. Job Expense Records
12. Time-sheet Records
13. Office Cost Budgeting

In analysing this list of operations, we can identify the type of software required, namely a flexible word-processing package, an accounting programme, together with a record-keeping programme, all must have the capability of producing a printed document.

B. Technical/Quantity Surveying Applications

As with Category A, a similar list has to be complied.

1. Standard Preambles
2. Standard Preliminaries
3. Tender Invitations
4. Tender Reports
5. Cost Indices
6. Fluctuation Records
7. Valuation Records
8. Bill of Quantity Production
9. Cash Flow Forecasting
10. Analysis of Tender
11. Storage of Cost Information
12. Cost Models
13. Manufacturers/Suppliers List
14. Legal Index

As with Category A, the list of operations is analysed, however, this time there is a noticable difference in the requirements of the programme, the emphasis being on the manipulation and storage of mathematical data. Also, the introduction of very specialist software packages such as programmes dealing with the N.E.D.O. Formula and Bills of Quantity Production.

The next step to isolate the common Programmes and then select the most suitable standard packages.

2. Software Selection

The identification procedure having been completed, we have now established the software requirements, the next step is the selection of the most suitable and adaptable software. This is probably the most crucial and decisive period during the setting up of the system, because of the prolification of software available, it is difficult to compare and chose the most suitable programmes. As one usually finds that each has it's own particular advantageous feature, and the choice apparently increases monthly. Ideally, ones choice could be based on short listing the most likely programme and testing each to establish the most suitable, however, this proves virtually impossible to do, mainly due to possible risk of having software copied and resulting in lost revenue by the publishers.

The alternative and usual path to take is,produce a short list, obtain as much technical information as possible, either from the technical press, suppliers and publishers or by contacting a fellow professional who is currently using the software on your list, then purchase the chosen programmes.

As with the process of identifying the operations, which are suit-able for computerization, a check list can also be prepared, forming the criterior on which suitability and comparisons can be made of standard software programmes.

1. Compatibility with Hardware
2. Capacity of Programme
3. Ease of Operation
4. Programme Flexibility
5. Up-dating of Software
6. Software Compatibility
7. Software Cost.

The importance and relevance of these criterior are outlined as follows.

1. Compatibility with Hardware

The decision taken on which type of computer system to purchase, directly narrows the choice of programmes available, in this part-icular example, with the Apple Computer,the first requirement is that the software must be able to run on Apple basic, however, the earlier assumption to buy the Z-80 Card, which allows CP/M to run on the Apple, vastly increases the choice, this additional feature is especially advantageous in chosing a word-processing programme. Another restriction is memory capacity, certain programmes require a minimum RAM of 48K, and some 64K, if one finds that the increased capacity is required, an add-on 16K RAM Card can be purchased for approximately £65.00, but normally the 48K RAM together with two disk drives will be sufficient.

2.   Capacity of Programme

Although this appears to be a minor consideration, at this time,
the capacity of a programme can become an important factor,
especially where record data is being stored. For example, a
standard programme called P.F.S. File, provides a data storage mode
and is capable of storing up to 1,000 forms of data, this is a good
capacity, where as a programme with only 200 form capacity would
prove inconvenient when carrying out a search or printing a report,
where changing disks would be required. The question of capacity
with reference to word-processing programmes can be critical when
chosing the programme, one obviously wishes to have as many pages on
a single disk as possible. Capacity is a constituent of operation
and has an overall bearing on the efficiency of the system operation.

3.   Ease of Operation

The ease with which programmes can be operated is extremely important,
and more so with the introduction of the computer into the office for
the first time. A major problem encountered is coping with large
numbers of commands for any one programme, this is further compounded
where a large number of programmes are used. This situation is
liable to arise where one may have as many as six programmes, some
using Apple basic and other CP/M, this tends to lead to two basic
problems, firstly, lack of familarization and thus efficiency in use
of the system and secondly, the arising confusion of commands between
the various programmes. In my own experience, operators easily mix-
up programme commands, leading to, in the worst event, erasing data
from disks. It is obvious that after a period of time, during which
the system is being constantly used, this confusion over commands
will disappear, but it is worth trying to obtain programmes with a
simple command structure, as the simpler the command structure, the
shorter the learning period becomes and all those using the system
become more comfortable with it's introduction and practice.

4.   Programme Flexibility

One of the requirements of a programme selection is the need for
flexibility, this is necessitated by the economic restraints. During
the preparation of the lists of operations to be computerized, we
have clearly illustrated the diversification of software requirements,
we therefore, have to select programmes that are capable of being
adapted and performing at least three or more of these operations.
Obviously with the degree of flexibility necessitated by the system,
the choice of software is once again reduced.

5.   Up-dating of Software

One of the greatest problems in the use of the micro-computer or any
computer, is contending with rapid advances in both Hardware and
Software. The situation can easily arise that once you have
purchased your software, in a matter of weeks a revised suite of
programmes is issued, so out-dating your recent purchase. It is
therefore, of the utmost importance that, while chosing software,
one ensure that your standard package allows for the supply of future

revisions and that the financial cost of this is minimal. It is worthless spending £200.00 on a programme, to find that at a later date you have to re-purchase the up-dated master diskette. Normally, the up-dated diskette is supplied for a nominal charge with most programme suites.

Another consideration is to confirm that where up-dating is likely to occur, data already stored is transferable onto the new format, otherwise exercises like searches will involve not only changing disks, but also re-loading the computer, reducing system efficiency and possibly leading to operational confusion.

The way to minimise these problems, is to purhcase software from established software houses, who have good reputations and customer service.

6. Software Compatibility

The compatability of your programmes with other software, although not a crucial factor, could later prove extremely useful, especially if the system expands, leading to a fully intregated computer system. The obvious advantage occurs in the preparation of reports, where for example, it would be advantageous to have both the word-processing programme and a financial forecasting programme linked. The Peachtree Software House have this capability, with their Magic Wand word-processing and a visicalc styled programme, the same applies to the P.F.S. Suite of programmes, P.F.S. File, Report and Graph, the latter has the added advantage of also linking into visicalc.

The disadvantage of over-emphasizing, this as a necessity is that one may be unnecessarily restricting choice of software, at the expense of obtaining a intregated system, which may not wholly be necessary.

7. Software Cost

Although an initial target of £1,500.00 was set, depending on the office's own finances, one may find it is not sufficient, especially if you wish to computerize more and more of the office practices in a shorter period of time, but for the purposes of this paper, the target reflects the aims. The advantages of keeping to a budget are, firstly, the selection of software tends to be more rigorous and ensures value for money, and secondly, it forces you to compare software prices, which do tend to vary, it is a good idea to establish a rapport with a local software agent, as this may be useful, if you wish to borrow new software packages at a later date, to determine their usefulness to your practice.

If one finds that you wish to expand your system but cannot afford the capital expenditure at once, the alternative is to have an annual software budget and thus build up a software library.

The selection of the software is now possible, for the various requirements of my own system. The following standard software packages have been purchased.

Title	Software House	Type of Programme	Price
Magic Wand	Peachtree	Word Processing	£ 250.00
P.F.S.File	Personal Computer	Data Storage	£ 55.00
P.F.S.Graph	Personal Computer	Report Writer	£ 55.00
P.F.S.Graph	Personal Computer	Plotter	£ 75.00
Visicalc	Personal Software	Methematical Sheet	£ 100.00
N.E.D.O.	Cyderpress	N.E.D.O. Formula	£ 450.00
N.B.S.	National Bldg. Spec.	Preambles/Prelims	£ 500.00
			£1,485.00

The only programmes which have not been chosen are an Accounts and
Payroll programme, together with a Bill of Quantities Production
programme, the reasons for this is twofold, firstly, these tend to
be difficult programmes to assess, and secondly, as they are
expensive programmes, they do not come within the overall cost
budget of £1,500.00.

3. Application of Software

The next step after having chosen the software, is to adapt these
standard packages to the individual applications within the system.
It is essential that an overall timetable and programme is produced
for this process.
.   When producing this, it is necessary to follow the rule, of
achieving computerization of as many operations as possible,as
quickly as possible, this has the advantage of providing encouragement to those both setting up and operating the system, and increases
the acceptance of the computer as a worthwhile office tool. One of
the major problems in introducing the computer into the office, is
the degree of suspicion with which it is treated, in order to overcome general apprehension, one should select a relevantly easy
programme to operate, which will show the benefits of computerization,
such as word-processing although in itself is a complex programme,
the degree of skill required for operatives is not high.
Preliminaries and Preambles are easily adaptable and the benefits of
computerization can be seen to the users almost immediately. Also
by commencing with word-processing valuable 'on-hands' experience is
obtained, as the command structure for this type of programme is
typical of most standard programmes.
    The period of nine months for completion of the computerization
process,takes into consideration the initial work required to adapt
the software,to the practice's specific needs, and this should be
carried out,without any distruption of the day-to-day running of the
practice. The best way to work within this restriction, is to allow
for so many hours per week, a realistic amount is 10 - 15 hours,
for this adaption,preferably after office hours, so avoid the usual
distruptions and interruptions.

Initally, the time is used, firstly, learning about each individual programme and secondly, applying the programmes to the specific application requirements. The input of data should be done during normal working hours, as a normal routine, this helps to normalize the presence and use of computer in the office.

One feature of having a computer-based office, is that it does require a degree of imposed standardization, as the majority of programmes used are dealing with the storage and manipulation of data, these obviously have to be presented in a special format, which is firstly, usuable by the particular programme and secondly, also by the practice's staff. It is at this stage that a crucial point is reached, namely how the data is to be collected and presented to the system. The main point to note, is that once the format is decided on it will prove very difficult at a later stage to re-design, without avoiding re-constituting system for that particular operation. In order to avoid this happening,or at a later date discovering the chosen format is of limited use and so having to duplicate formats and input, the format has to be very carefully thought out, with not only the short term aims being considered, but also the possible alternative use of the data which is to be stored. The general approach to take, is to design formats that allow data to be continually added to, thus avoiding changing the previously delevoped format and stored data. In the future, all standard programmes probably will incorporate a facility enabling the alteration of format, which does not destroy the existing stored data, in fact, P.F.S. File has recently introduced this facility.

When designing the standard input forms, it is essential to keep them as simple as possible, the importance of this is twofold, firstly, the more complex a form,the likelyhood of input error is increased, secondly, the period of checking and correcting data, may destroy the credilibity of the system. In a well-designed system, the format should be simple to complete, clear to read and a stage of verification should be allowed for. If one remembers that the computer application will only be as good as the data supplied, a successful system can be obtained.

To enable the completion of data input forms and ultimately the efficiency of the system, the full understanding and co-operation of the staff is essential, this is best achieved by the provision of an office manual, dealing with the system. The manual serves two purposes, firstly, to instruct and illustrate how input data is to be presented and the correct completion of data forms. Secondly, it provides a type of beginners introduction to computing, the advantage of the manual is that staff members, can at their own pace adapt to the use of the computer and they should also be encouraged to use the terminals whenever possible.

4. Execution of System

Once the scope and framework of the programmes have been established, the next stage is to run the system in the actual office environment. It is at this stage that one finds out exactly how good was the earlier groundwork.

As pointed out earlier, it is necessary, no matter how small the practice to have a defined computer group, in the smaller practice this will form only part of their full-time work, as the amount of computerization increases, so it may prove advantageous to have a full-time researcher/operator. The initial computer staff should be two, composed of a quantity surveyor of reasonable seniority and a member of the general office staff.

In the execution of the system, the critical ingredient for it's success or failure is the quality and reliability of the data. There are two areas were this is relevant, firstly, the collection of data and secondly, the actual input of data by the operator.

1. Data Collection

The quality of this is dependent on two items, the sources and method. For data which has it's source from within the office, one has to except it's quality as being good, for example, the accuracy of time-sheets, expense sheets, these are difficult to check, unless gross errors are committed. For data which comes from external sources, it may be possible to cross-check this by comparison of sources, e.g. price data, this can be checked by comparing various retail/trade sources. The importance of good data is essential especially where it is used in formulating various outputs, where errors may be compounded.

2. Input Control

A system of verification has to be devised, the simplest method is to have the input done by one operator and then verified by a second operator. Generally verification of micro-computer data is an inherent problem, as opposed to main frame computers on which the above method is practiced.

The alternative method of verification using a single operator, is to take advantage of the monitor or printer. The data having been stored, the verification can be done by either using the monitor to display the data or a hard-copy of data and checking the stored data against the original input forms.

The constant up-dating of data is of paramount importance to the working of the system, where possible a routine of up-dating data should be established. When dealing with records such as staff time sheets, expense sheets, etc., each has it's own defined period of collection, whether it is weekly, monthly or daily, is of no consequence, it must be done on a regular basis, to maintain the system quality.

One of the greatest dangers of a data storage with a computer-based system, is accidental destruction of data, there are two safe-guards open to the user, firstly, by keeping copies of all disks, so in the event of a disk being damaged, there is a back-up copy. Secondly, by keeping a hard-copy of all stored data. In the majority of applications,both a back-up disk and a hard-copy should to be kept.

The final stage of execution is the presentation of data and/or output, in the majority of cases, a tabulated format is the most useful, either for purely recording data or for comparison. In some instances, graphic presentation is desirable, the P.F.S. Graph programme used in this system has this capability. An advantage of the system that has been described, is the ease with which stored data can be manipulated for presentation, this is due to the programmes used.

One of the benefits of good presentation and use of flexible programmes, is to have the ability to present computer-based data in a form that the reader is unaware of it's computer source.

Conclusion

As stated in the Introduction, the aim of this paper is to provide a framework, with which the computer can be introduced to the professional office, using standard software programmes. The paper has shown that there are several important steps which must be taken.

1. The identification of operations, which can be readily adapted to computerization.
2. The methods by which standard software programmes are selected
3. The adaption of selected programmes to suit the office requirement.
4. The application of the programmes.
5. The successful running of the system as a whole.
6. The costs involved in purchasing software.
7. The planning of the system has to consider the long-term development.

The paper has not covered all the various points necessary for the establishment of an office system, nor, I am sure, answered everyone's questions, but I trust that it has provided a guide and framework which my fellow professionals will be able to use, as an aid in setting up their own system.

One of the basic problems quantity surveyors have when dealing with computers, is that there are few purpose-designed Q.S. systems, which are either suitable for every practice of affordable.

I hope this paper shows with a little basic knowledge, the benefits of the mirco-computer are within most practice's capabilities and finances.

COMMUNICATIVE BILLS OF QUANTITIES - DO THEY INFLUENCE COSTS?

ROBIN M SMITH, Napier College

1. *Introduction*

Cost research is generally dependent on data produced by the
contractor which quantity surveyors then analyse and manipulate.
This work is extremely useful and interesting but temperamentally
I have always found myself attracted to a more fundamental role,
that of the quantity surveyor actively contributing to the
construction process, influencing the contractor's management
methods and ultimately the cost of the building. This desire seems
to complement one of our fundamental professional principles, that
of ensuring value for money, and the logical vehicle for exerting
this influence is the bill of quantities.

In 1965 Higgin and Jessop[1] in their report on "Communications in
the Building Industry" stated, "It was not until we concluded that
the bill of quantities is usually a hypothetical construct and not
necessarily a fully accurate description of reality, despite its
detail, that we began to understand the communication process it
serves". They concluded that the bill of quantities in its
traditional form has little function beyond its contractual purposes.
They were of the opinion that redesigned documentation techniques
could be set up with the needs of construction planning and control
in mind so that analyses done for "contract documentation" would be
of much greater direct and immediate use to the construction phase.

This paper looks fairly briefly and simply at more than 10 years
of practical experience in pursuing and developing this concept.

2. *Background*

In technical and professional journals 10 or 12 years ago a lot of
column space was devoted to discussing operational bills.
Distinguished experts argued over the principles and practicalities.
Much of this debate had been initiated by BRE (British Research
Establishment) papers putting forward proposals for operational
bills and later for bills of quantities in operational format.[2,3]
These ideas by Skoyles, Forbes and Bishop struck a responsive chord
with the small team of quantity surveyors with whom I worked in
the /.....

the Scottish office of the NBA (National Building Agency) who had
been exploring similar concepts but in a less organised way. Part
of the Agency's remit at that time was to improve productivity and
efficiency in the building industry. The Architects believed that
this could best be achieved by rationalising building design and
construction. The design team's intentions had then to be conveyed
to the construction team via drawings, specification and bill of
quantities. The drawings are the vehicle for conveying this
information to the "building" arm of the construction team, but
in the financial basis of the BQ we felt that there was potential
for communicating to the contractor's management team. This feeling
lead to the development of the hypothesis that a BQ specially
structured to communicate project information to the contractor
would result in improved management and efficiency, and hence lower
costs.

An early attempt at communicative documents - at that stage we
were still feeling our way - had been to try out an operational bill
for a small housing contract.

A staggering input of man hours that could never have been
afforded by a private practice ensured that the contract was
reasonably successful, but it required a far higher degree of
co-operation and trust than is usual between professional quantity
surveyors and contractor. It was obvious that the contracting
industry was not ready for such radical innovatory changes, partly
through lack of familiarity with the concept and documentation, but
largely because the contractor's cost data, especially for estimating,
is not geared to this system. The logical decision then was to try
something akin to BRE's bill of quantities in operational format,
labelling them "Activity Bills".

An activity, so titled to avoid confusion with operational bills,
is a parcel of work carried out by one man or gang of men at one
time without interruption.

The activities were illustrated on a sequence diagram bound in
with the bill.

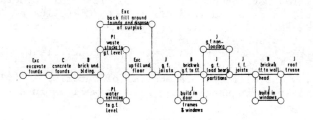

Figure 1. Typical Sequence Diagram

The bill was prepared using the activities as a sub-heading
within the normal work (trade) sections, to give a brief but
comprehensive picture of that section of the work which was then

itemised in accordance with the current SMM. E.g."First lift
brickwork; from DPC to first floor joist level; include building in
lintels, windows and door frames". So to the recipient in the
contractor's office the bill appeared to be fairly normal, not very
different from the usual familiar document they knew and loved!

It was believed by the design team that the "activity" was the key
which would allow the mass of information processed by the quantity
surveyor and locked in the normal bill of quantities to be released
for the benefit of the contractor.[4]

3. *The Billing System*

3.1. Bill Preparation

The bills were prepared by the surveyors under the usual pressures
of late information, details that would not work, and an immoveable
date for issuing the contract documents. The work was carried out
at the appropriate RICS scale fees. Initially, the extra time
involved in the bill preparation was funded by Grant-in-aid but the
jobs quite quickly became self-supporting although little profit was
made! Like any other practice we then had to live with the bills
through the post contract and final account stages.

One point should be made clear at this stage. To avoid
contractual difficulties it was emphasised in the preamble that the
additional information in the bill was for guidance only and the
contractor could choose to use it in any way he pleased or not at all.
In practice no user disagreed significantly with the design team's
analysis.

Before billing started the job was analysed and broken down into
activities. The first two bills were prepared manually but it
quickly became apparent that production would be easier using a
computer, the benefits being felt not so much in the bill production
itself but in facilitating the post contract analyses.

The computer system chosen was the CLASP (now LAMSAC) bill of
quantities program. To suit our needs the programme was extended
and adapted in a number of ways. Firstly, we persuaded the CLASP
Bill of Quantities Development Group to develop their program so
that bills could be produced in a CST format (Consolidated Schedule
Technique). We adapted the existing "Features" analysis to produce
activities and then as we further clarified our ideas we extended
this to allow for the separate analysis of Labour, Plant, Material
and Overheads within the activity. This facility was still further
refined to allow the different categories of material and plant to
be coded and analysed in these categories as well as in activities.

Code	Item	Description	Cont Qty	Unit	Rate £	Extension £0 – 00	L% £0 – 0'	P% £0 – 00	Mat's £0 – 00	M... £0 – 0
G25		BRICKWORK IN APEX *****************								
FF		PRECAST CONCRETE ***************								
FF2		NORMAL: MIX 1.2.4 20 MM AGGREGATE -----------------------------------								
FF3Y		PADSTONES: BEDDING IN CEMENT LIME MORTAR (1.1.6)								
FGC4	85	440 MM X 100 MM X 215 MM; TWO MORTICES 20 MM X 25 MM X 75 MM DEEP FOR RAGBOLTS; RUNNING WITH CEMENT MORTAR	72	NO						
GC		BRICKWORK *********								
GCC		COMMON BRICKS, BS. 3921. ORDINARY QUALITY IN CEMENT-LIME MORTAR (1.1.6) -------------------------------------								
GC5A		HALF BRICK THICK								
GC62	81	SKINS OF HOLLOW WALLS	780	M2						
G21L		ROUGH CUTTING								
G24N	82	GENERALLY	48	M2						
		0021/H1/G25 33 TO COLLECTION								

Figure 2.

Figure 2. illustrates the bill layout. The A4 page is used in landscape format which permits the introduction of Labour, Plant, Material and Overhead columns. The information on these columns is of course confidential to the contractor and is not completed on the copy of the bill submitted with the tender. A disadvantage of this layout is that fewer items can be printed on each page and so the bill is about 15% thicker than a comparable A4 portrait format.

The taking-off process is straightforward, subject only to the slight constraint involved in identifying activities. The bill items are coded from the LAMSAC library which uses a form of Fletcher and Moore standard phraseology.

Activities are coded using a simple 'ad hoc' system and elements are also identified, preferably with an easily recognizable code like BCIS.

3.2. Post Contract

To run his activity analysis the contractor sends the fully priced bill to a computer bureau who punch only the item reference numbers and the figures in the Labour, Plant, Material and Overheads columns. A sample of the output is illustrated in Figure 3.

LAMSAC BQ PROGRAM - NBA CONTRACTOR'S ANALYSIS

FOUR PERSON NORTH HOUSE

GENERAL EXCAVATIONS AND DISPOSAL

B-REF	QUANTITY	UM	TOTAL		LABOUR		MATERIAL		PLANT		OVERHEADS	
			£	£	£	£	£	£	£	£	£	£
0152	49.13	M2	0.04	1.96					0.04	1.96		
0153	9.83	M3	0.03	0.29					0.03	0.29		
		TOTAL		2.25		0.00		0.00		2.25		0.00

TRENCH EXCAVATION AND DISPOSAL

B-REF	QUANTITY	UM	TOTAL		LABOUR		MATERIAL		PLANT		OVERHEADS	
			£	£	£	£	£	£	£	£	£	£
0154	5.86	M3	3.40	19.92	1.85	10.84			1.24	7.26	0.31	1.81
0155	1.06	M3	3.50	3.71	1.91	2.02			1.27	1.34	0.32	0.32
0156	0.68	M3	2.15	1.46	1.17	0.79			0.78	0.53	0.20	0.13
0157	6.24	M3	0.68	4.24	0.00	0.00			0.62	3.86	0.06	0.37
0158	13.27	M2	0.18	2.38	0.16	2.12			0.00	0.00	0.02	0.26
0159	0.75	IT	100.00	0.75	91.00	0.68			0.00	0.00	9.00	0.06
0160	20.19	M2	0.30	6.05	0.27	5.45			0.00	0.00	0.03	0.60
0162	3.68	M2	0.30	1.10	0.27	0.99			0.00	0.00	0.03	0.11
		TOTAL		39.61		22.89		0.00		12.99		3.67

CONCRETE FOUNDATIONS

B-REF	QUANTITY	UM	TOTAL		LABOUR		MATERIAL		PLANT		OVERHEADS	
			£	£	£	£	£	£	£	£	£	£
0163	0.20	M3	22.25	4.45	6.31	1.26	13.92	2.78			2.02	0.40
0164	2.29	M3	20.33	46.55	3.64	8.33	14.84	33.98			1.85	4.23
		TOTAL		51.00		9.59		36.76		0.00		4.63

Figure 3.

This is a monetary analysis but the labour total can quickly be converted to man hours.

It is obvious on practical grounds that if the bill is presented in an activity format then an elemental layout is impossible. However, it is equally obvious that elemental data is a fundamental requirement of the professional quantity surveyor's cost planning function. The document can therefore be re-run to produce an elemental bill or more simply an elemental analysis.

Where the CST facility is used the analyses, both activity and elemental, can be obtained for individual units (house types or blocks), then multiplied up for the total units per type, and finally brought together as a grand total.

The quantity surveyor can also run an activity analysis, using only the item totals, and the resultant output establishes convenient cost centres for interim valuation.

A small sample of these analysis for a large housing contract is shown in Figure 4.

ELEMENT 2.G - INTERNAL WALLS AND PARTITIONS

B-REF	QUANTITY	UM	TOTAL	
551	5.40	M	£.500	£2.700
552	9.60	M	£.720	£6.912
		TOTAL		£9.612

ELEMENT 2.J - PARTY WALL

B-REF	QUANTITY	UM	TOTAL	
132	22.75	M2	£9.660	£219.765
215	22.75	M2	£9.660	£219.765
281	20.14	M2	£9.660	£194.552
314	10.96	M2	£9.660	£105.873
318	2.26	M2	£2.240	£5.062
		TOTAL		£745.017

ELEMENT 3.B - FLOOR FINISHES

B-REF	QUANTITY	UM	TOTAL	
803	95.10	M2	£.930	£88.443
114A	95.10	M2	£.680	£64.658
1151	95.10	M2	£2.420	£230.142
1152	47.55	M2	£2.000	£95.100
1166	142.65	M2	£1.650	£235.372
1169	12.00	NO	£.020	£.240
1170	6.00	NO	£.020	£.120
1171	3.00	NO	£.100	£.300
1194	143.65	M2	£.000	£.000
		TOTAL		£714.385

5 PERSON MID HOUSE   NUMBER OFF 29

GENERAL EXCAVATIONS & DISPOSAL

NO HEADING FOR CATEGORY

B-REF	QUANTITY	UM	TOTAL	
1	7.68	M3	£1.950	£14.976
2	2.00	M3	£6.000	£12.000
3	7.68	M3	£2.130	£16.358
		TOTAL		£43.334

EXCAVATIONS FOR FOUNDATIONS

NO HEADING FOR CATEGORY

B-REF	QUANTITY	UM	TOTAL	
4	8.85	M3	£2.260	£20.001
5	2.00	M3	£6.000	£12.000
6	3.29	M3	£1.530	£5.033
7	5.56	M3	£2.530	£14.066
8	5.95	M3	£1.140	£10.089
9	9.84	M2	£.070	£.688
10	1.00	IT	£.000	£.000
11	2.00	HR	£1.250	£2.500
12	8.69	M2	£.380	£3.302
13	21.06	M2	£.380	£8.002
		TOTAL		£75.681

UPFILL & BOTTOMING

NO HEADING FOR CATEGORY

B-REF	QUANTITY	UM	TOTAL	
19	33.17	M2	£.950	£31.511
21	33.17	M2	£.630	£20.897
22	33.17	M2	£.300	£9.951
		TOTAL		£62.359

CONCRETE FOUNDATIONS

NO HEADING FOR CATEGORY

B-REF	QUANTITY	UM	TOTAL	
23	9.84	M2	£.930	£9.151
24	2.96	M3	£21.420	£53.403
25	32.80	M	£.500	£16.400
26	0.14	M2	£6.140	£.859
27	0.06	M	£1.310	£.078
		TOTAL		£89.891

Figure 4.
Quantity Surveyor's Elemental and Activity Analyses.

4. *Assumed Benefits*

It was assumed that benefits would flow to practically every
department in the contractor's organisation. They can be summarised
briefly as:

4.1. Estimating
The CST format avoids repetition of the same item and the additional
pricing columns allow Labour, Plant, Material and Overheads
information to be built up easily and clearly on the working bill.
Bill codes can be cross referenced to a library of estimating costs.

4.2. Planning
The sequence diagram is a guide to the arrangement of the contents of
the bill. The builder can decide what activities he intends to follow,
change the order if desired, or group similar consecutive activities
together without affecting the data which can be extracted from the
bills. The analysis can readily be converted into man-hours which,
when related to available labour, will indicate duration.

4.3. Bonusing
The bonus surveyor is concerned with the work content of an activity
in each block or unit. He will apply his own time constants to
arrive at the target time.

Where the make up of the bill is in activities he can check that there is a reasonable relationship between the budgeted time and his targets.

4.4. Buying and ordering

To make information about cutting and waste more clearly available additional information is provided on sheet sizes, joist lengths, etc. For example, joists can be listed separately for each length enabling decisions to be made about cutting on site or buying timber pre-cut. This concept also aids the design team in rationalising design and reducing unnecessary variety. The close relationship of the activities to the programme enables orders to be accurately pre-planned.

4.5. Cost and Control

The total cost of each activity (or groups of activities) becomes the valuation cost centres for stage payments and can be set out on "tick lists" or similar recording systems.

Interim valuations then prepared by the client's surveyor in company with the contractor's surveyor give the builder the value of his work with which costs can be compared. Thus the valuation is in units which have been used for planning, bonusing and site management and the contractor can collect his costs for labour, plant and materials in a way which can be compared with the valuation. Many contractors use computer programs to facilitate these tasks.

4.6. Sub-contracting

The activity, by definition, can never be partially sub-contracted. The sub-contractor's work always consists of one or more activities which can readily be detached from the bill for copying (each activity starts on a new page).

5. Does it Work?

For the first two or three contracts meetings were arranged with the tendering contractors and the philosophy and anticipated benefits of this type of approach were explained. Subsequently, all the bills of quantities were issued "cold" with only an explanatory note at the beginning of the document. No systematic attempt was made to measure the bill's effectiveness in conveying design rationalisation or to test the usefulness of the management information it contained.

I, therefore, proposed that this would be a suitable area for research, a suggestion accepted by Heriot-Watt University. The evaluation process is, therefore, being carried out under the aegis of the University as part of a research thesis.

5.1. Survey of Contractors

The first stage was a survey of contracting firms who had had some experience of this type of bill of quantities. A pilot study of three firms, followed by discussions with a statistician determined that questionnaires would be the most effective method of assessment. The pilot scheme also revealed that a personal visit with the questionnaires would guarantee a high return of valid forms and would minimise misunderstandings.

Part of the questionnaire is illustrated in Figure 5.

PLEASE TICK THE APPROPRIATE BOX

	very unhelpful	slightly unhelpful	makes no difference	slightly helpful	very helpful
6. Location information noted beside items.	☐	☐	☐	☐	☐
7. The bill prepared and set out in consolidated schedule technique (C.S.T.) format.	☐	☐	☐	☐	☐
8. The ability to break down the C.S.T format and reprint bills for each block or house type.	☐	☐	☐	☐	☐
9. The measurement of sheet materials expanded to itemise the full and different cut sizes of sheet separately.	☐	☐	☐	☐	☐
10. Structural timbers itemised separately for each cutting length.	☐	☐	☐	☐	☐
11. Comprehensive specification descriptions within each item (as required).	☐	☐	☐	☐	☐
12. The ability to use Activities as cost centres for valuations.	☐	☐	☐	☐	☐
13. The ability to obtain labour values from each Activity to build up or check bonus targets.	☐	☐	☐	☐	☐
14. The ability to obtain the data needed for planning and programming (by analysing the amount of labour and material in each Activity).	☐	☐	☐	☐	☐
15. The ability to obtain total quantities of each category of material for buying purposes.	☐	☐	☐	☐	☐
16. The ability to monitor costs by comparing budgetted (Bill) values of each Activity with the actual cost of the work as carried out.	☐	☐	☐	☐	☐
17. The ability to prepare a cash flow diagram by relating the net value of each Activity to the programme.	☐	☐	☐	☐	☐

Figure 5.

Each form initially establishes the variable factors which it is believed would affect the respondent's opinion - his department, age, qualifications, size of company, and so on - and then it seeks a graded response to a series of simple, unambiguous questions about the additional information and facilities contained in the bills. Organising and co-ordinating this part of the exercise was complicated and in the end just over two years were spent "on the road", a whirlwind programme of one visit about every six weeks! Twenty one contractors in the south of Scotland and north-east of England were visited and 81 questionnaires were completed.

5.2. Results

The data thus gathered has now been tabulated and summarised. It will be analysed on the SPSS (Statistical Program for the Social Sciences) program, a popular and comprehensive integrated system of computer programs designed for the analysis of social science data. SPSS contains procedures for the usual descriptive statistics and frequency distributions as a first stage which can then be used in simple correlation and analysis of variance routines.

Results of this computer analysis are not yet available but an inspection of the data has revealed certain general trends and results. Figure 6 summarises the total of the five graded responses.

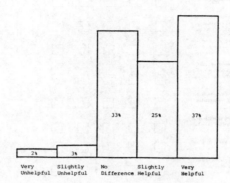

		33%	25%	37%
2%	3%			
Very Unhelpful	Slightly Unhelpful	No Difference	Slightly Helpful	Very Helpful

Figure 6.

Few respondents found the additional information unhelpful, apart from the Fletcher and Moore phraseology which was disliked by 22%, the CST format which 15% thought unhelpful, and somewhat surprisingly the comprehensive specification information in the bill item which did not appeal to 10%.

The "no difference" column as expected attracted a substantial 33% but it should be noted that this response was boosted by the peripheral bill users like buyers who use the document in a fairly limited way.

The majority of users (62%) found the enhanced bill of quantities to be helpful to some degree. At this point in time few results or relationships involving the variables can be assessed with any degree of confidence. One preconceived expectation, however, was proved to be wrong. It was expected that the most favourable response would be from younger, qualified staff but there is no evidence to support this view.

The most obvious backing is from the larger type of contracting organisation who from actual experience probably most appreciates the need for efficient internal co-ordination and dissemination of management information.

Some supplementary questions were introduced in an attempt to test reactions to further possible enhancements. Again, most of these suggestions were favourably received except the proposal to convert the bill to an elemental format. This received the "thumbs down" from most of the recipients. 32% thought it would be very unhelpful, 26% slightly unhelpful and 25% felt it would make no difference. Only 17% thought it would be helpful.

6. *Conclusion*

It has to be recognised that in our practical and conservative profession and industry there are very few genuinely new discoveries. Progress is made in small steps, one development leading gradually to another. Older Scottish quantity surveyors, for example, will recognise a similarity between the "new" systems outlined in this report and their method of taking-off in accordance with the Scottish Mode of Measurement. What came out of the BRE work was a timely stimulus, focussing attention on an important area of our professional work. Their research structured the material in a proper form, providing a sound framework for further development. In turn the Scottish NBA QS team built on this, altering and adapting the system from their practical experience. Their contribution was nothing particularly innovatory, it was their commitment in doggedly pursuing their objectives and persevering in using their system for all the jobs produced in the Edinburgh Office for 10 years.

Has all this work been worthwhile?

The questionnaire test results show a generally favourable response to the communicative bill but it would be overstating the case to suggest that contractors were positively clamouring for this type of document. The results should rather be read as a green light to the profession, encouraging them to move forward. But the pace and direction must be set by the quantity surveyors themselves, although in consultation with our contracting colleagues.

Last, but not least, do these bills result in cost savings? The answer here is a qualified, cautious, "yes". The qualification is because it was only when contractors tendered for a second, similar job that apparent savings became apparent. The caution is because there were not enough repeats for any degree of confidence. However, the NBA/SDD study at Pitcoudie[5] clearly showed the benefits both of rationalised design and the improved methods of communication. The tender for a contract of 283 houses was 11% below cost limits and the contractor never denied that it was a profitable contract.

References

1. Higgin, G and Jessop, N. (1965), Communications in the Building
    Industry, Tavistock Publications.

2. Bishop, D (1966), Operational Bills and Cost Communication.
    Architects' Journal, Vol. 139, No 828, pp158, 160,2

3. Skoyles, E.R. (1968), Introducing Bills of Quantities
    (Operational Format). Building Research Station CP62/68

4. National Building Agency (Edinburgh) (1971),
    Repetitive House Building - A study of design, communications
    and building management.

5. National Building Agency (Edinburgh)(1978),
    Monitoring documentation on housing sites - Pitcoudie
    Report to the Scottish Development Department

CAPITAL COST ESTIMATING AND THE METHOD OF PRESENTING INFORMATION
FOR PRICING OF CONSTRUCTION WORK

CLIFF HARDCASTLE, Newcastle upon Tyne Polytechnic

Introduction

In the majority of the manufacturing industries the interrelated
processes of objectives management and design management are usually
carried out within the same organisation resulting in information
flow and communication linkages at such a level and in such a form
as satisfies the accounting and estimating criteria under which the
firm operates.

In the construction industry a different situation exists. The
production processes are undertaken by various parties. The
separation of design and construction in particular results in a
breakdown in the information flow between the parties involved.[1]
This breakdown in the information flow inevitably produces a
situation in which the iterative processes of design management do
not operate efficiently[2] as the appraisal operation cannot be
carried out without suitable material for use in the measurement
operation.

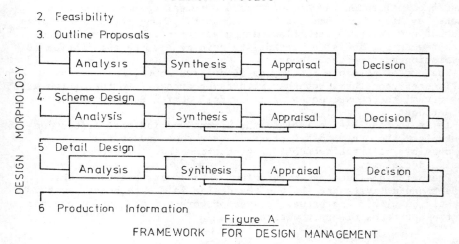

DESIGN PROCESS

2. Feasibility

3. Outline Proposals

Analysis — Synthesis — Appraisal — Decision

4. Scheme Design

Analysis — Synthesis — Appraisal — Decision

5 Detail Design

Analysis — Synthesis — Appraisal — Decision

6 Production Information

DESIGN MORPHOLOGY

Figure A

FRAMEWORK FOR DESIGN MANAGEMENT

Although various performance features will require appraisal at each design stage, this paper is particularly concerned with the quality and form of the initial capital cost information which is at present available to the designer and therefore to the client at the early stage of a design.

The paper draws on research work originally carried out in the context of building conversions but which is now being extended to new building work.

Cost or Price

Comment has been made above on the need for a realistic and accurate method of capital cost estimating. It is presumed that such an estimating system would allow the determination of the cost to a client of certain proposed works, but this is not simply production cost but production cost ± mark up.

Production cost is itself an amalgam of direct cost, indirect cost and overheads.

Therefore the information utilized by the client's cost advisers should impart knowledge of these costs.

Although the relative importance of each type of cost will vary between projects it can be assumed that the direct cost is the primary cost feature of any project.

Indirect costs are those which relate to the overall duration of the project. They are affected by the rate of completion of individual activities in as much as these cause variations in project duration.

Overhead costs include management and head office overheads.

Cost Influences

In order to determine the client cost of a project the above costs must be determined, as must the anticipated profit mark up of the contractor. The client cost is, therefore, subject to a number of possible influences many of which are external to those features quoted by him in his brief to the designer. Among these are aspects such as contractors' cash flow manipulations[3], human error, the market economy, social acceptability pricing[4] and the application of absorption costing techniques[5].

The sum of possible cost types identified above, and the influences upon them, together with those external to the design will not reveal the consequences of individual design decisions upon client cost.

If it is possible to isolate the direct and indirect production costs from the other cost factors of overheads and profits, it may, be possible to establish cost of production and, therefore, of design decisions.

The principle factors which affect these costs are those which affect productivity and include the design in its technical sense, repetition within and between stages of work and the interdependance of the operatios of construction.

Thus it is desirable for the design team to have knowledge of production costs. The cost types are known and understood as are the major influences upon these costs. There is at present, however, no interpretative document which can translate and transmit this information from the production team to the design team.

Cost Estimating New Work and Conversion Work

Despite the problems of estimating cost, the quantity surveying profession in the U.K. has evolved very successful methods of 'cost estimating'. The term 'cost' is at this point used to convey the meaning of production cost ± mark up. The logic underlying this success has been investigated in order to understand the paradox and perhaps indicate how the logic could be used to evolve a system for conversion and rehabilitation work, an area in which traditional approaches to cost planning have not been very successful.

From this investigation the models in Figure C and Figure E were developed.

The building environment results from the input of labour, material and plant, which in turn produces elements which produce the characteristics of the environment. The elements are not only those which because of uniqueness of location and utilization can be identified with an individual building, but also those which service these elements such that the environment is sustained.

Where the functional unit is a complete building, the cost factors 1 and 3 (fig. C) will, except in highly unusual and obvious cases, contain over 75% of the total cost. The cost factor 2 is found to be highly variable but of considerably less cost importance.

A situation exists, therefore, in which the functional elements of the functional unit, and those support elements which can be related to building parameters, assume a cost importance which far outweighs that of the support elements which cannot be related to building parameters. It is, therefore, the former cost factors upon which current cost estimating techniques are concentrated. It is seen that the cost function about which the cost estimator will have most information is that which determines the material cost M_{CE}.

Also it can be further demonstrated that labour costs are a function of those aspects which determine material cost.

Total Cost of Functional Unit = The sum of the material, labour and plant costs of the functional elements and the support elements.

$$C_T = (M_{CE} + L_{CE} + P_{CE}) + (M_{CS} + L_{CS} + P_{CS})$$

Functional Elements   Support Elements

and

$$M_{CE} + L_{CE} = f(Q_{ME}, R_{ME}, T_{ME})$$

where Q = quantity
R = quality
T = type.

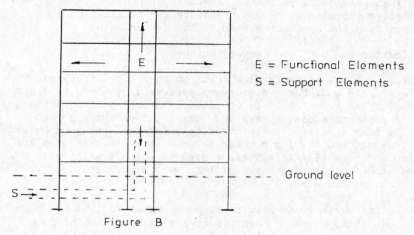

E = Functional Elements
S = Support Elements

Ground level

Figure B
The Functional Unit in New Work

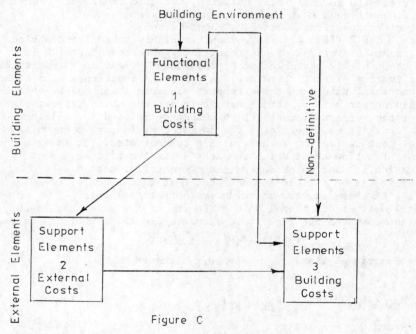

Figure C
ELEMENT LEVEL AND COST LEVEL

376

The plant function is similar but complicated by the addition of a factor W_{ME} which is the location of the material within the unit. Thus the cost model of the functional unit is:-

$$C_T = (f(Q_{ME}, R_{ME}, T_{ME}) + f(Q_{ME}, R_{ME}, T_{ME}, W_{ME})) +$$

$$(f(Q_{MS}, R_{MS}, T_{MS}) + f(Q_{MS}, R_{MS}, T_{MS}, W_{MS}))$$

$$C_T = (f_1 + f_2) + (f_3 + f_4)$$

In the construction of new buildings the cost important function is f_1 i.e. that which determines labour and material costs for the erection of the functional element.

In conversion work the problem is more complex although the basic model remains the same. The cost estimator will still have to devise a method for determination of a total cost which will include the costs of materials, labour and plant of the elements E and S.

In cost type (1) are the elements, the cost of which can be determined at the level of individual spaces as a result of their uniqueness of location and utilization.

The extent, quantity and specification of the support elements necessary will produce a cost which can only be evaluated at the building level. These are identified as cost type (2).

In cost type (3) the costs incurred are those which result from a building level requirement which can be satisfied by an element which can be identified, and therefore cost estimated, at the space level. Cost types (4) and (5) are not considered in this work as they are elements which have been identified as being of a fixed condition and therefore not convertible.

A Project Analysed

Using the previous model a conversion project was statistically analysed in each cost type. A brief summary is given here.

The conversion was of an early 20th century warehouse which was altered to house two university departments to include 179 activity spaces.

The analysis has four sections the first of which forms the basis from which the others are consecutively developed.

The sections are:-

(1) Item Analysis
(2) Element Analysis
(3) Space Analysis
(4) Space Element/Building Element Comparison

The analysis at the item level has shown that any attempt to develop an approach to cost planning based upon the space as a unit of description is difficult at present as a result of the vast quantity of information which would be generated from an analysis of a traditional bill of quantities, i.e. the mean number of items necessary to describe a space was 94 with a standard deviation of 54.

THE PRODUCT

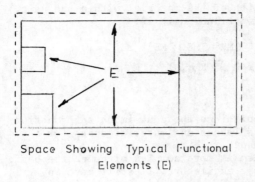

Space Showing Typical Functional
Elements (E)

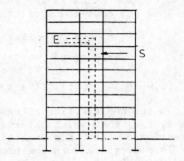

Space within building environ-
-ment showing functional
elements and support elements

Figure D

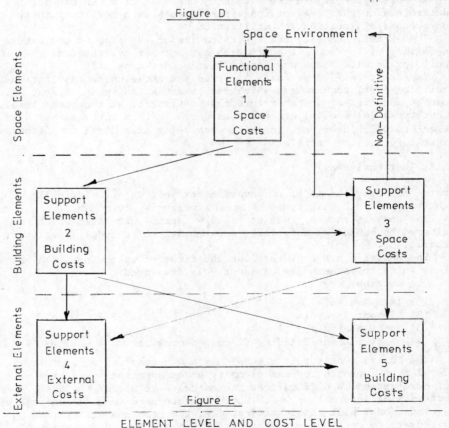

Figure E

ELEMENT LEVEL AND COST LEVEL

The extensive use of provisional sums also creates problems in the development of a cost planning system at the space level for the percentage influence of these items shows greater fluctuation at the space level than occurs at the total project level.

Investigations into the possibility of reducing the number of items necessary to describe a space while maintaining accuracy, have indicated that it is possible to identify over 80% of the cost of a space with a number of items which approaches 24% of that which is currently used.

A study of the variation in cost of similar elements among spaces has revealed that, with particular elements, the cost per unit of element shows little variation among spaces, even though it is thought necessary to describe these elements in great detail.

The analysis at the element level revealed that the relative importance of the cost of an element to the cost of a space is not constant for any particular element.

It is suggested, therefore, that in the development of a cost estimating system one of the criteria for inclusion in the system should be the possible cost of an element and not its possible cost importance or what is an acceptable level of cost ignorance and not what is an acceptable level of inaccuracy.

The second criterion is that of the effect upon other levels of cost, (i.e. building element/building cost, building element/space cost) which the inclusion of an element in a space may have.

The relative cost importance of elements among spaces remains a good indicator of the need for the inclusion of particular element types within a cost estimating system and this survey has shown that, in any cost estimating system only very few elements could possibly be excluded.

The analysis of the relative cost importance of elements among spaces has served to indicate that space cost profiles exist. Although the definition of these spaces has not been based upon scientific assessment, (the titles are those of the architect), spaces designed for similar functions do show similar cost profiles.

The comparison of cost types (1), (2) and (3) indicates that a proportion of the cost of a conversion cannot be allocated to individual spaces e.g. structural work, distribution systems. This aspect can be overcome by use of;

(a) threshold analysis applied to the capacity of building cost and space cost levels and, ultimately, to the other element cost types, (4) and (5), in the model,

(b) the identification of the location of elements which produce demands on the consequent element cost types.

The problems of estimating the costs of the cost type (3), i.e. building element/space costs, are similar to those of the cost estimating of distribution elements, although without the function of location in the cost model. There remains, however, the need for the cost estimator to be able to assess the capacity of existing plant and equipment and the load upon that plant and equipment generated by the needs of the spaces. This information, together with a knowledge of cost of plant and equipment could then be used in a threshold analysis.

As a space unit of accommodation can be understood to have a similar relationship with its immediate environment as a complete building unit of accommodation has with its immediate environment, it is suggested that the cost of the development can be determined by the same method as is used for complete building units i.e. the application of costs to a $y = mx + c$ equation.

The analysis carried out has served to show that a cost can be produced using the equation $y = mx + c$.

It has also served to indicate which element costs can be interpreted as a function of superficial area and which cannot. Furthermore, it has revealed the importance of building element costs relative to space element costs.

It has been shown that a unit superficial area cost of perimeters can be linked to the prospective function of a space in the same manner as those element types listed previously.

From the results of the analysis, it is suggested that an index system should be developed based upon the POP[6] ratio and size of a space. Such indices need not reflect every possible size and shape of space but should be delineated according to possible effect upon superficial area cost of variations in size and shape. A possible list of indices is given figure F.

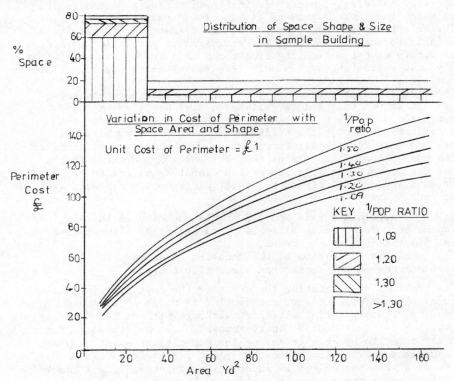

It has not been possible to link all element costs to the superficial area of space and these must therefore be considered as costs
type C in the equation. An Example of Inputs to Operation of the
Cost System
(Using the substitute identifier, space function)

    y = Mx + C

M Costs	C Costs
Floor Elements	Doors
Ceiling Elements	Windows
Wall Elements	Fittings
Lighting	Ventilation (Local)
Power	Hot Water Installation
Heating	Sanitary and Cleansing Installations
Ventilation	Communication Installation
(Central)	Building Level/Space Costs

As the system would operate at present there would exist, for
each space function and size, an M and C value each of which contains
the summed costs of the necessary element types. These M and C
values would differ from those of spaces to house other functions by
virtue of differences in element types required and differences in
element specification. Consider figure G.

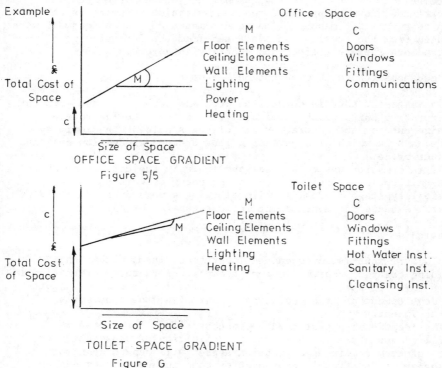

OFFICE SPACE GRADIENT

Figure 5/5

TOILET SPACE GRADIENT

Figure G

If the y = mx + c approach to the cost estimating of spaces is accepted in principle there exists a system capable of estimating possibly only 50% of the cost of a building. The remaining costs which are a function of the support elements cannot logically be related to space parameters although their occurrence will be determined by space needs. In the situation where a new distribution system is required throughout a building it may be argued however that the element tends towards a crystalloid formation and, therefore can have its cost related to floor area in the same way as space elements.

The quantity and specification of such elements will depend, not only upon the number of service outlets which they are to serve and upon the quantity of output required, but also upon the relative position of these outlets to each other and to the unit which provides the media to be conveyed (i.e. building element/space cost). Prior to the development of a cost planning system which uses the space as the unit of planning at the feasibility stage changes in methods of communication of prices must take place.

Results have shown that it is possible to develop a logical system of cost planning at the space level. This has been proven for the complex situation of conversion work, using information obtained from a bill of quantities which was not conducive to such an exercise. It is therefore suggested that such a system could operate for new work provided the communication document is structured in a suitable manner. The requirements of such a document were deduced from the previous analysis and from the interviewing of contractors representatives and quantity surveyors.

Consequences for the Communication Document

It is suggested that in conversion work and perhaps in new work the interprative unit, which permits a link between the information requirements of the contractor and client should be the activity space, a unit which also permits a link between production cost and product price.

If information was to be made available in the correct format, this unit could be used in the cost planning of work at the feasibility stage of design while allowing a more detailed build up later. Aspects for considering include the following:

Location

The lack of information upon the location of items within the building creates problems in any attempt to structure information in space units or space elements for cost planning, and also inhibits the development of true operational costing by the contracting side of the industry.

The locational aspect need not necessarily relate to every item in a bill of quantities. The specific definition of those items or types of item requiring a location aspect will depend upon the influence of location upon production cost and upon other design variables.

The separation of information relating to building cost elements and space cost elements, either by use of a locational aspect or some other feature, is also a necessary inclusion.

Indirect Production Costs

The influence of these cost types is particularly increased when cost planning at the space level. This relative increase suggests that a more definite identification of these costs would be helpful.

The variation of such costs could then be related to features of the work such as location and choice of material, access, storage etc.

Overheads and Profit

The cost of overheads and the profit mark up are not functions of the design of the building and as such should not be included in the item rates where they become unknown variables in the price estimating equation of the quantity surveyor.

Item Numbers

The suggestion to use the activity space as a unit of location will result in a great increase in the number of items necessary to describe the elements which can be interpreted as space costs. To allow the development of a bill of quantities with a locational aspect, it is, required that the number of items describing the work which is to be carried out should be reduced and more use made of the principle of item coverage.

Prime Cost and Provisional Sums

If ultimately cost planning is to develop in the manner suggested it is imperative that the number of P.C. and Provisional sums which are generally found to occur in a bill of quantities are reduced and fuller descriptions included in the main sections of the bill.

This would ease the allocation of item costs to individual spaces and facilitate the development of space costs.

Attempts have been made in the past to cost estimate certain service element types using regression analysis techniques. These techniques have shown the need for separate identification of the space elements and building level elements.

Component and Assembly Information

There is a difference of opinion between quantity surveyors and contractors' representatives as to the separation within a bill of quantities of the above types of information. It is considered that it would improve the accuracy of estimating.

It is concluded that a logical and real approach to the capital cost estimating at the space level cannot be developed until radical amendments are made to the methods of presenting information for pricing. If the amendments were made as suggested then building work would be more easily and accurately capital cost estimated at the early stages of design, for the price information could be related where necessary to production features as well as to a functional unit at a scale of great significance to the client.

Research is currently progressing into the identification of a definition level suitable for the contractors operational costing which will link with the space as a cost planning unit.

References

1.   Tavistock Institute: Interdependance and Uncertainty: Tavistock Publications 1966.
2.   Building Performance Research Unit, Building Performance, London, 1972.
3.   Denton, R., Cash Flow Analysis – An Approach to Tendering, MSc thesis, University of Manchester Institute of Science and Technology. 1975.
4.   Fine, B., "Tendering Strategy," Building, 25th October, 1974.
5.   Brown, W. and Jaques, E., Product Analysis Pricing, London, 1964.
6.   ABACUS, University of Strathclyde.

THE EFFECTS OF INITIAL STANDARDS OF CONVERSION ON THE 30 YEAR
LIFE OF REHABILITATED BUILDINGS

M.S. ROMANS
J.G. LITTLER
G.W. AITKEN
B.A. TABRIZI

Building Unit, Polytechnic of Central London

1. *Introduction*

Details concerning *conversion* costs of 300 properties, and data
extracted from 30,000 *maintenance* job sheets and 3,000 ledger
entries were analysed, in order to determine whether or not higher
expenditure on the initial conversion of a rehabilitated property
would result in less ongoing maintenance. The properties belong to
the Notting Hill Housing Trust who cooperated in the Project.

2. *Procedure*

Complete lists of the parameters examined are shown in Table 1
(conversion stage) and Table 2 (maintenance).

Table 1. Parameters Examined at Conversion Stage

Capital expenditure on conversion
Private consultant retained for the particular conversion
Property size: total floor area
Scope of the conversion scheme: extensive or minimising major works?
Quality of design and specification
Funding authority
Age of building*
Time since conversion
Bed-spaces/unit
Floor area/person

*In the event all the buildings were of similar age.

Table 2. Parameters Examined During Maintenance

Street code
House number in the street
Flat number in the house
House status: converted, unconverted (house, hostel, commercial, new build)
Category of the particular maintenance job as defined by the research team:
   Roofs
   Mechanical and electrical
   External fabric
   Fixtures and fittings
   Finishes
   External plumbing and drains
   Dampness and timber rot
   Joinery repairs
   Miscellaneous
Price of the particular job
Date of the job
Tenancy factor*
Expenditure category of the job as defined by the Housing Association:

Minor repairs	Welfare	Cyclical maintenance
Major repairs	Development	
Anti-squat measures	Relets	
Management	Tenant to pay	

*Examined in Case Studies, not in the Data Base

The parameters of Tables 1 and 2 were examined for the Trust's properties, but filters were applied as follows to select the sample:

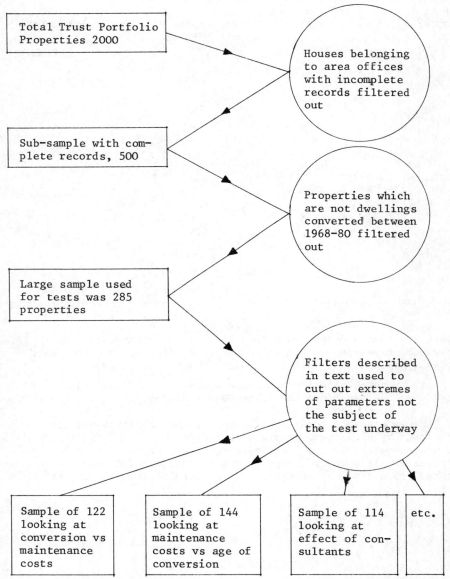

Total Trust Portfolio Properties 2000

Houses belonging to area offices with incomplete records filtered out

Sub-sample with complete records, 500

Properties which are not dwellings converted between 1968-80 filtered out

Large sample used for tests was 285 properties

Filters described in text used to cut out extremes of parameters not the subject of the test underway

Sample of 122 looking at conversion vs maintenance costs

Sample of 144 looking at maintenance costs vs age of conversion

Sample of 114 looking at effect of consultants

etc.

Analysis concentrated on the correlation between conversion costs and subsequent maintenance costs and on the relationship between maintenance costs and factors at the conversion stage which might be thought to be of significance in future maintenance.

3. *Selected results of analysis of the data base*

3.1 Are maintenance costs related to level of expenditure at conversion when the whole sample is considered?

Figure 1 shows a scatter plot of conversion costs for the "large sample" of 285 properties considered and the 7,500 maintenance jobs occurring to those properties between 1968 and 1979.

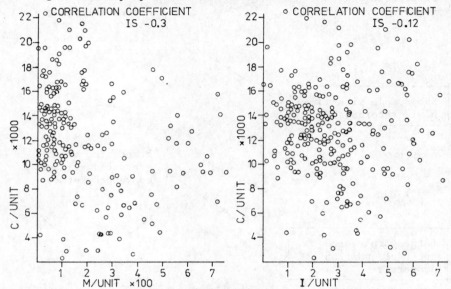

Figure 1. Cost and incidence (I jobs/y) of maintenance: Conversion costs (£C) and maintenance costs (£M) are all expressed per year per dwelling unit (following conversion) at 1980 prices.

There is no correlation between conversion costs and maintenance costs when one considers the large sample of 285 properties.

3.2 Are maintenance costs related to expenditure at conversion stage if a filtered sample is selected?

122 houses were selected from the large sample by filtering for extremes of all factors except conversion costs. The results in Figure 2 also show no correlation of conversion and maintenance costs.

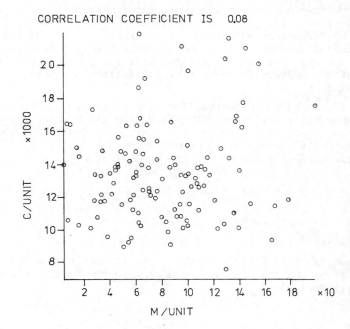

Figure 2.   Maintenance versus conversion costs for a homogeneous
           sample

3.3   Are maintenance costs related to conversion costs when
      conversion costs above or below average are considered?

	Below average conversion cost	Above average conversion cost
For the large sample:		
Maintenance cost/unit	£175	£103
Maintenance jobs/year	2.8	2.4

Again when all extreme factors were filtered out (except extremes
of conversion costs), there was no relationship of maintenance to
conversion costs:

	Low conversion cost	High conversion cost
Number of houses	64	58
Cost of maintenance per unit	£ 82	£ 78
Maintenance jobs/ unit. year	2.3	2.3

3.4   Are maintenance costs related to particular features of the
      conversion process?

The list of factors considered at the conversion stage (Table 1) was
used to define sub-sets of the whole sample. Whilst maintenance
costs do not depend on the aggregated conversion costs, particular
aspects of the conversion process were isolated for further
investigation.

   (a)   Effects of consultants – The sample was divided up on the
basis of the six consultants A–F retained by the Housing Trust at
the conversion stage, and whilst tests were carried out on the
whole sample (285 houses or 750 dwelling units) in this and future
Tables reference is made only to the samples structured to eliminate
extremes.

Consultant	A	B	C	D	E	F
Maintenance cost/unit (£/y)	96	24	96	108	65	124
Incidence/unit	1.9	1.2	2.4	2.6	1.9	2.4

   The results above include costs of cyclical maintenance and works
to the external fabric. These elements were removed, giving the
following figures:

Consultant	A	B	C	D	E	F
Cost/unit	96	24	96	101	65	87
Incidence/unit	1.9	1.2	2.4	2.5	1.9	2.4

   In all the tests, maintenance was broken down into categories,
and this is particularly valuable when discussing the effect of the
consultant:

Consultant code	A	B	C	D	E	F
Roof	96	24	96	108	65	124
M & E	24	0	12	12	4	9
Ext. fabric	25	13	35	34	16	32
Fix/fittings	4	5	5	8	8	6
Finishes	9	0	12	10	6	12
Soil/drain	1	1	1	4	2	2
Damp/decay	2	0	0	6	1	1
Joinery	7	5	14	10	6	6
Miscellaneous	7	0	7	9	16	10
Unclear	8	0	3	2	2	20

   It appears then that the <u>high</u> maintenance costs shown in properties converted under the guidance of consultants F and D are
spread throughout the whole range of jobs and under all "tests".

(b) Funding authority - *Three* sources of funding have been
available to the Trust for this area of London for use at conversion
stage.

When conversions of very long standing were excluded, most of the
properties converted with funds from one authority were deleted.

Structured sample of 114 houses:

Funding authority	A	B
Maintenance/unit	72	110
Incidence/unit	2.0	2.5

The research team feels from familiarity with the documents concerning conversions funded by these two authorities, that the difference might arise from the authorities monitoring procedures.

(c) Scope of conversion - Direct information on the extent of
the conversion scheme for a particular property, was not collected.
The following characteristics were used to select a sample of 30
houses which shared: high conversion cost and high property size and
low floor area per bed-space (intensive conversion); or the converse
(average degree of conversion).

Scope of conversion	High	Average
No houses	30	168
Maintenance/unit	138	106
Incidence/unit	2-6	2-4

It is possible that intensive conversion leads to higher costs of
maintenance. Typical examples are internal bathrooms with their
added problems.

(d) Size of property - Properties were divided into those of
<250 m², 250-350 and >350 m². The costs of maintenance of the <u>units</u>
within the property are shown versus the size band; results follow
for the structured sample.

Floor area	<250	250-350	>350
No houses	37	45	24
Maintenance/unit	68	81	86

The size of the property appears not to affect subsequent maintenance costs for the units created within it.

3.5 Are maintenance costs related to particular factors since
     conversion?

(a) Time since conversion - In Figure 3a properties are classified by the time since conversion.

CORRELATION COEFFICIENT IS 0.67

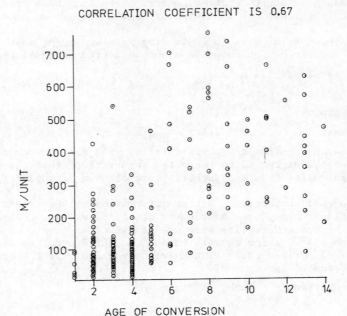

Figure 3a. Maintenance versus age of conversion

The cost of maintenance/unit.year rises with time since
conversion. A structured sample of 144 houses in which extremes in
other determinants (size, consultants, occupancy, conversion costs,
etc.) have been eliminated also shows such a dependance, see
Figure 3b.

There are grave doubts about interpreting these particular
figures. For example, they depend on the indexation adopted and the
larger sample includes properties converted before the 1974 Housing
Act.

(b) Designed occupant density - No relationship was found
between the number of bed-spaces/unit and overall maintenance costs
or indeed maintenance costs in <u>any</u> of the categories of Table 1.

3.6 Overall pattern of maintenance costs
The most important implication is the dominance of mechanical and
electrical work.

The limited number of plots for new build and old properties
belonging to the Trust suggest similar patterns of maintenance.

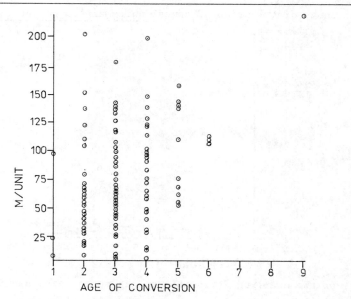

Figure 3b. Maintenance versus age of conversion

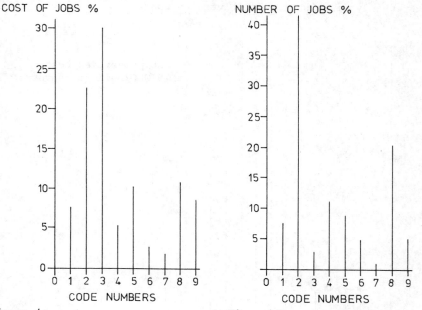

Figure 4a.   Figure 4b.

1. Roofs 2. M & E 3. External Fabric 4. Fixtures and Fittings
5. Finishes 6. Soil and drains 7. Damp and rot 8. Joinery
9. Miscellaneous

3.7 Frequency distribution of annual-unit-maintenance costs

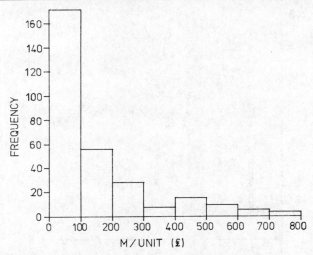

Figure 5.

For most of the dwelling units maintenance costs less than £100/y (1980 prices).

3.8 Frequency distribution of conversion costs

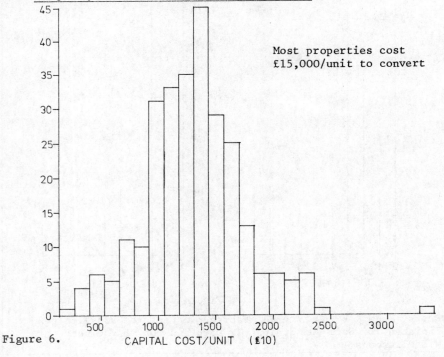

Most properties cost
£15,000/unit to convert

Figure 6.

3.9 New build versus conversions
Limited data suggests that new built dwellings experienced slightly
more maintenance costs than converted ones.

3.10 Maintenance which should not arise
An attempt was made to divide maintenance jobs into two categories.
Jobs carried out on items newly installed at conversion stage (e.g.
central heating); or jobs carried out on items totally renewed at
conversion stage (e.g. roofs, plumbing, electrics and damp proofing).
  This classification suggests that more than 50% of the maintenance
arises from new installations or elements totally renewed at the
conversion stage.

4. *Information revealed by case studies*

Some of the considerations listed in Table 1 were not easily
amenable to statistical analysis of the records available.
  Some considerations in Table 2 were also not susceptible to
analysis from the data base.
  These factors were examined during 20 'case studies' in which the
Development Files (conversion stage) and the Property Files (maintenance stage) were analysed manually.

4.1 Conversion stage
Builders - Maintenance costs were higher for properties converted
by builders who went bankrupt after conversion had been finished.
Design criteria - Close attention was paid to the quality of consultant's work. Several indicators were used: clarity and precision
of specifications for roofs, plumbing and heating; clarity or indeed
presence of site minutes; % of cost yardstick; % overspent and so on.
  The relationship between high "quality" and low incidence of subsequent maintenance (from the data base) was strong.

4.2 Maintenance stage
Tenancy factors - No evidence was found that tenancy had a strong
effect on maintenance costs, except that the repair work at relet
stage, was rather variable, but still small.
Maintenance management - After initial repair, jobs recurred in a
number of properties. Job recurrence could be due to: poor
diagnosis, poor workmanship or poor supervision.

FACTORS AFFECTING MAINTENANCE COSTS IN LOCAL AUTHORITY HOUSING

ROY HOLMES AND CHRISTOPHER DROOP, Bristol Polytechnic

1. Introduction

Over the last three years the authors have been involved in studies
on the incidence and cost of maintenance on local authority dwellings. The aim has been to assess the affect that various factors
have on the maintenance demand and ultimately on cost. A number of
factors have been researched including: Age of Stock, Storey Height,
Method of Construction, Social Status of Estate, Labour Force and
Policy. These factors were examined to test the hypothesis that the
amount spent on maintenance is 'budget' orientated rather than 'need'
orientated.
   The hypothesis emerged after it became clear that officers involved with maintenance cost control were influenced by two major
factors:
   (a) the need to establish a relationship between the housing
revenue account and maintenance expenditure,
   (b) the lack of suitable data from which maintenance needs and
cost parameters could be predicted.
   The studies so far have involved several local authorities with
stocks of between 12 000 and 41 000 dwellings, from whose records
large samples of data have been gathered. Detailed cost data for
periods of seven to ten years have been used in the research and the
work is continuing.

2. Age

In global terms the cost of maintenance increases with age; this has
been clearly demonstrated by Skinner[1] where costs are shown to
increase significantly during the first 20 years and then assume a
slow-growth pattern. From Skinner's work it would seem reasonable to
suggest that maintenance costs continue to rise for the economic life
of the dwelling.
   The rise in costs against age is confirmed in simplistic terms by
comparing cost/age histograms for two different authorities, figure
1. in which dwellings in three age groups, inter-war, 1945-64, and

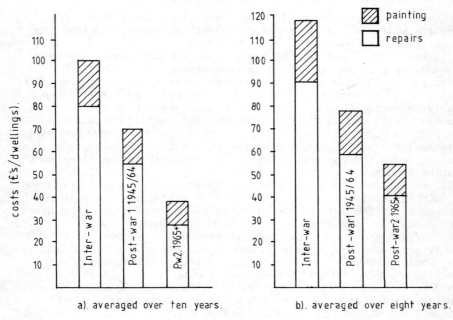

a). averaged over ten years.   b). averaged over eight years.

FIG 1. Maintenance costs for two authorities at 1980 prices.

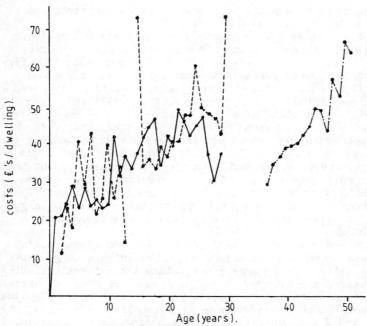

FIG 2. Repair costs against age (1971/2 to 1977/8).

1965 onwards, are compared. In both cases it is clear that costs
correlate with age, and in terms of overall maintenance costs, the
younger dwellings subsidise the older stock.

Figures 2 and 3 support the general concept that costs increase
with age. The graphs are based on seven years data at 1977/78
prices for 41 000 dwellings. Differences between the two graphs, for
repairs and for painting (external and internal decorations), can
clearly be seen. Painting costs show a much greater variation, and
this can partly be put down to the fact that individual painting jobs
take much longer than repair jobs, and there are fewer of them; small
variations in job incidence, therefore, have a much greater effect on
painting costs than on repair costs. The cyclical nature of external
painting also has its effect, in that policy decisions affecting the
work done on each cycle can have considerable effects on the costs.
The painting cycle is normally set at five years, and the peak in
costs in the fifth year of the non-traditional dwellings' life shows
this. A smaller peak can also be seen at ten years of age.

Both figures 2 and 3 show costs increasing on average quite
clearly with age, and the non-linear relationship backs up Skinner's
work. Costs, although rising quickly to begin with, appear to be
levelling out. The rate of the initial increase, as well as the level
at which they settle, are different for repairs and painting. Whereas
painting costs appear to level out after about ten years, repair
costs are increasing over a longer timespan, and seem to be rising to
a much higher level.

A valid hypothesis to explain this must consider the mechanism by
which costs are thought to rise with age.

It will be appreciated that any (strictly hypothetical) building
entirely composed of elements which are supposed to need repairing or
replacing only every sixty years will not require much attention for
the first twenty or thirty years, but after sixty years the annual
repair bill will be substantial. Alternatively, a building composed
largely of components lasting less than five years will have
substantial maintenance costs after only two or three years of life;
however, in this case it can also be appreciated that after five or
seven years the average cost would not increase any more, because all
the components would be failing fairly regularly, every five years or
so. These are extreme examples, but they serve to make the point that
any maintenance environment with a large proportion of short-lived
components will show costs that rise quickly and level out quickly
(apart from random fluctuations), while an environment with a large
proportion of long-lived components will show a slow rise in costs
over a long period.

When painting costs are considered separately from repair costs
(as they are here) they constitute an environment with a large proportion of short-lived maintenance items, since the external painting
cycle is five years, and internal or relet decoration might on average be done once in every ten or fifteen years to each house. Repairs
on the other hand, include many components which last thirty or forty
years or more, and which contribute a large proportion of the cost.

3. Element Groupings

Having considered the trends of cost/age the next step was to examine
the composition of the costs to highlight, if possible, those elements of construction that had received an increasing percentage of
the maintenance resource. It became clear that certain costs, for
example structural repairs (mainly long-lived components), only began
to rise after 15-20 years and costs for elements with short-lived
components, such as plumbing, rose quickly to begin with but levelled
off after about twenty years.

The two most costly element groupings were external painting and
structural finishes and fixings. A comparison of costs for broad
categories against time (Table 1) shows that finishings and fixings
are a major contributor to maintenance costs in both inter-war and
post-war dwellings. The increase for age is greater for the younger
dwellings. For post-war dwellings, plumbing has become an important
contributor. This is probably due to the increased sophistication of
hot water systems.

With non-traditional dwellings the main problem is structural,
with a more than three-fold increase over fifteen years. This could
be indicative of some greater structural problem yet to be faced, and
it would not be surprising to see major capital works on some non-
traditional stock during the next ten years. Alternatively some
authorities may deem such stock to have reached the end of their
useful life.

Figure 4 shows, for each element grouping, the costs per dwelling
for inter-war, post-war and non-traditional stock. As stated above
it can be seen that external painting and finishes and fixings rank
higher than the other groupings. There is less painting on the
younger stock but more could be done to eliminate this grouping
altogether; many of the non-traditional dwellings are colour washed
externally. What is interesting is that for both plumbing and heating & lighting the assumed reduction in costs due to age, does not
occur. This was confirmed, particularly for heating and lighting, in
studies carried out with other authorities. It may be assumed that a
change either in design or in standards in these groupings has
brought about benefits to the tenants without a corresponding cost
benefit, in terms of reduced maintenance, to the client. The multiplicity of systems for both hot water and heating requires further
study and rationalisation.

4. Dwelling Type

Many local authorities are unaware of the maintenance costs for
specific dwelling types. Some have followed the basic classification
set out in an early working-party report on maintenance(2) which
separates costs for pre-war, post-war 1, and post-war 2 dwellings
according to height (1-2 storey, 3-4 storey and 5+ storeys). This
method of costing has been the basis for returns on maintenance to
the Chartered Institute of Public Financing and Accountancy (CIPFA).
Many authorities have not separated the dwelling types within those

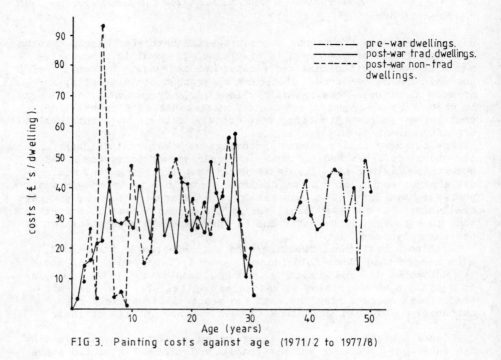

FIG 3. Painting costs against age (1971/2 to 1977/8)

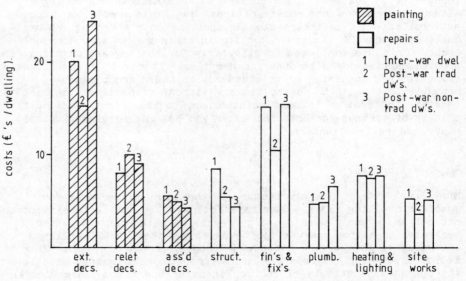

FIG 4. Costs of each element grouping (1968/9 to 1977/8).

broad groupings and therefore are unable to assess the performance of specific dwelling types. However, where specific costings had been recorded we were able to compare the performance, in terms of cost, of twenty-two different dwelling types: this covered a total stock of 41 000 dwellings.

Figure 5 shows costs for each dwelling type averaged over a ten year period; the first thing to notice is that differences in painting costs are irrespective of age. The second point of note is that painting costs in this authority are in the same order as repair costs for most of the dwelling types. Subsequent research with other authorities shows that percentage painting costs in figure 5 are unusually high (42 per cent of maintenance) and this may be due to other factors (eg policy rather than dwelling type.)

Post-war dwellings are cheaper to repair than inter-war stock (21 per cent cheaper) but non-traditional are only marginally cheaper (6 per cent) to repair than inter-war stock. From figure 5 it can be seen that non-traditional dwellings vary more in their costs than traditional dwellings. Some non-traditional dwellings demonstrate high repair costs, higher than inter-war stock; these include Airey, Cornish Unit (3-4 storey), Unity, Woolaway, Reema and Costain. Some Airey houses have subsequently given concern about their structural stability (3) and this raises questions about other designs which have employed small sectioned pre-cast columns and concrete panels; these would include Unity, Woolaway and Cornish Units.

Painting accounts for 57 per cent of total maintenance in the BISF houses. This is due to the externally painted steel panels at first floor level. Little is known about the state of the steel frame within the cavity and the repair costs on these houses give no indication whether a very high maintenance commitment is pending.

There is little evidence to support the view that size of dwelling should be taken into account in the maintenance equation.

5.   Social Status of Estates

In the early days of the research programme it was assumed that the social status of estates would affect maintenance costs. That assumption was confirmed but not in the way that had first been expected; it was assumed that those areas or estates enjoying a good reputation would require less maintenance than those estates which were less desirable. The assumption was wrong. In fact Area 1, with the highest reputation, had the highest maintenance cost per dwelling and Area 2, which was much less desirable, had lower maintenance costs. These costs are shown in figure 6 and are confirmed by a detailed time sheet study figure 7. Area 3 contains estates which enjoy an average reputation amongst the tenants.

Some of the particularly high or low costs can be simply explained. For instance, the low cost of site works in Area 5 is because most of the dwellings are multi-storey flats. The low cost of structure in Area 8 is due to the low average age of the dwellings. Areas 1 and 2 however have similar aged low-rise stock. There are two possible explanations for the difference in maintenance costs in

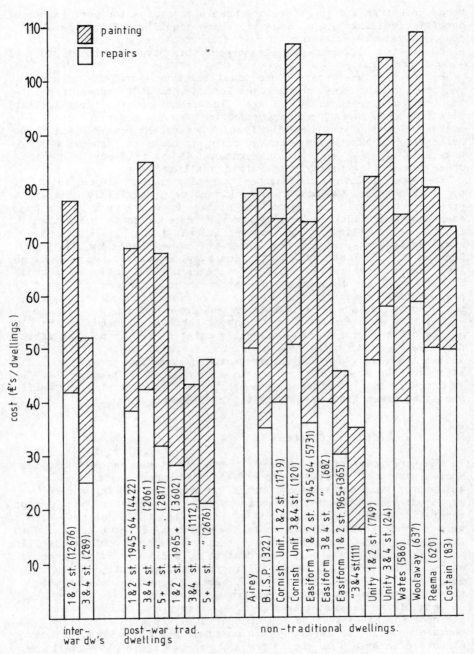

FIG 5. Costs of each dwelling type (1968/9 to 1977/8). (Figures in brackets show the numbers of each type at 31/3/78)

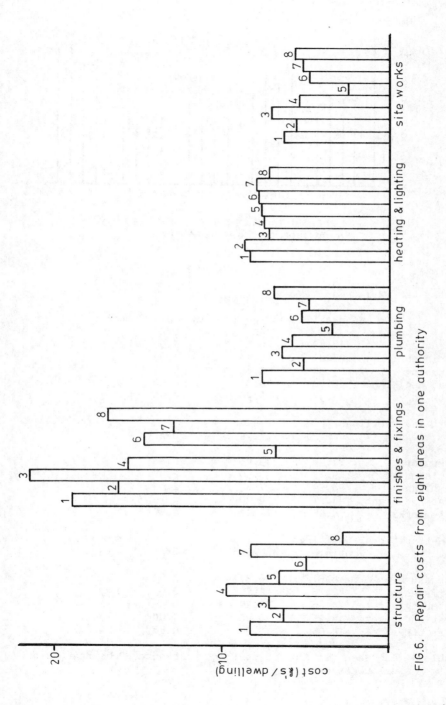

FIG.6. Repair costs from eight areas in one authority

cost (£'s/dwelling)

structure   finishes & fixings   plumbing   heating & lighting   site works

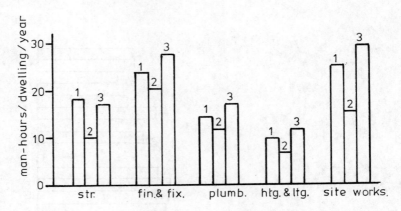

FIG. 7. Time spent in three areas on each repair type
(from the timesheet sample)

Inter-war Dwellings

	% cost at age 37 A	% cost at age 51 B	Increase factor B/A
Structure	18.89	20.65	1.09
Finishes	27.81	44.95	1.62
Plumbing	11.58	9.69	0.84
Heat and Light	22.28	15.31	0.69
Site Works	18.97	9.40	0.50

Post-war Dwellings

	New	at age 29	B/A
Structure	21.97	15.05	0.68
Finishes	19.11	30.50	2.07
Plumbing	8.29	17.02	2.05
Heat and Light	41.27	14.30	0.35
Site Works	8.05	14.11	1.75

Non-traditional Dwellings

	at age 15	at age 30	B/A
Structure	4.51	16.37	3.63
Finishes	42.75	41.79	0.93
Plumbing	21.66	12.49	0.58
Heat and Light	16.95	16.90	1.00
Site Works	14.05	12.34	0.88

Table 1 Increase factors for cost/age (estimated from regression
calculations)

Areas 1 and 2. First, the response of the authority to attend to work may vary according to their perception of a satisfactory service. If the response is good, which might be the case for the best estates, the tenants will be motivated to ask for more to be done: tenants who are sceptical about the authority's response time will be less inclined to ask for some jobs to be done. Secondly, on estates which lack a high social status many tenants will simply not bother to report work; this will result in a backlog of maintenance, a reduction in standards but lower annual costs. Much of the work will not be done until a change of tenancy occurs thus, artificially lowering the annual costs.

6. Labour Force

When the labour costs for each Area were considered it became clear that since, on average, 75 per cent of all repair costs (excluding overheads) were attributable to labour, the size of labour force could be an important factor. The labour force was checked for size in each area for each year that maintenance costs were available, having omitted work to acquired and temporary dwellings. It was found that the costs correlated with the number of men employed. In other words it was of little consequence how many dwellings there were in an area, since demand for maintenance can be manipulated by the time taken to respond to a maintenance request. It might be expected that the authority would expand the labour force in an area to satisfy demand. To do so would, however, be to invite increased expenditure since an increase in labour force would not only increase tenant satisfaction but also increase tenant demands for further work.

In seeking to satisfy recent legislation concerning profit on capital employed many authorities have separated their DLO's from the housing department and introduced a client/contractor concept. This may have the effect of increasing maintenance costs because DLO's must now demonstrate that a profit has been made and so will attempt to do as much work as possible. The result will be an increase in the labour force to meet demand rather than the use of a time buffer to minimise the labour resource.

To illustrate the point of manpower levels, further imagine two areas of housing each with a depot having the same number of men but one area having twice as many dwellings: the total cost of maintenance would be roughly the same (if outside contractors were not used) since the wage bill is the largest component and that is dependant on the number of men employed. That would result in costs per dwelling, for the largest area, being half that of the other area.

Providing that tenants are willing to accept a longer response time to their requests (which will cause them to demand less maintenance) and some have little option, and providing that the standards do not fall below a prescribed minimum, the smaller the labour force the lower the maintenance costs will be. Some tenants, dissatisfied with long response times, resort to doing the work themselves or to having it done privately; this reduces the costs still further.

7. Policy

There are many variations in the policies of local authorities particularly in those areas where discretion can be exercised for financial and social reasons. The most important of these discretionary
areas are:
  (a) External painting where the cyclical period can be extended
or reduced,
  (b) Internal decoration where the responsibility can be placed on
the tenant,
  (c) Planned maintenance where the periods or elements can be
modified,
  (d) Tenant requested repairs where the response time can changed,
  (e) Discretion given to labour force over what needs to be done.

  All maintenance organisations are affected by discretion in terms
of interpreting policy. The interpretation of policy varies as it
passes down through management and often the outcome is dictated by
the operative rather than top management(4).
  For instance, it was found that, in the case of roof repairs,
houses in one area cost twice as much to repair as identical houses
in another area, even though the number of repairs were in the same
order. The operatives in one area applied quite a different policy
in their method of solving the problems, though in fact they were no
more successful; but over a period of seventeen years their approach
to the problem doubled the cost.
  Some authorities operate a 10 per cent prior inspection policy
before the work is passed to the operative. This means that the
operatives have discretion on the other 90 per cent of the tenant
requested repairs, and many operatives prefer to replace than to
repair. There is no doubt that the discretion exercised by the
operative can greatly influence the final cost.

8. Conclusion

A simple model of the overall maintenance system might be as follows.
The overall budget for maintenance is determined by the number of men
employed by the DLO, or alternatively, the number of men is determined by the budget - in reality, there is probably a balance of the
two; in any case, the overall cost of maintenance is not determined
by the type or age of the properties, and not even directly by the
number of properties. The number of properties, however, does have
some influence on the number of men employed by the DLO.
  Within the overall budget for maintenance, money is absorbed by
houses "needing" maintenance done on them. The "need" for maintenance
is not necessarily actual; although determined to some extent by
factors such as age, or dwelling type, much of it is aggravated or
tempered by the expectations and desires of the tenants and of the

maintenance team. Indeed, all maintenance is subject to a human factor, in that somebody has to decide to ring up the maintenance office to report a fault, or somebody has to decide on the painting cycle period, and so on. Thus it is neither the age nor dwelling type that determines maintenance costs, but rather the people using or maintaining them; it is they who "absorb" the money from the budget, supposedly on the building's behalf but often influenced by their own expectations and desires in response to factors such as age and dwelling type.

It must be said, however, that such a system is not necessarily a bad thing. Public housing organisations exist, after all, to provide a service to a substantial proportion of the population, and it should be argued that one important element of a good maintenance system is to be seen to be satisfying the expectations and desires of tenants, by responding efficiently and effectively to their requests for maintenance, (as well as to provide adequately maintained property).

References

1. Skinner, N.P. (1981), House Condition, Standards and
   Maintenance Housing Review, Vol 30 (4), 106-109
2. HMSO, (1964), The Report of the Working Party on the Costing
   of Management and Maintenance for Local Authority Housing.
3. Holmes, R. and Droop, C.S.S. (1982), Problems of Airey Houses
   Analysed, Municipal Journal, 16 April.
4. Holmes, R. and Marvin, H. (1980), Maintenance costs and policy,
   Housing, Vol 16 (1), Jan. 17-19.

AN INVESTIGATION INTO HOSPITAL MAINTENANCE EXPENDITURE IN THE NORTH
WEST REGIONAL HEALTH AUTHORITY

KEITH SLATER, Sheffield City Polytechnic

1. Introduction

This paper summarises the investigations carried out during a three
month field-study with the North West Regional Health Authority
during 1980 to enquire into spending patterns on the maintenance of
the hospital building stock, and is based mainly on costs for the
year 1978-79.

 The total expenditure for all types of maintenance expenditure in
the Region for 1978-79 was £20,163,546 (the latest date for which
figures were available at the time of the research) and this represented 6.5% of the total hospital revenue expenditure for that year.
Over the past two decades or so the expenditure on maintenance has
been increasing annually, although as a percentage of the total revenue expenditure it has actually been decreasing steadily.

 Costs from the eleven Areas in the Region were collected and then
analysed, beginning with broad costs and extending through degrees of
detail until individual element costs for hospitals were examined.

 Due to the concise nature of this paper it is not possible to show
cost data in detail and, apart from the data shown in figures 1 to 3,
the information has been generally excluded.

 Some basic questions were addressed to the data along the following lines:

 (1) Are larger buildings cheaper to maintain than smaller ones
per unit of accommodation?
 (2) Are older buildings cheaper or more expensive to maintain
than newer ones?
 (3) What influence has location and exposure on costs?
 (4) What influence has use and function on maintenance costs?
 (5) What influence has a planned preventive maintenance policy on
costs?
 (6) Is the estate being maintained at a satisfactory level?

2.   Area building maintenance costs

Building maintenance costs in two Areas within the Region were anal-ysed in detail (referred to as Area 'A' and Area 'B') in an effort to discover any existing trends in expenditure patterns. (See figures 1-3).

An analysis of total building maintenance costs (excluding engineering maintenance) for NWRHA, Area 'A' for year ending 31st March, 1979, per hospital.

1 Number of beds in each hospital	2 Number of $100m^3$ cost units	3 Number of $100m^2$ floor area units	4 Total Cost per bed	5 Total Cost per $100m^3$ unit	6 Total Cost per $100m^2$ unit
737	2223	652	246.900	81.860	279.090
395	596	151	167.610	111.080	438.440
145	294	90	86.890	42.850	139.990
61	131	28	159.970	74.490	348.500
48	128	44	143.130	53.670	156.140
53	108	33	218.360	107.160	350.700
38	83	25	139.210	63.740	211.600
31	59	20	482.960	253.760	748.600
1289	3144	886	183.320	75.160	266.700
486	920	305	216.300	114.260	344.670
384	773	225	212.220	105.420	362.190
123	303	104	310.980	126.240	367.790
115	271	75	215.990	91.660	331.190
110	113	34	191.840	186.740	620.650
70	101	32	104.460	72.400	228.500
47	62	18	38.538	29.210	100.610
41	52	18	337.170	265.850	768.000
24	36	14	544.250	362.830	933.000
23	31	10	185.090	137.320	425.700
1544	3661	447	120.460	50.800	416.080
436	1316	353	368.420	129.540	455.040
385	1240	331	372.110	115.540	432.820
196	351	66	79.210	44.230	235.240
125	293	72	259.720	110.000	450.900
57	140	19	103.400	42.100	310.210
52	90	22	101.330	65.490	267.910
24	80	11	295.880	88.760	645.550
29	37	8	76.350	59.840	276.750
1270	2277	508	166.620	92.940	416.560
709	1457	512	182.010	88.570	252.040
163	575	158	122.090	34.610	124.960
292	300	88	99.240	96.590	329.300
78	99	31	198.230	156.180	498.770
44	99	35	54.800	24.350	68.890
43	60	19	321.630	230.500	727.890
421	1334	282	230.850	72.850	344.640
321	880	237	204.340	74.540	276.760
61	127	30	234.280	112.530	476.370
41	90	21	70.730	32.220	138.100
35	55	12	6.890	4.380	20.080
112	110	22	37.320	38.000	190.000
22	65	7	672.680	227.680	2114.140
62	119	36	149.600	77.940	257.640

Figure 1.

Variable	Mean	Std. Dev.	Std. Error	Maximum	Minimum	Range
1	249.814	360.826	55.025	1544.000	22.000	1522.000
2	564.721	852.607	130.021	3661.000	31.000	3630.000
3	141.651	200.314	30.548	886.000	7.000	879.000
4	202.635	136.180	20.767	672.680	6.890	665.790
5	102.927	72.924	11.121	362.830	4.380	358.449
6	398.805	331.863	50.609	2114.140	20.080	2094.060

Figure 2.

Area 'B' (three districts) total basic building maintenance costs (1978-79).

(1) Number of $100m^3$ cost units	(2) Number of $100m^2$ floor area units	(3) Cost per $100m^3$ unit	(4) Cost per $100m^2$ unit
2933	788	131.18	488.26
1203	222	70.15	380.15
699	159	137.80	605.79
440	110	76.83	307.30
264	75	65.80	231.63
263	71	122.63	454.24
2581	642	91.71	368.71
1210	367	58.78	193.80
394	111	95.74	339.83
283	69	41.85	171.64
246	54	34.24	156.00
50	20	9.52	23.80
3033	1205	84.54	212.78
2153	722	95.11	283.61
967	428	76.49	172.81
324	76	47.35	201.84

The number of beds have been ignored since two hospitals in the area only have day visitors.

Variable	Mean	Std. Dev.	Std. Error	Maximum	Minimum	Range
3	77.482	35.274	8.819	137.800	9.520	128.280
4	287.011	147.414	36.854	605.790	23.800	581.990

Figure 3.

The data showed that costs are generally of a 'random' nature and do not appear to be significantly influenced by the size of the hospital, in terms of the number of beds, volume of floor area.

3.   Cost related to the size of hospitals performing the same function.

Whilst the size of a hospital did not seem to have any influence on building maintenance costs over an entire Area, it was not clear whether or not this was also true for hospitals performing the same medical function, for example acute and mainly acute hospitals.   In order to discover whether the way a hospital is used, in terms of function, had any influence on costs a study of fifteen acute and mainly acute hospitals in Area 'A' was made.   Once again, unit costs did not relate to size, in terms of area or volume, and no detectable pattern could be seen.

Next, the total maintenance costs related to the number of beds (and, therefore, patients)' was considered.   This examination of maintenance costs per bed did not reveal any clear pattern to the scatter either, except that the correlation between costs and bed numbers appeared to improve slightly when the smaller hospitals were ignored.

The pattern is certainly too vague to enable predictions to be made.

4.   The danger of 'unique' figures occurring for a single year.

The difference in spending between the Areas and Districts within the Region were often very large.   This could have been caused by unusual figures appearing for one particular year - which did not appear the year before.   The 1978 figures were compared with the 1979 figures, but the same large and unexplained differences appeared again.   It was found that the same wide variations exist in annual figures for the previous five years.

5.   Comparison of Regional building and engineering maintenance expenditure.

The next avenue of investigation was to separate building and engineering costs to see if one group was more closely grouped around the mean than the other.   Costs were compared over two years.

The results of this analysis follow:

	Building Maintenance		Eng. Maintenance	
	1978	1979	1978	1979
Mean	£643.83	£ 728.15	£331.07	£384.80
St. Dev.	£164.41	£ 213.26	£ 82.16	£ 94.18
Range	£386.93 to £893.20	£ 483.20 to £1414.79	£192.65 to £482.51	£232.05 to £607.37

It can be shown that there is little difference between the variability of building and engineering maintenance costs when expressed as a percentage of the mean costs.

6. Analysis of building maintenance by broad element groupings.

Building maintenance costs were analysed under the main primary element headings (e.g. internal decoration, main structure, finishes and
fittings) in order to determine whether some elements had costs which
were very closely grouped around the mean. Any such close groupings
should indicate that these elements are more cost predictable than
others. Once again no clear close banding was observed and costs
varied considerably through the Region.

7. Individual element costs.

The next stage in the study was to focus the investigation on the
'fine detail' and twenty nine individual element costs were recorded
for fifteen teaching hospitals.
    At this level of detail it is to be expected that there should be
quite large differences in annual spending on individual cost elements from one hospital to another. For example, one hospital may
have a large internal decorating programme running, whilst another
hospital may have an external decorating programme at the same time
but hardly any internal painting.
    It was noted that the average figures for the fifteen hospitals
studied are reasonably close to the National Average Costs for 1973-
74. (The only year for which National figures were obtainable at the
time of the research).

8. Planned preventive maintenance (PPM).

The PPM was examined next in an effort to detect what influence this
might have on general maintenance costs.
    The planned preventive maintenance picture on a regional basis was
somewhat fragmented. Some Areas and Districts allowed for this type
of work; others ignored it. Some carried out PPM in the engineering
area only.
    It is difficult to determine just what effect, if any, PPM has on
overall maintenance costs.
    An examination of expenditure showed that it is extremely diffi-
cult to detect patterns of general maintenance expenditure linked
with PPM expenditure. For example, costs for the two Areas spending
the most money on general maintenance for 1979 were studied; one had
a planned preventive maintenance programme, the other had not.

9. Comparison of the largest and smallest.

When extremes were examined, i.e. the largest and the smallest hospital in each Area or District, it was apparent that no pattern existed
between maintenance costs (per unit) and size. It may be thought
that there could be a trend for smaller hospitals to produce different costs than the very largest ones. But once again, costs fluctuate considerably and seem to be unrelated to size.
    In one District studied the lowest unit cost was for the largest
hospital and the highest cost for the third largest hospital in the
rankings. (The % increase of the highest over the lowest cost in

this case was 491%).

A similar pattern emerges with total engineering maintenance costs
as appeared with Building maintenance costs when individual unit
costs per hospital are examined, with a considerable range in the
costs.

10. Maintenance costs and age.

It is usually believed that repairs and maintenance work to old
buildings are generally more costly than for modern buildings, due
mainly to the types of construction encountered, e.g. moulded architraves, built-up skirtings, large sash-windows, moulded staircases,
decorative plaster work, moulded brick courses, etc. etc.

Labour and material costs are often particularly high when matching up to existing work has to be done.

Whilst the age of a building can be a factor in determining
maintenance expenditure, it will also depend on the standard of
maintenance which the building has received over the years.

In 1972 the Directorate of Quantity Surveying of the Property Services Agency published a guide to costs-in-use entitled, "Costs-in-
Use. A guide to data and techniques". Part 2 of this guide deals
with maintenance costs and several building types were studied. For
each type a regression equation was calculated and a graph was shown.
With one exception all of the studies have shown that there is a
close relationship between maintenance costs and the age of the
building. The exception is the BRS paper on hospitals. This one
study did not fit the consistent pattern of costs increasing with
age. Because the complete hospital was taken as a unit, this pattern
could well have been masked by the varying ages and construction of
buildings within the hospital complex.

It is not particularly surprising that hospitals do not show a
close relationship between maintenance costs and age. Some information collected on the one large general infirmary, as an example,
showed that it is virtually impossible to look at age/maintenance
costs on a hospital basis. Some parts were quite recently constructed, whilst other parts were built as long ago as 1859. The age
scatter of the individual parts of the hospital interferes with any
simple deductions.

Maintenance costs are not recorded for each individual part of
hospitals and, therefore, it is not at the present time possible to
show what relationship exists between maintenance costs and age of
buildings. If costs can be recorded in sufficient detail to enable
maintenance costs to be isolated for each part of the hospital, then
a pattern similar to other building types may well emerge.

11. Maintenance costs and location.

DHSS research (National Hospital Service Building Maintenance Cost
Data, 1973) shows that pure building maintenance costs are influenced
by location. The figures indicate that industrial areas are likely
to be around 41% more costly than maintenance carried out in rural
areas.

It was not possible to make any detailed comparison of costs with other Regional Health Authorities, except that partial comparison has been possible with the Devon Area Health Authority. (A case study for the Devon AHA was produced and published by the RICS BMCIS.)

The DHSS research suggests that there will be a difference of about 34% when an industrial area is compared with a coastal area. In fact, a comparison of NWRHA with the Devon AHA shows a difference of 50.4%, i.e. the NWRHA maintenance costs are on average 50.4% above the average for the Devon AHA. (Based on 1978/79 figures.)

12. Reliability of maintenance estimates.

Actual costs of maintenance were compared with the previously estimated costs of maintenance, for the year ending 31st March 1978. It was observed that in 14 cases out of the 18 the estimates were greater than the actual amounts spent, sometimes by over 300%.

The estimates appear to include large contingency sums and may well be inflated considerably in order to allow for 'fierce pruning' of budgets at a later stage. Works Departments usually have to compete with other interests for the limited funds available and it would not be surprising to find that part of the 'tactical negotiation' technique is simply to paint a blacker picture than is actually the case when staking a claim.

13. The effectiveness of maintenance spending.

It is easy to show how much money is being spent annually on various aspects of maintenance but it is quite another matter to be able to say whether this expenditure is adequate for maintaining the estate in a satisfactory state.

A short questionnaire was sent to all the twenty Works Officers in the Region inviting them to respond to some questions on maintenance policies, etc. Thirteen replies received by mid July, 1980 (the end of the field study) indicated that the Works officers were not satisfied with the condition of the estate. Ten replies show that the value of backlog maintenance was actually increasing. Ten Works Officers put an approximate estimated value on maintenance backlog work although this was a very subjective judgement.

14. Conclusions

14.1 Expenditure patterns on maintenance work.
As various sets of data were scrutinised it soon became apparent that there were no clear reasons why money was spent in the way it was. In other words, the expected patterns did not emerge. There appeared to be no link between the size of a building and its maintenance costs. There was no link of any significance between age and maintenance costs, as previously mentioned. There did not appear to be any detectable relationship between function and maintenance costs, and so on.

There was little difference between the variability of building and engineering maintenance costs when expressed as a percentage of the mean costs. The way costs were 'bunched' around the mean was about the same for both areas.

It was relatively easy to show where the differences occurred, i.e. between individual hospitals or between Areas and Districts. It was, however, quite a different matter to explain why costs differed so much.

Much of the cost data was plotted on a computer linked graph plotter in order first to visually check the linear (or curvilinear) regression lines and then to calculate the correlation coefficients. The correlation coefficients were so low in every case as to suggest that there was no obvious pattern to the scatter. (In fact, this was already quite obvious from the plots.) A visual appraisal showed that there were no curvilinear relationships emerging. (The computer program did not allow curvilinear lines to be drawn automatically.)

More regular results began to appear when the data was examined via a standard computer programme 'Statpack' to determine how closely costs matched a normal distribution. It now seems to be reasonably clear that most expenditure is of a random nature in terms of distribution. This does not necessarily mean that such factors as volume, area, function, location, age, etc., do not influence maintenance costs, it is possible that they do (and arguably probable that they do). All that can be concluded at this stage is that the influence of the mentioned variables is not so far detectable in any clear way. A deeper and more thorough study than is possible here may reveal hitherto undetected patterns and links.

14.2 Performance standards.
One key factor which could not be investigated during the case study was that of maintenance standards and a vital question had to be left unanswered, namely, "What performance/maintenance standards have been set in each of the Areas and Districts?" It will be necessary to know much more about these standards and how uniform they are throughout the estate before any final judgements can be made about spending patterns and levels.

It is possible that different performance requirements, maintenance requirements and maintenance standards exist in the Areas and Districts. If this is so, then it would help to explain the differences in maintenance costs throughout the Region. More research could be aimed at, for example, discovering whether or not some Works Officers are attempting to satisfy a very high maintenance standard, higher than that which would be strictly necessary to simply prevent deterioration.

14.3 Maintenance cost data: interpretation and use.
It is by no means a simple matter to estimate accurately the capital cost of new construction but the estimating of capital costs is a much more predictable business than estimating maintenance expenditure. There are numerous factors which, in varying degrees, influence maintenance spending, e.g.:

Maintenance standards required.
Previous maintenance standards attained.
Degree of wear and tear experienced.
The total failure of a component.
Accidental damage.
Vandalism.
Obsolescence of equipment, components, etc.
Location and exposure.
Function and use of building.
Original standard of construction.
The 'Knock on' effect of delayed maintenance.

The 'equation' becomes very complex and it is extremely difficult to predict what influence each variable is likely to have from year to year.

Because of the variables and because of the rather general way in which maintenance records are kept (without background facts) it is difficult to establish accurate figures for specific buildings in advance, for maintenance work.

Does available maintenance expenditure information give a useful picture? This study of maintenance expenditure suggests that the limited block of money allocated annually for maintenance purposes is spent in accordance with the pressures applied to it. Engineering elements tend to have a high maintenance priority since the use and function of a hospital is so closely connected with the engineering provision, e.g. electrical services, heating, lighting and plumbing. Other, low priority, aspects of maintenance may be ignored and delayed with eventual costly consequences, out of all proportion to the cost of normal, regular maintenance. Therefore, what the figures reveal is the expenditure on maintenance, but not necessarily the true cost of maintenance policies. In effect, the amount of maintenance work done is a compromise between the need for maintenance and the ability of the NWRHA to pay for it.

A statement in 'Building Maintenance Statistics', May, 1971 (DOE) reads: "It is by no means axiomatic that maintenance statistics serve a useful purpose." Perhaps the committee on Building Maintenance spoke with more wisdom than is generally realised.

14.4 The relationship between maintenance needs and money available.
A suspicion began to form that expenditure was not necessarily generated by need but by the availability of money and that the use of this money was influenced by decisions made by Works staff who had to commit limited annual funds to the work of greatest priority.

When annual figures of building maintenance expenditure are examined they reveal that expenditure is sometimes restricted, for example, in the year ending April, 1978 there was a 'squeeze' on finances allocated. The broad conclusion derived from the data and from discussions with professional and technical staff is that money is often limited for various 'political' reasons and strategic financial planning is not always based on the actual maintenance needs of the Estate.

It is not obvious why there should be such wide fluctuations in allocations from year to year apart from this latter suggestion being

at least partly true. There may be other reasons which cause costs to
fluctuate but they are not evident from the data so far available.

EXPLORATIONS TOWARDS AN OPTIMAL INSPECTION POLICY FOR THE FAILURE OF BUILDING COMPONENTS

CHRISTINE GROVER
RICHARD GROVER, Portsmouth Polytechnic

Introduction

In recent years there has been increasing interest in improving the efficiency of building maintenance policies. Fundamental to the achievement of an efficient maintenance policy is the choice of an appropriate inspection policy. This paper focuses attention on some of the principal issues involved and sets out a framework within which an optimal inspection policy may be developed. The framework is shown diagramatically in figure 1. The term optimal is used in a restricted sense to refer to the policy that best meets the objectives of the decision maker within the existing constraints. The choice of an inspection policy depends upon two main factors; the characteristics of deterioration and failure in buildings, and the values placed by the interested parties upon the costs and uncertainties involved in any particular course of action. A model of failure rates with respect to age is developed using the Weibull distribution so that the probabilities of a fault being present and detected in a given component at any one time may be estimated and the consequences of delay in remedying it determined. The costs of inspection and delay are examined and related to strategies for dealing with risk and uncertainty. The paper begins though with a review of the role of planned inspection in building maintenance policies.

The Role of Planned Inspections in Building Maintenance Policies

Under a planned inspection policy the function of inspection is to determine whether and when maintenance or replacement takes place. There are however building maintenance strategies that do not involve planned inspections. These include replacing or maintaining as and when failure occurs and planned regular replacement or maintenance. The latter may be carried out periodically, by blocks, or randomly if the planned activity happens to be scheduled for an inconvenient time and is postponed. Such strategies may prove optimal under certain circumstances (1). This may arise when the costs from delaying maintenance are insignificant because, for example, failure does not prevent the building from functioning efficiently in the short term

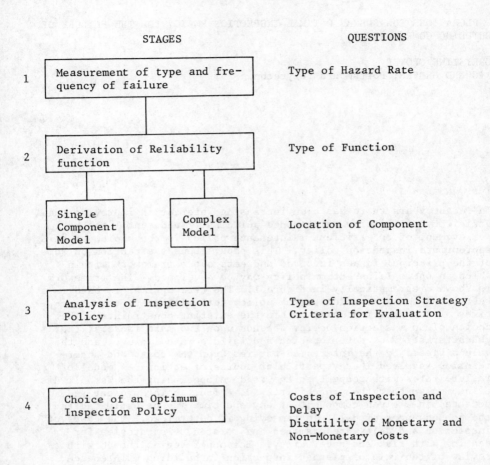

Figure 1.   Framework for the Selection of an Optimum Inspection
Policy.

or where it does not trigger failure elsewhere in the system. Alternatively, when inspection costs are similar to or greater than the costs of replacement or maintenance, planned inspections may add to total costs without commensurate benefit. Under the strategies of regular planned maintenance or maintenance on failure, the role assigned to inspections is a subordinate one confined to determining the extent of the maintenance work to be undertaken but not its timing. Thus in a periodic painting cycle, inspection may be used to determine whether existing paintwork should be burnt off or defective woodwork replaced, but not whether repainting should take place at that time. It is likely however that the selection of either of these strategies has, in most cases, not resulted from a careful comparison with the alternatives and, in consequence, may prove suboptimal. The framework for an optimal inspection policy facilitates comparison with alternative strategies by isolating the criteria upon which an evaluation should be based.

There are certain conditions favourable to the use of planned inspection policies. In some cases the state of the system may be established with certainty only through inspection. A system may have failed but the failure is not apparent as, for example, in the case of an alarm system whose inability to function may not be realised until it is put to the test. A feature of many systems is that their condition degenerates with age but the precise degree of deterioration and, consequently, the quality of their performance, may not be known without inspection. For some systems the failure rates at each stage in the life cycle may be known but the actual incidence of failure is stochastic. Inspection may be required to establish which individual cases have failed. Whether a planned inspection policy is optimal in situations in which inspection is required to establish the state of the system depends upon the value of the information gained by inspection relative to the inspection costs.

Inspection can sometimes enable failure to be detected at an earlier stage than would otherwise be the case in circumstances in which failure and its detection are not simultaneous. Where failure is an absorbing state, the consequences become progressively worse. When inspection facilitates the early diagnosis of failure, remedial action may be able to limit the spread of damage throughout the system, thus minimising the consequential costs of failure. The early detection of failure may permit preventative maintenance to be carried out prior to a breakdown and this may prove cheaper than emergency repairs. A deteriorating situation identified by inspection may subsequently be intensively monitored so that the point at which the system breaks down can be predicted. One of the functions of inspections is to enable a lead time to be gained. This can be defined as the time interval between the point at which early detection occurs as the result of an inspection and that when failure would have become apparent in the absence of inspection (2). In figure 2 the system moves from normal functioning (S_0) to a state of undetected failure (S_1) at time t. Failure is discovered in the absence of inspection at time x_d. In figure 3 inspection successfully identifies the failure at time x_i. The period of previously

FIGURE 2

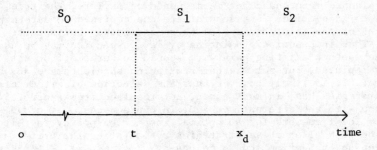

Failure and its detection without an inspection policy.

S_0 = no failure

S_1 = failure present but undetected in absence of inspection.

S_2 = failure present but known in absence of inspection.

FIGURE 3

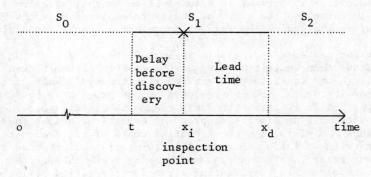

Failure and its detection with an inspection policy.

undetected failure is divided into the lead time gained and the period of delay between failure and its discovery by inspection. The division between these two is a function of the frequency of inspection and the efficiency with which it is able to detect failure. The value of an inspection policy under these circumstances depends upon the costs of inspection relative to the benefits to be derived from the early detection of failure. The latter depends in turn upon whether inspection can be followed by effective remedial action and the penalties resulting from any delay in reacting to a failure.

Modelling Failure Rates in Building Components

Although failure in many building components is influenced by a variety of factors, reliability theory suggests that it is possible to model it using a function displaying the probabilities of failure at each stage in the life cycle. There are several distributions that could be used to represent the function but the most promising appears to be the Weibull. This is a family of functions named after Walodi Weibull, a Swedish physicist, who used it to represent the distribution of breaking strengths of materials (3), and subsequently applied it to a range of other problems (4). It has since been widely used in maintenance and inspection problems in a variety of fields, including building maintenance (5). Although it has normally been regarded as an empirical distribution, similar equations appear to be generated by certain processes of nucleation and growth (6).

The Weibull frequency distribution is of the form

$$f(t) = \frac{\beta}{\alpha} \left(\frac{t - \gamma}{\alpha} \right)^{\beta - 1} \exp - \left(\frac{t - \gamma}{\alpha} \right)^{\beta} \tag{1}$$

the cumulative frequency distribution

$$F(t) = 1 - \exp - \left(\frac{t - \gamma}{\alpha} \right)^{\beta} \tag{2}$$

and the hazard or failure rate

$$r(t) = \frac{\beta}{\alpha} \left(\frac{t - \gamma}{\alpha} \right)^{\beta - 1} \tag{3}$$

where the parameters α, β, and γ represent the scale, shape, and location respectively (7). The function is a versatile one as figure 4 illustrates. When $\beta > 1$ there is an increasing hazard rate meaning that the incidence of failure increases with age. When $\beta = 1$ the function is exponential so that the hazard rate is a constant proportion. For $\beta < 1$ the hazard rate decreases with age and when $\beta = 3.6$ the function approximates to a normal distribution.

The life of a component is often made up of each of the three hazard rates so as to produce the characteristic bath-tub curve (figure 5). This can be modelled by a two-fold mixed Weibull distribution where

Probability (%)

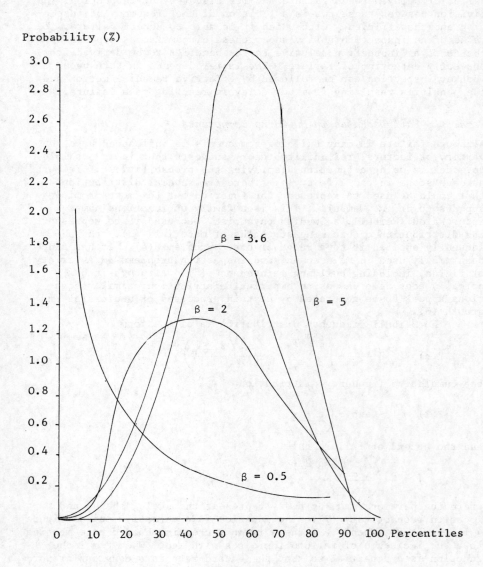

Figure 4   The Weibull Distribution

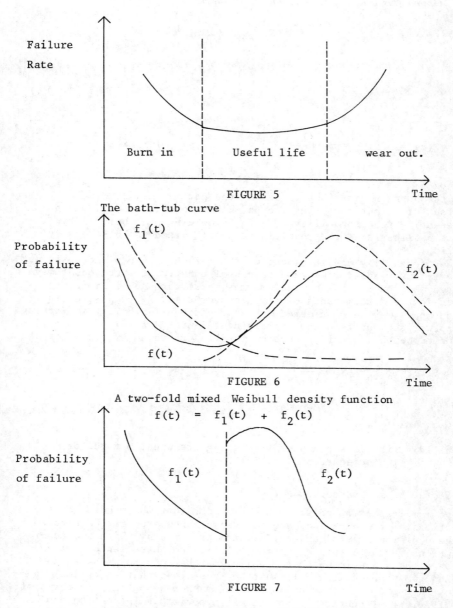

Failure Rate

Burn in | Useful life | wear out.

FIGURE 5
The bath-tub curve

Probability of failure

$f_1(t)$

$f_2(t)$

$f(t)$

FIGURE 6

A two-fold mixed Weibull density function

$$f(t) = f_1(t) + f_2(t)$$

Probability of failure

$f_1(t)$

$f_2(t)$

FIGURE 7

A two component composite Weibull density function.

$$f(t) = p \ \frac{\beta_1}{\alpha_1} \left(\frac{t}{\alpha_1}\right)^{\beta_1 - 1} \quad \exp \ -\left(\frac{t}{\alpha_1}\right)^{\beta_1} +$$

$$(1-p) \ \frac{\beta_2}{\alpha_2} \left(\frac{t - \gamma}{\alpha_2}\right)^{\beta_2 - 1} \quad \exp \ -\left(\frac{t - \gamma}{\alpha_2}\right)^{\beta_2}$$

$$\tag{4}$$

and

$$F(t) = p \left[1 - \exp \ -\left(\frac{t}{\alpha_1}\right)^{\beta_1} \right] \quad + \quad (1-p) \left[1 - \exp \ -\left(\frac{t - \gamma}{\alpha_2}\right)^{\beta_2} \right]$$

$$\tag{5}$$

with the burn in distribution having the parameters α_1, β_1, the wear out distribution having the parameters α_2, β_2, and γ, and p representing the probability of a failure being of a burn in type. Figure 6 illustrates the resulting two-fold mixed Weibull density function.

The probability density function is of future events and so is not directly observable. The experiences of cohorts of a component may be used as proxies though problems may arise through significant shifts in the determinants of failure over time. A composite distribution model may be considered, thus providing flexibility in fitting an appropriate function and explaining failure data (7). Such a composite Weibull density function is illustrated in figure 7 for which

$$F(t) = \begin{cases} 1 - \exp \ -\left(\frac{t}{\alpha_1}\right)^{\beta_1} & \text{for } o \leqslant t \leqslant \delta \\\\ 1 - \exp \ -\left(\frac{t}{\alpha_2}\right)^{\beta_2} & \text{for } \delta \leqslant t \leqslant \infty \end{cases} \tag{6}$$

Weibull distributions can be fitted using graphical methods or maximum likelihood equations (8) (9).

The Choice of an Inspection Policy

An optimum inspection policy was defined as that which best met the objectives of the decision makers. We must now explore the nature of the decision makers' objectives, the criteria by which the contributions of different inspection policies to these can be evaluated, and the sense in which any policy may be said to be best in meeting these objectives. It must be recognised that an optimum policy in the specific sense used here may not be optimal in terms of the efficient allocation of resources in society. Inspection and maintenance policies generate spillover effects for neighbouring properties. These arise because the market value of a property is determined not only by the characteristics and condition of the property itself, but also by the attributes of the neighbourhood in which it is located. For example, Wilkinson has estimated that over 30 per cent of the variation in house prices in Leeds is due to

factors primarily associated with aspects of the locality in which
the house is situated as distinct from the structure and amenities of
the dwelling unit itself (10). A property owner must take account
of the condition of surrounding properties in deciding whether to
maintain or improve his own. This can lead to a situation in which
the rational property owner may fail to keep his property in an
adequate state of repair so that he may extract the maximum beneficial spillovers from neighbouring properties whilst minimising his
own contribution to the neighbourhood's attributes so as to incur the
least cost to himself. (11). If such a strategy is widely adopted
it may lead to certain localities becoming blighted and a sub-
optimal resource use for society (12).

Inspection policies must be capable of satisfying the varying
objectives of a number of potential decision makers. It would seem
likely, for example, that the objectives of a tenant on a full
repairing and insuring lease will differ from those of an owner
occupier. Thurley has demonstrated how the objectives of a housing
maintenance programme can vary according to its dominating influences. Thus an occupant-based system places its priorities upon the
responses to the occupants' needs. It therefore stresses the importance of speedy reactions to occupant initiated demands, the quality
of finishings, and good client relations. Resources tend to be
geared to meeting peak demands. By contrast an owner-based system
emphasises the impact upon the value of the property and has the
objective of the minimisation of costs. A workforce-based system
tends to result in the work accomplished reflecting the skills and
resources of the labour force. Whilst an insurance-based system
emphasises that choices should be based upon a rational assessment of
structural needs (13). The objectives of the policy cannot therefore be divorced from their social context. Whilst building maintenance may be "undertaken in order to keep, restore or improve every
facility ..... to a currently acceptable standard, and to sustain the
utility and value of the facility" (14), it is doubtful that there
can be any concensus as to what is an acceptable standard in view of
the differences between potential decision makers as to the utility
to be derived from any particular course of action. Azzaro's
comments on maintenance apply with equal force to inspection policies:

> "there is no single standard of condition which will be equally
> applicable to all the different types of building and circum-
> stances likely to be met in practice" (15).

Whilst we may presume that all the potential decision makers experience a disutility from costs, are adverse to taking risks, and
prefer present to future gains, the choice of an inspection policy
will vary according to the precise values attributed to these as this
will determine the point at which the trade-offs between alternatives
become acceptable. Whilst we may establish the framework within
which the choice of an inspection policy takes place, there is no
universally applicable solution to the problem.

Four principal types of inspection policy may be postulated (16)(17)
    (i) Periodic. Inspection occurs at equal intervals.
    (ii) Sequential. The time interval between inspections is given

by $d_i = wd_{i-1}$ $i = 2,3...$ (7)

For an increasing hazard rate choose $w < 1$ and for a decreasing one choose $w > 1$, whilst $w = 1$ gives the periodic policy.

(iii) Constant Hazard or X_p. Inspection times are chosen so that the probability of failure between any two inspections is constant.

(iv) Shifted Periodic. Periodic policies have the attraction of being simple to implement but will only be optimal when the failure rate is constant. When used in other circumstances it is clearly of minimal use to inspect in the earlier years when there is an increasing hazard function as failures are unlikely to occur before a time t_w. In this case we choose

$x_i = t_w + (i-1)d$. $i = 1,2,3...$ (8)

The reader is referred to (18) and (19) on the evaluation of a warranty assurance which may give an estimate of t_w.

The technical evaluation of an appropriate inspection policy is based upon three criteria:

(i) mean delay between failure and its detection

(ii) the probability of false positives and negatives from the inspections

(iii) mean number of inspections required to detect a failure.

The mean delay is the expected time between failure and inspection

$$E(x_i - t) = \sum_{i=1}^{\infty} \int_{x_{i-1}}^{x_i} (x_i - t) \, f(t) dt$$

$$= \sum_{i=1}^{\infty} \left\{ x_i \int_{x_{i-1}}^{x_i} f(t) dt - \int_{x_{i-1}}^{x_i} tf(t) dt \right\}$$

$$= \sum_{i=1}^{\infty} x_i \left[F(x_i) - F(x_{i-1}) \right] - E(T)$$ (9)

where (x_i) are inspected times $i = 1,2,3....$ and $E(T)$ is the expected time to failure. For a Weibull distribution

$$E(T) = \alpha \, \Gamma \left(1 + \frac{1}{\beta} \right)$$ (10)

where $\Gamma(x)$ is the gamma function (20). The equation holds true only when we can be sure that failure will be discovered at the first inspection after its occurrence. The fault may be missed at this or subsequent inspections. Failure to detect the fault gives a false negative. Similarly inspection may record a failure incorrectly where none exists. This is called a false positive. False negatives give rise to longer delays and, hence, increase costs due to further deterioration of the system and additional inspections being necessary to discover the fault. False positives cause the cost of further investigations and unnecessary remedial work to be incurred. Shahani (16) gives the mean delay given a false negative probability of b as

$$E(x_n - t) = \sum_{n=1}^{\infty} x_n \sum_{i=1}^{n} \left[F(x_i) - F(x_{i-1}) \right]$$

$$(1-b)b^{n-i} - E(t) \tag{11}$$

where x_n is the inspection that reveals failure. The mean number of false positives is given by

$$E(F) = a \sum_{i=1}^{\infty} (1 - F(x_i)) \tag{12}$$

where a is the probability of a false positive, and the mean number of inspections by

$$E(N) = \sum_{i=0}^{\infty} (1 - F(x_i)) + \frac{b}{1-b} \tag{13}$$

From Shahani, the constant hazard policy gives

$$E(Delay) = \begin{cases} p(1-b) \sum_{i=1}^{\infty} (\frac{b^i - q^i}{b-q})_F{}^{-1} (1 - q^i) - E(t) & \text{if } b \neq q \\ p(1-b) \sum_{i=1}^{\infty} ib^i F^{-1} (1-q^i) - E(t) & \text{if } b = q \end{cases} \tag{14}$$

Optimisation involves selecting the inspection policy that best meets the decision makers' objectives. Inspections are costly but so may be the consequences of undetected failure. An optimum inspection policy minimises the disutility from the various costs involved, a process of balancing the costs from inspection against those from delay.

Barlow & Proschan (21) present a cost model to facilitate optimisation that assumes:

(i) failure is discovered only on inspection
(ii) inspection takes negligible time and does not cause further deterioration of the system
(iii) the system does not fail during inspection
(iv) each inspection involves a fixed cost (C_1)
(v) the time between failure and its discovery involves a cost of C_2 per unit of time
(vi) inspections cease on the discovery of the failure.

For a periodic policy the cost is

$$C_1 \left[N + 1 \right] + C_2 D \tag{15}$$

where N is the number of checks prior to failure and D is the interval between failure and its discovery. When the time of failure has a failure distribution F(t) then

$$E(C) = \int_0^{\infty} \left\{ C_1 (E(N)+1) + C_2 E(D) \right\} dF(t) \tag{16}$$

When the inspection policy is sequential then

$$E(C) = \sum_{i=1}^{\infty} \int_{x_{i-1}}^{x_i} \left[C_1 (i+1) + C_2(x_i - t) \right] \, d\, F(t) \qquad (17)$$

Barlow & Proschan present a recursive formula

$$x_{i+1} = x_i + \frac{F(x_i) - F(x_{i-1})}{f(x_i)} - \frac{C_1}{C_2} \qquad i = 1,2,3.... \qquad (18)$$

where once an optimal x_i is known then x_{i+1} can be calculated. This process involves substantial computational effort to select x_i.

Nakagawa & Yasui (22) give an equation for calculating an optimal periodic inspection time for a Weibull distribution and a cost ratio C_1/C_2. Assuming $E(D) = T/2$ then an approximate checking time is

$$T_1{}^* = (2\, \alpha\, \Gamma(1 + 1/\beta)\, C_1/C_2)^{\frac{1}{2}} \qquad (19)$$

Munford & Shahani (23) give a useful nomogram where an optimal p in the constant hazard policy can be chosen given the Weibull parameters and the ratio of cost C_1/C_2.

Cost functions of the form used in these models suffer from two main deficiencies. Firstly, their restrictive assumptions make it unlikely that they are realistic when applied to building inspections. Secondly they treat the objectives function of the decision makers as being a function of costs rather than the disutility experienced from the costs. They assume monetary values to be directly comparable and so do not pay adequate regard to the timing of cash flows or to different responses to the risks involved.

Barlow & Proschan assume that there is a fixed cost per inspection so that the average and marginal inspection costs are equal. If there are economies of scale in inspections then such a relationship will not hold true. Economies of scale are likely to arise through the existence of fixed costs so that the unit costs decline as these are spread over a greater number of inspections. They may include the costs of setting up an inspection system, such as a computer data base, or costs which remain constant over a range of inspections, such as premises for the inspection staff. The economies of scale may be reinforced by a learning curve effect through which the unit costs of inspection are reduced through repetition (24). The variable costs of each inspection will comprise items such as labour costs and materials expended, as well as any degredation of the system or disruption to services caused by inspection. It seems unlikely that inspection costs will be identical for each inspection even in the absence of economies of scale due, for example, to differences in travelling time to inspections and accessibility within buildings (25). As repair time distributions appear to follow a log normal distribution (21), it may be that variable inspection costs follow a similar pattern.

The costs of delay (C_2) seem likely, as in the Barlow & Proschan model, to be a function of the period of delay. There is no reason, though, to suppose that the relationship is linear. The costs of delay are likely to include:

    (i) further damage to the system including the failure of re-
        lated parts and the possibility of higher costs of remedial
        work

   (ii) a loss of productivity from the activities within the
        premises which may in turn result in a loss of rent.

  (iii) landlord/tenant conflict which may lead to higher manage-
        ment costs, compensation for breaches of convenants, and
        damage to the corporate image.

The precise relationship between C_2 and time will depend upon the
probabilities of particular outcomes.

    The existing cost models disregard the timing of the costs. Given
that the decision makers have a preference for present over future
returns, the costs must be converted into time equivalents through
the use of an appropriate discount rate before the minimum cost can
be computed. Essentially a present value is computed that is the
average discounted cost for each point in time weighted by the probabilities of particular outcomes (26). It should not be assumed that
disutility increases in strict proportion to the monetary value of
the costs so that a cost minimisation approach does not necessarily
result in the minimisation of the disutilities. For example, certain
consequences of failure may have a very low probability of occurrence so that their cost is given a low weighting in the computation
of the total cost. The consequences may be regarded as so horrific
that they may be given a disutility out of all proportion to the
likelihood of their coming about. In such cases the weighting for
the utility calculation differs markedly from that based on monetary
values. We must also bear in mind that monetary values may not
always be a good guide to the true costs involved as, for example,
where these are intangible or the product of imperfect competition.
It may be possible to use utility computations directly in the optimisation process (27). However there are strategies that enable
utilities to be incorporated where their precise values are unknown
or where the probabilities of the various outcomes are uncertain.
For example, we may use a minimax approach to ensure the selection of
the course that leads to the least bad of the worst possible outcomes from each eventuality (28).

    Strategies are available to deal with complex systems comprising a
number of components. If the parts within a complex system are independent, then each may be treated as a simple system. Where the
components are interdependent, opportunistic policies may be used.
The maintenance of a single uninspected part depends upon the state
of one or more continuously inspected parts. Such policies prove
optimal when the cost of joint maintenance is less than the sum of
maintaining each part separately.

Conclusion

This paper has set out a framework within which the best inspection
policy for the decision maker is selected. The framework may also be
used to determine whether a maintenance strategy such as replacement
on failure or block replacement should be used instead of one of the

inspection policies. The circumstances favourable and otherwise to an inspection policy have been outlined and a method of determining the costs and consequences suggested. This has involved the use of reliability functions to generate the consequences of delay prior to failure being detected. The utility of early warning may then be compared with the disutility from inspection costs. In proposing this framework we have posed a number of questions for which answers cannot be given at present. Our framework is therefore an exploration towards an optimum inspection policy for building components.

Acknowledgement

We are grateful to Dr. A.K. Shahani for introducing us to the Weibull distribution and its uses.

References

1. D.B. James & M.F. Green,
    'The Use of Decision Models in Maintenance Work', in E.J.
    Gibson (ed), Developments in Building Maintenance 1 (1979).
2. P.C. Prorok,
    'The Theory of Periodic Screening 1: Lead Time and the
    Proportion Detected', Adv Appl Prob, 8 (1976).
3. W. Weibull,
    'A Statistical Theory of the Strength of Materials',
    Ingeniörsventenskapsakemiens (1939).
4. W. Weibull,
    'A Statistical Distribution of Wide Applicability', J. Appl
    Mechanics, 18 (1951).
5. A.H. Christer,
    'Economic Cycle Periods for Maintenance Painting', Opl Res Q,
    27 (1976).
6. J.H. Gittus,
    'On a Class of Distribution Functions', Appl Statistics, 16
    (1967).
7. N. Mann, R. Schafer & N. Singpurwalla,
    Methods of Statistical Analysis of Reliability and Life Data
    (1974).
8. N.L. Johnson & S. Kotz,
    Continuous Univariate Distributions, 1 (1970).
9. J. Kao,
    'A Graphical Estimation of Mixed Weibull Parameters in Life-
    Testing of Electron Tubes', Technometric, 1 (1959).
10. R.K. Wilkinson,
    'House Prices and the Measurement of Externalities', Econ J,
    83 (1973).
11. O.A. Davis & A.B. Whinston,
    'The Economics of Urban Renewal', Law & Contemporary Problems,
    26 (1961).
12. J. Rothenberg,
    Economic Evaluation of Urban Renewal, Brookings Institution,
    Washington (1967).

13. K. Thurley,
    'Improving the Organisation of Maintenance in Married Quarters
    Estates', in E.J. Gibson (ed), Developments in Building
    Maintenance 1 (1979).
14. Dept. of the Environment,
    Report of the Committee on Building Maintenance (1972).
15. D.W. Azzaro,
    'The Economics of Maintenance', in E.D. Mills (ed), Building
    Maintenance and Preservation (1980).
16. A.K. Shahani,
    'Choice of a Process Inspection Scheme: Some Basic Considera-
    tions', in H.J. Lenz et al, Frontiers in Statistical Quality
    Control (1981).
17. A.G. Munford,
    'Comparison Among Certain Inspection Policies', Management
    Science, 27 (1981).
18. N.R. Mann & S.C. Saunders,
    'On Evaluation of Warranty Assurance When Life has a Weibull
    Distribution', Biometrika, 56 (1969).
19. N.R. Mann,
    'Ordered Sample Observations from a Weibull Population'
    I.E.E.E. Transactions on Reliability (1970).
20. M.R. Spiegel,
    Mathematical Handbook of Formulas and Tables (1968).
21. R.E. Barlow & F. Proschan
    Mathematical Theory of Reliability (1965).
22. T. Nakagawa & K. Yasui,
    'Approximate Calculation of Inspection Policy with Weibull
    Failure Times', I.E.E.E. Transactions on Reliability (1979).
23. A.G. Munford & A.K. Shahani,
    'An Inspection Policy for the Weibull Case', Opl Res Q, 24
    (1973).
24. A Review of Monopolies and Mergers Policy,
    Appendix C, Cmnd 7198 (1978).
25. E.G. Lovejoy,
    'Safety and Security in Accessibility for Maintenance', in
    E.D. Mills (ed), Building Maintenance and Preservation (1980).
26. J.M. Samuels & F.M. Wilkes,
    Management of Company Finance, 3rd edn (1980), ch 12.
27. G.C.A. Dickson,
    'An Empirical Examination of the Willingness of Managers to
    use Utility Theory', J Management Studies, 18 (1981).
28. M. Peston & A. Coddington,
    The Elementary Ideas of Game Theory, C.A.S. Occasional Paper,
    no 6 (1967).

LIFE CYCLE COSTING - A COMPREHENSIVE APPROACH

ROBERT FULLER, Building Design Partnership

INTRODUCTION

Life cycle costing is commonly interpreted as the operating cost of
buildings over the forecast useful life attributed to them.   The
wider and more correct view considers the full range of costs arising
from the provision of accommodation to fulfil a particular user
requirement and which accrue from initial provision of the facilities
right through the anticipated life span and include initial capital
costs, subsequent operating costs and replacement or renewal at
intermediate stages of the life cycle.   All these factors require
evaluation on a common basis and should be included in the decision
making process at development stage.

An even wider and more comprehensive approach to life cycle costing
can be taken and that is the approach which also considers the
organisational costs of the users of a building and assesses the
effectiveness and economic efficiency of the facilities related to
the functions of the user body.   When considering cost effective
policies at the design and planning stages it can easily be
demonstrated that there is direct correlation between design decisions
and the operating costs of the facility users. As these latter costs
are on a considerably larger scale (generally speaking) than the life
cycle costs of the facilities themselves, then the more comprehensive
analysis has considerable significance in the early decision making
process.

The form of the analysis may well prove to be similar for ALL types
of building use and in each case it is necessary to establish the
relative importance of each main category of costs. To arrive at
this point it is first desirable to review the nature of the full
range of costs arising from the provision of accommodation and the
methods available for their comparative evaluation.

LIFE CYCLE COSTS

The range of costs which accrue from the initial provision of the
facilities right through the anticipated useful life to termination
may be considered under a number of major categories.

Initial Capital Cost

This covers the acquisition of land, (and existing premises if new
building is not proposed), construction of the required premises,
the furnishing and equipping of the premises and all the legal and
design fees, loan charges and incidental costs relating to these
operations.

Annual Operating Cost

This covers the annual costs of services utilities, repair and
maintenance (including periodic redecoration) of the fabric,
fittings and services installations, rates and insurance, staff
costs related to building maintenance and other similar general
overheads.

Periodic Replacements

This covers the periodic replacement or renewal (rather than
interim repair) of certain parts of the building fabric, furnishing
or services installations and represents a capital expenditure at
different points in time during the life of the building (e.g. felt
roof coverings or heating equipment).

Modernisation Schemes

At certain points in the life of a building, user requirements have
changed over a preceding period to the extent that considerable
modification is needed within the building shell to maintain
efficiency of operation, but without the necessity for total
replacement of the facilities. As with periodic replacements,
modernisation is a capital cost arising at a specific time.

USER COSTS

The details of any user costs will vary very considerably according
to the type of activity. There will, however, be three main areas
under which significant costs will occur which may be divided up
between labour costs, material costs and equipment costs. Of these,
labour and material costs will be continuous or recurrent expenditure,
whilst equipment costs are more likely to be the subject of periodic
capital investment.

Equipment costs are sometimes included in the capital costs of
buildings as an initial life cycle cost but where this equipment is
particularly related to the activity or operation of the building
user then it is more properly considered as a User Cost. Thus

manufacturing plant, specialised laboratory equipment and other
commercial plant right down to copying machines in offices would be
treated separately from the building. It is of interest to look
briefly at the user costs arising under the nine main categories of
building type. Whilst there is further wide divergence within these
categories, the broader divisions considered will give at least an
indication of the nature of variability.

Transportation (e.g. Bus Stations, Airport Termini, etc.)

In this category, operating staff salaries (with 'on' costs) will be
a quite significant item of 'user' cost on the labour side. In this
context also, user is probably less suitable as a descriptive term
than building owner as the reference is to the operator rather than
the public. Material costs arising will be the 'software' for
carrying on the operation - paper for tickets, timetables, posters,
information leaflets, administration, etc., and all the normal
adjuncts to routine office performance. Specialised equipment might
comprise such things as ticket dispensing machines, computer
equipment, electronic office equipment, catering equipment and
similar items. With this type of operation it is difficult to know
where to draw the line and whether one should include or exclude the
costs of operation of the transportation network which lies outside
the physical confines of the particular building. Where the planning
of the termini may, in itself, limit the use which can be made of
these other facilities then the total cost consideration and
analysis might be the widest possible one.

Industrial (e.g. Factories, Warehouses, etc.)

There is a wide divergence between costs of production units, such
as factories, and storage units such as warehouses. In the former
case, labour costs in the form of supervisory and production workforce (other than automated situations) will be high as will also
material costs represented by the raw materials for processing and
equipment represented by the production machinery. In the case of
warehousing, labour will be represented by the staffing costs of a
small operative workforce with relatively small material costs and
equipment costs probably restricted to goods handling equipment
(forklift trucks, etc.).

Commercial (e.g. Shops, offices, etc.)

Again there is a wide divergence between the cost patterns emerging
with shops on the one hand having a relatively low staffing cost but
very high materials cost and offices on the other hand having a very
significant labour cost in the salaries and wages of staff in a
fairly intensively occupied building, but with a lower material cost
in the supplies (mainly stationary) that they process or utilise.

Entertainment and Recreational (e.g. Cinemas, Swimming Pools, etc.)

These buildings will, in the main, have relatively low user costs
related to the operational staff, materials and equipment needed for
the specific activity. In the case of theatres, wider production
costs of presentations might be relevant however.

Public Service (e.g. Libraries, Police Stations, Fire Stations, etc.)

These can be expected to be similar in user cost terms to the Entertainment and Recreational buildings above, though staffing levels and patterns of use may be more intensive.

Religious (e.g. Churches)

These will generally have low user costs, and the nature of the activity is such that planning or design decisions are unlikely to affect these user costs in any marked degree.

Health and Welfare (e.g. Clinics, Hospitals, etc.)

Hospitals particularly in this category have very high user costs both in terms of staffing costs for twenty four hour operation, material costs for medical supplies, catering and general administration and equipment. It is significant that 'cost-in-use' studies have been directed extensively towards hospitals in recent years though this has been more in terms of energy cost than total operating cost.

Educational (e.g. Schools, Colleges, Universities, etc.)

Generally in an intermediate category of user cost, educational activities have teaching staff costs, materials costs related to teaching activities and some equipment costs also related to teaching though this latter cost varies with the sophistication of the institution.

Residential (e.g. Houses, Flats, etc.)

There is no user cost in this category which is of significance.

RELATIVE IMPORTANCE

When seeking economies, it is good practice to commence with consideration of the largest items of expenditure and work downwards. It is pointless to seek large scale improvement in financial performance by adjustments in relatively low cost items. Therefore it is necessary to define the relative importance of the different cost areas identified above before the scale of any improvement can be properly evaluated. To do this, it is necessary to make the assessment on a common basis and as all the costs are either capital or recurrent it will be necessary to convert one or the other. It is not significant as to which course of conversion is used, what is important is that all costs are reduced to the same basis which is either net present value or annual equivalent cost. The relevant processes for conversion (and for evaluation) were clearly established by Dr P A Stone[1] as long ago as 1967 and are in common usage. It is therefore unnecessary to deal in detail with the technical processes of conversion though one or two aspects, particularly in relation to user costs, are worthy of comment.

In all conversion calculations, the two factors which are of most significance are the rate of interest assumed and the term set for the expected building life. The latter will also affect the extent of future costs forecast. The precise application of these will vary from case to case and will be further affected by allowances for taxation and for inflation, the rate of interest and taxation allowances being specifically related to the precise nature of the building owners financial operation.

In the past it has been customary to allow sixty years as the normal life span of a building constructed to any degree of permanence. This has served quite well, generally, and would certainly seem adequate as a notional maximum.Beyond sixty years, further operating costs become extremely difficult to forecast and, when discounted to present value, of little financial significance. It is also quite problematical whether after sixty years the building will still be suitable for the purpose for which it was erected or to which it may subsequently have been converted. Whilst it is a fact that buildings continue a useful life beyond sixty years, in some cases for several hundreds of years, the reasons for this are rarely purely commercial ones and the process of evaluation and consideration must therefore be on a different basis. Periods of less than sixty years may, however, be quite relevant in some circumstances. The growth of electronic systems (computerisation, communications, controls) and their miniaturisation, together with automation in manufacturing production, are rapidly changing the commercial world. The American market has already demonstrated that commercial properties on prime sites can become obsolete and subject to replacement in a mere ten years. Thus realistic allowances must be made when evaluating time scales, and future costs and the effects of changing patterns of use clearly defined. In the relevant cases, forecast life cycles should be reduced or additional allowance made for in-built flexibility to accommodate rapidly changing patterns of use.

As indicated in the range of user costs described above, the relative importance of these and other life-cycle costs will vary from situation to situation. One example is shown below taken from a study of life cycle costs of university institutions[2].

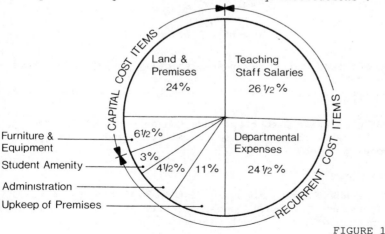

FIGURE 1

In this example the costs have been averaged across a number of similar institutions and converted to an average annual cost in each category. In this case, periodic replacements are not identified separately but included with Departmental Expenses or Upkeep of Premises whilst the forecast allowances for modernisation costs are included in the total amortised cost for Land and Premises. The above figures have been derived on the basis of a sixty year life of the buildings and a discounting rate of 7%. The analysis indicates that the current user costs are as great as the combined capital and recurrent expenditure in the life cycle costs.

A further example of model life-cycle costs, only this time excluding the effects of the User Costs, is shown in Fig. 2. This shows more clearly the incidence of periodic replacement of both furniture and equipment and some short term building adaptation. This model is taken from a study of laboratory buildings[3].

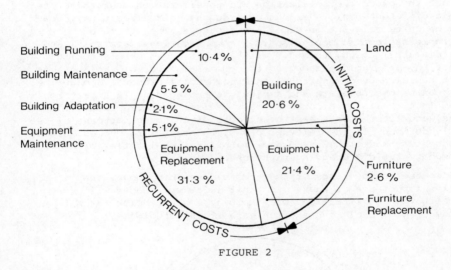

FIGURE 2

It has been predicted, though never statistically proved in practice, that at the very top end of the scale - offices - the user costs would account for over ninety percent of the total combined expenditures.

The process of evaluation used to produce the above models can be easily applied to any other definable organisation and a separate individual model obtained.

OPERATING EFFICIENCY

Introducing user costs into life-cycle cost analysis creates a new and larger dimension. The test of economic efficiency in life-cycle costs which include only the direct capital and recurrent building

related costs, is the level of rate of expenditure for a given range
of facilities. The test of economic efficiency in the wider life-
cycle cost system is somewhat different. In this situation it becomes
necessary to equate the total expenditures with the scale of the user
operation, so that the flow of resources used (measured in financial
terms) can be said to support a certain level of economic activity
over a period of time.

There are two aspects of this activity pattern. One is the scale and
nature of the activities: the other is their level of efficiency in
operation - in commercial terms, their profitability. When dealing
with development decisions, it is not necessary to consider both
aspects. It may reasonably be assumed that the user organisation
will have tested the potential profitability or level of efficiency
of its operations prior to proceeding to the provision of the
necessary facilities. In these circumstances, it is therefore only
necessary to ensure that the desired activity pattern can be
achieved.

In the example of life-cycle costs for universities quoted above the
activity pattern was taken as the extent of teaching and research
activity carried on and measured by the total staff/student contact
hours, staff and student research time and student private study time.
Separate definitions of activity patterns would be required for each
type of user organisation.

The measures of economic efficiency can be obtained from the
comparison of alternative solutions, the optimum solution being that
which provides for the required activity pattern at the lowest total
cost. The comparison would be carried out by constructing the cost
model described above for each of the alternatives proposed.

This comparison method was used in the university study[4] and
produced some interesting results. Two alternative planning
solutions for a university development were worked out, both based
on the same basic academic plan and population statistics. In the
first case, traditional planning systems based on departmental
teaching arrangements and accepted space standards were adopted.
In the second case, inter-departmental teaching, intensive space
utilisation and sophisticated time-tabling methods were assumed.
The result of the evaluation of the anticipated life-cycle costs of
the two schemes, expressed in annual cost terms, is shown in the
table below.

TABLE 1

Category of Cost	Annual Equivalent Cost per Student		Proportion of Total Cost	
	Alternative 1 £	Alternative 2 £	Alternative 1 £	Alternative 2 £
Recurrent Costs				
Administration	91.40	91.40	7.23	7.94
Teaching Staff Salaries	391.50	350.00	30.96	30.39
Departmental Expenses	267.00	280.00	21.11	24.31
Student Amenity	47.45	47.45	3.75	4.12
Upkeep of Premises	162.00	140.00	12.81	12.16
Capital Costs				
Land and Premises	238.75	191.84	18.88	16.66
Furniture and Equipment	66.50	50.95	5.26	4.42
TOTAL	1,264.60	1,151.64	100.00	100.00

CONCLUSIONS

The importance of models lies, not in the identification of a precise scale of economy that can be achieved, but in the demonstration of a wide financial basis of comparison for planning and management alternatives. It is apparent from the example shown above, that planning decisions can have significant effects on user costs and that as the significance of these user costs grows relative to other life-cycle costs, the more important does it become that such influences are properly identified and taken into account.

References

1. P A Stone (1967).   Building Design Evaluation: Costs-in-Use. E & F Spon Ltd.

2. R A Fuller (1978).   An Analysis of the Economic Structure of British Universities with special reference to Development Planning.   Ph.D. Thesis.   London University.

3. Laboratory Investigation Unit of the DES (1969).
   An Approach to Laboratory Building. A Paper for Discussion.   Paper No 1.   London.

4. R A Fuller. op.cit.

SOFT LANDSCAPING IN HOUSING LAYOUTS

HECTOR GOW, Dundee College of Technology

Introduction

The increasing use of landscape planting in the housing environment is
now recognised and accepted as a necessary component of the total
development.

In this regard it can be compared to the introduction of central
heating which brought with that decision the concomitant need for
annual servicing and maintenance. The components of external works
do likewise and soft landscaping especially has to be recognised as
an annual maintenance commitment if the design philosophies that led
to its development are to be allowed to produce the desired effect in
aesthetic, design and human terms.

In the present economic climate it must also be recognised that
design decisions should be closely aligned to a maintenance regime
that balances the need for a minimum commitment in upkeep with a
maximising of effect in overall design terms. This is more easily
said than achieved in practice and it falls to the design and
maintenance teams to liaise closely on the components and practices
which will closely relate design aims with the most efficient
maintenance practices.

Case Studies

In order to show these problems in a practical situation the following case studies review three essentially different design solutions
in housing developments and analyses the capital needs and maintenance costs of each. No conclusions are attempted for it must remain
with the individual authority and its officers to set the priorities
and standards within their own areas. To some extent however the
studies do illustrate that the differences between the design
solutions, given a reasonable level of landscaping, are not as great
as might have been expected.

The table illustrates the analyses structured in terms of costs
per house between the three layouts.

TABLE: COMPARISON OF COSTS-IN-USE BY LAYOUT/HOUSE/YEAR

CASE STUDIES	1-VDAL	2-PVSL	3-PVJUL
Initial Costs	2345.00	2511.00	3148.00
Annual Equivalent @ 10% [1]	258.35	276.64	346.84
Maintenance Costs	113.90	65.86	101.93
	372.25	342.50	448.77
Costs/House/Annum	18.61	17.13	22.44

It can be seen from these figures that a variety of expenditure in soft landscaping is possible for both traditional and newer layouts. Case Study 2, to produce the lowest cost per annum over the 25 year period, had no grass areas whatsoever. No designer could continue such a design solution indefinately. It illustrates however that a traditional layout will or could have a cost similar to that produced by the newer solution appearing on the housing scene.

Interestingly grass costs in maintenance terms can only be brought down to the shrub maintenance costs by reducing the actual manpower operations involved.

This would require hard surface edging to all grass areas which would increase capital costs or the shrubs could be allowed to overhang the grass areas[2]. This is the essential message of soft landscaping; efficiency can be matched to a balanced design solution if the maintenance operations are evaluated in conjunction with the design. With this philosophy it can be shown that reasonable costs can still produce better and more acceptable landscaped environments.

References

(1) Stone, P.A., Building Design Evaluation Costs-in-Use, Spons 1980
(2) Gow, H.A., Section Nine: External Works, SLASH Group 1981
(3) Long, R. et al, Comparative Layout Studies, SLASH Group 1979

CASE STUDY NO.1: VEHICULAR DOMINATED ACCESS LAYOUT [3]

CAPITAL COSTS OF DESIGN

Shrub Planting	($211m^2$ @ 5.00)	1055.00
Trees	(3No @ 27.00)	81.00
Grass	($418m^2$ @ 0.50)	209.00
Toddlers Play ($277m^2$)	(£1000)	1000.00	2345.00

MAINTENANCE COSTS OF YEAR 2-5

Shrub Planting*	($211m^2$ @ 11 p/m^2)	(23.21 x 4)	92.84	92.84
Trees	(3No @ 2.00)	(6.00 x 4)	24.00	24.00
Grass**(1)	($418m^2$ @ 13 p/m^2)	(55.02 x 4)	220.08	
Grass**(2)	($418m^2$ @ 7.5 p/m^2)	(31.60 x 4)		126.40
Toddlers Play ($277m^2$)	(@ 3% Cap. Cost)	(30.00 x 4)	120.00	120.00
			456.92	363.24

MAINTENANCE COSTS OF YEARS 6-25

Shrub Planting*	($211m^2$ @ 7 p/m^2)	(14.77 x 20)	295.40	295.40
Trees	(3No @ 2.00)	(6.00 x 20)	120.00	120.00
Grass**(1)	($418m^2$ @ 13 p/m^2)	(55.02 x 20)	1100.40	
Grass**(2)	($418m^2$ @ 7.5 p/m^2)	(31.60 x 20)		632.00
Toddlers Play ($277m^2$)	(@ 3% Cap. Cost)	(30.00 x 20)	600.00	600.00
			2276.80	1647.40

*Based on the minimum level of maintenance indicated in the planning section of this sub-section.

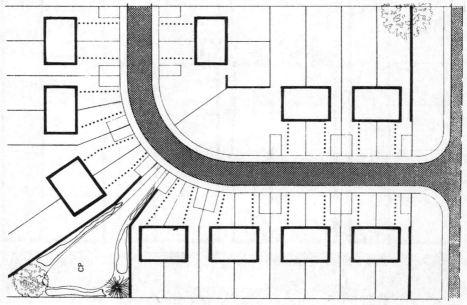

CASE STUDY NO.2: PEDESTRIAN/VEHICULAR SEGREGATED LAYOUT

CAPITAL COSTS OF DESIGN

Shrub Planting	(259m^2 @ 5.00)	1295.00
Trees	(8No @ 27.00)	216.00
Grass	(0 @ 0.50)	
Toddlers Play (408m^2)(£1000)	1000.00	2511.00

MAINTENANCE COSTS OF YEARS 2-5

Shrub Planting	(259m^2 @ 11 p/m^2)	(28.40 x 4)	113.96
Trees	(8No @ 2.00)	(16.00 x 4)	64.00
Grass	(- @ -)	(-)	-
Toddlers Play (408m^2)(3% Cap. Costs)	(30.00 x 4)	120.00
			297.96

MAINTENANCE COSTS OF YEARS 6-25

Shrub Planting	(259m^2 @ 7 p/m^2)	(18.13 x 20)	362.60
Trees	(8No @ 2.00)	(16.00 x 20)	320.00
Grass	(- @ -)	(-)	-
Toddlers Play	(3% Cap. Costs)	(30.00 x 20)	600.00
			1282.60

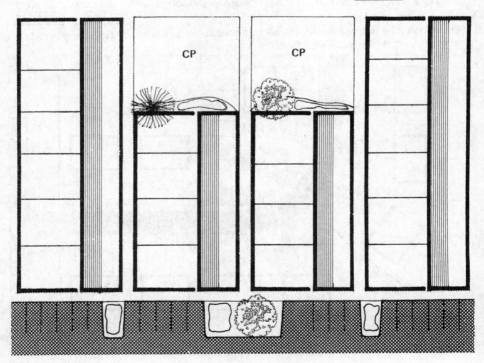

CASE STUDY NO.3: PEDESTRIAN/VEHICULAR JOINT-USE LAYOUT

CAPITAL COSTS OF DESIGN

Shrub Planting	$(363m^2$ @ 5.00)	1815.00	
Trees	(3No @ 27.00)	81.00	
Grass	$(504m^2$ @ 0.50)	252.00	
Toddlers Play $(180m^2)$	(£1000)	1000.00	3148.00

MAINTENANCE COSTS OF YEARS 2-5

Shrub Planting	$(363m^2$ @ 11 p/m^2)	(39.93 x 4)	159.72	**159.72**
Trees	(3No @ 2.00)	(6.00 x 4)	24.00	24.00
Grass**(1)	$(504m^2$ @ 13 p/m^2)	(68.70 x 4)	274.52	
Grass(2)	$(504m^2$ @ 7.5 p/m^2)	(38.10 x 4)		152.40
Toddlers Play $(180m^2)$	(3% Cap. Costs)	(30.00 x 4)	120.00	120.00
			578.24	456.12

MAINTENANCE COSTS OF YEARS 6-25

Shrub Planting	$(363m^2$ @ 7 p/m^2)	(25.41 x 20)	508.20	508.20
Trees	(3No @ 2.00)	(6.00 x 20)	120.00	120.00
Grass**(1)	$(504m^2$ @ 13 p/m^2)	(68.70 x 20)	1374.00	
Grass**(2)	$(504m^2$ @ 7.5 p/m^2)	(38.10 x 20)		762.00
Toddlers Play $(180m^2)$	(3% Cap. Costs)	(30.00 x 20)	600.00	600.00
			2602.20	1990.20

451

NOTES

(a) GRASS MAINTENANCE REGIME**   (1)   (2)

1.   No. of cuts per annum   12 x (12)$*\dfrac{(12\times0.63\times418)}{100m^2}$ 31.60   31.60

2.   Preliminary stone picking 1 x (O)*(0.73/100m^2)   3.05   -
3.   Shear edging (41m)   4 x (O)*(4.83 x 4)   19.32   -
4.   Spade edging (41m)   2 x (O)*(2.05 x 2)   4.10   -
                                                        ——————   ——————
                                                        58.07   31.60=
                                                                      $_2$
Less stone picking let in capital works contract   3.05   7.5p/m$_2$
(This produces a rate of approx. 13 p/m^2)   ——————
                                                        55.02   = 13p/m^2

(b) SHRUB MAINTENANCE REGIME

1.   Residual Herbicide (211 x 1 @ 5 p/m$_2^2$)⎫   (1)*=7 p/m^2
2.   Hand Grub/Fork   (211 x 4 @ 1 p/m$_2^2$)⎬= 11 p/m^2 (1)* during
3.   Contact Herbicide   (211 x 2 @ 1 p/m^2)⎭ in years   (1)* years 6-25
                                                    2-5

Note: Shrubs may require replacing during the life of the scheme outlined.
(No)* Indicates regime during years 6-25.

COMPARISON OF SHRUB AND GRASS COSTS FROM CASE STUDY NO.1

Shrub - Capital Cost = £5.00/m^2 Annual Equivalent =   0.55p
      - Maintenance Costs year 1 - 5 =   92.84
      - Maintenance Costs years 6 -25 = 295.40
                                        ——————
                                        388.24 = Annual Cost=0.08p
         (Note: This represents £1.84/m^2 ————
         over a 211m^2 shrub area)   0.63

Grass - Capital Cost = 0.50/m^2 Annual Equivalent =   0.06
      - Maintenance Costs year 1 - 5 = 220.08
      - Maintenance Costs years 6 -25 =1100.40
                                       ———————
                                       1320.48 = Annual Cost=0.13
         (Note: This represents £3.16/m^2 ————
         over a 418m^2 grass area)   0.19

      - Alternative deleting specified operations 3 and 4
      - Maintenance costs year 1 - 5 =   126.40   0.06
      - Maintenance costs years6 -25 =   632.00   0.08
                                        ——————   ————
                                        758.40   0.14
         (Note: This represents £1.81/m^2
         over a 418m^2 grass area)

The essential differences are that while shrubs have a capital cost
almost ten times that of grass areas there is a commitment to almost
twice the expenditure on grass maintenance per annum as there need be
on shrubs. Costs-in-Use Studies reduce the weighting from a 10:1
ratio to a 3:1 ratio against the shrub option. Reducing the grass
maintenance to simply cutting brings it to the same costs as the shrub
maintenance. However shrubbery costs could be reduced to 25% of the
minimum/

minimum grass maintenance cost by deleting operation 1 (residual
herbicide) and accepting that operation 2 (hand grub/fork) is restricted to 20% of the shrub bed. The lesson in landscape terms is
that a proper balance in the design is needed between these two
different but equally essential components of ground cover in order
to satisfy the aesthetic design needs of most situations.

A STRUCTURED REGIME FOR SHRUB AREAS

YEAR 1*Stone Cleaning (1)
    Residual Herbicide (1)
    or
    Contact Herbicide (3)
    Hand Grub and Weed (4)(15%)
    Shrub Replacement (1)(5%)
    Fence Maintenance (1)

YEAR 2 Residual Herbicide (1)
    or
    Contact Herbicide (3)
    Hand Grub and Weed (2)(15%)
    Shrub Replacement (1)(5%)
    Fence Maintenance (10%)

YEAR 3 Residual Herbicide (1)
    or
    Contact Herbicide (2)
    Hand Grub and Weed (2)(15%)
    Prune Shrubs (15%)
    Shrub Replacements (5%)
    Fence Maintenance (Removal)

YEAR 4 Residual Herbicide (1)
    or
    Contact Herbicide (2)
    Hand Grub and Weed (2)(15%)
    Prune Shrubs (15%)(?)
    Shrub Replacement (5%)

YEAR 5 Residual Herbicide (1)
    or
    Contact Herbicide (2)
    Hand Grub and Weed (2)
    Prune Shrubs (15%)(?)
    Shrub Replacement (5%)

YEARS
6 - 25 Residual Herbicide (1)
    Prune Shrubs (15%)(?)
    Shrub Replacement (-)
    Hand Grub and Weed (2)(15%)

NOTE:
YEAR 1* Denotes Capital Contract operations on newly handed over
      estates as well as year one on new layouts to existing
      estates.
  (3) The number in brackets represents the times per annum the
      operation should be performed.
  (%) Refers to the amount of shrubs actually pruned and is in
      keeping with the modern practice to cultivate a natural or
      semi-natural ground cover requiring less attention than
      traditional horticultural practice would suggest.
  (?) Refers to the selection of shrubs which once shaped in their
      formative years should be allowed to develop as the climate
      and soil dictates. The form, species and situation will
      govern the need to prune in later years.

The author is indebted to Kevin Brame, Principal Landscape Architect,
Glenrothes Development Corporation, for pursuing the work upon which
this article is based, and his encouragement and help without which
it would not have seen print.

EFFECT OF INFLATION AND RATE OF INTEREST ON COST-IN-USE CALCULATIONS

EDWARD A. PRZYBYLSKI, Central London Polytechnic

1. Introduction

Those in the construction industry responsible for the evaluation
of design decisions in monetary terms, have come to realise over
recent years that decisions on the basis of initial cost alone are
no longer adequate.
The running costs of a building throughout its lifespan have
increased to such an extent that they considerably outweigh the
initial cost of the building.
Construction economists are familiar with techniques used to
convert the capital cost of a building, to an annual equivalent or
to convert future costs to a present day value. This conversion
uses the concept of the time value of money; namely that future
payments are discounted to their present value.

2. The time value of money

Given that cashflow is the important decision variable and that the
object of the construction economist can be interpreted as minimising
the flow of cash on the maintenance, operation and alteration of
a building throughout its life, an important statement must be
explained and justified.
"Money which occurs at different points in time cannot be compared
directly".

Thus, just as it would be nonsensical to add £3 + $5 + DM7 to
equal 15 in exactly the same way adding £2 which arises now, with
£5 that arises in 12 months' time, with £6 that arises in six
years' time, to total £13, would be nonsensical. Money which arises
at different points of time, just like money in different currencies
cannot be directly compared. In order to compare such a flow of
cash (e.g. to total it), it must be converted to a "common"
point in time, just as to be able to add £3 + $5 + DM7 the amounts
must be converted to a common currency.

The fundamental principle for the concept of the time value of money
is that individuals possess a personal "rate of time preference"
such that cash which arises in the near future is preferred to cash
that arises in the far future. Thus an individual who is offered
a choice between receiving £100 "now" or £100 in "12 months' time"
will not reason that, as the two amounts of cash are the same, he
does not mind which alternative is chosen (i.e. he will not be
indifferent between the two alternatives). He is likely to prefer
the £100 "now". This choice does not fundamentally have anything
to do with uncertainty in the sense that he might doubt whether the
offer of £100 in 12 months' time will actually transpire, he is
likely to prefer the £100 now, even if both alternatives were
certain.

3. Inflation and capital markets

The foregoing analysis has nothing to do with the presence of
inflation; nor has it anything to do with the fact that individuals
can earn a rate of interest on their money by placing it on
deposit in the capital market.

   However, although the idea of the time value of money would still
exist if capital markets did not exist, and so you could not earn
a rate of interest on your money, capital markets do have an
important part to play. This comes from the assumption that not
only will individuals have a time value of money, but that they
are rational, i.e. they always prefer more money to less.

   As an example of this process it will be assumed initially that a
capital market does not exist. Suppose that an individual believes
that his time value of money is 6% per annum. That is if he were
offered a choice between £100 now or a sum of money in 12 months'
time, he would elect to have the £100 now as long as the sum of
money offered in one year's time was less than £106. If he was
offered the choice of £100 now or £106 in a year's time he would
be truly indifferent; whilst if the sum offered in a year's time
was greater than £106 he would prefer the alternative to the £100
now.
Thus an individual's 'time value of money' can be viewed as a rate
of exchange through time. The fact that the individual in this
example had an exchange rate of 6% arises from his own personal
circumstances. Other individuals are likely to have other
exchange rates: the older the individual the higher it is likely
to be; the poorer the individual the higher it is likely to be, and
vice versa in both cases.

   If it is assumed that a capital market now exists in which money
can be invested to earn a 10% annual rate of interest, the situation
changes. When a capital market did not exist, if the individual
was offered a choice between £100 now or £108 in 12 months'
time, the £108 would be chosen. This is because he is being offered
more than the additional £6 (i.e. 6%) necessary to persuade him to

forego the cash now in favour of a greater amount of cash in 12
months' time. However, given the same choice, but now with the
possibility of earning a 10% rate of interest, the individual will
change his decision and choose the £100 now. The reason for this
change in decision is that this money can be placed on deposit to
produce £110 in 12 months' time, which the individual will rationally
prefer to the alternative of only £108. Thus the capital market
interest rate can directly affect an individual's decision.

4. Interest/Discount rate

The choice of interest rate for discounting must be closely related
to the decision of whether or not inflation is ignored. When
dealing with periods as long as the life of a building it is clear
that inflation can have a significant effect. However, as the
purpose of cost-in-use calculations is to assist in making informed
choices between alternative courses of action, rather than actually
to predict the sum which should be set aside in a sinking fund to
provide for future expenditure, then the problem is diminished.
It is assumed that general inflation not only increases the cost
of future expenditure but also increases 'pro rata' the general
revenues from which the expenditure is met.
If inflation is ignored in costs then it must also be ignored in the
choice of discount rate. In other words the discount rate should
represent the real cost of finance. This real cost can be calculated
as the actual cost of money less the amount included for inflation.
There are many factors which affect the precise calculation of the
real cost of finance, however discount rates which are too high
have often been used in calculations. The effect of using such
high discount rates is to favour projects with low initial costs
and high running costs.

In recent years there have been large increases in interest rates
on the capital market and also an increase in the capital necessary
to finance business at higher prices. Even in the circumstances
of a laissez faire national economy with a minimum of government
intervention, interest rates have been used to achieve some measure
of control over the economy. Traditionally when the economy has
become overheated, deflationary action has been taken by increasing
interest rates. Similarly interest rates have been lowered when
it has been deemed advisable to expand business activity. During
a long period of rising prices, and particularly when during the
1970's the pace of the rise has accelerated greatly, the lenders
of money became aware that the purchasing power of the interest,
and of the principal when ultimately repaid, falls year after year.
Consequently they seek compensation in the rate of interest.

If a loan is made at 10% interest for a period over which
inflation is estimated at 6% per annum the pure rate of interest
is 3.77% only and the remaining 6.23% is for inflation
compensation. The formula for the calculation of pure interest
(R), where G is the gross interest and F the rate of

inflation, is

$$R = \left[(1 + G)/(1 + F)\right] - 1$$

Therefore where 10% is the gross rate of interest and 6% the rate of inflation, the rate of pure interest is

$$\left[(1 + 0.10)/(1 + 0.6)\right] - 1$$

$$= (1.10/1.06) - 1 = 1.0377 - 1 = 0.0377$$

   In simple terms, at the end of the year the borrower has the principal of £1 owing to him plus interest of 0.10, making a total of £1.10.
But the purchasing power of £1 which he originally lent is now represented by £1.06. Therefore, he laid out £1.06 in year end pounds and now has £1.10 in year end pounds, a gain of 3.77%. In these circumstances interest rates will remain at a level historically high due to the inclusion of the element of inflation compensation until the danger of inflation is reduced. The danger which appears to exist in cost-in-use calculations is that whilst inflation is ignored in future expenditure it is not possible to ignore it in the discount rate.

5. The cost of capital

Present Value appraisal requires the use of a discount rate. This should properly reflect the organisation's opportunity cost (or time value) of cash. In a situation of a perfect capital market, this opportunity cost is reflected by the market interest rate.
However in a perfect capital market there is not simply a single market interest rate but an almost infinite range of interest rates, each of which is related to a specific level of risk. Investors are assumed to be generally risk averse (i.e. risk is disliked) and to be persuaded to take on risk an investor has to be offered additional expected return by way of a reward or compensation for so doing. Thus the higher the risk of any particular appraisal the higher will be the perfect capital market interest rate. Under these circumstances an organisation has got to decide which of this infinite range of interest rates it uses as the discount rate for its appraisals.

   The solution to the problem would appear to be obvious: if the opportunity cost of cash increases with increasing risk because investors are risk averse, then the discount rate used to appraise the desirability of any particular solution for a project should be the interest rate that reflects the risk of the project (i.e. should reflect the uncertainty that surrounds its future cashflows). This is the logic behind the approach which combines the expected returns on an organisation's existing debt and equity capital into

a weighted average in order to reflect the capital market interest
rate (or rate of return) appropriate for the organisation's
existing risk level, and then uses this as the appraisal discount
rate. One of the assumptions here being that the investment
proposal has a level of risk similar to the organisation's existing
risk level.

6.   Taxation

Taxation and its impact on project cashflows has an important
bearing on appraisal calculations. As the object of financial
appraisal is to minimise the flow of future expenditure, the
organisation is only interested in the actual cashflow of a project.
Thus project cashflows should take into account tax charges that
they bear and also any tax reliefs which are attached to the project
such as capital allowances.

   With industrial buildings some relief can be obtained on the initial
cost, through depreciation allowances, investment, initial and
cash allowances, their actual form and impact varying from time
to time. Amounts spent on maintenance and repairs, heating and
lighting, and other running expenses are classified as business
expenses and are deductible from profits in the case of all types
of buildings. It is postulated that the current regulations and
levels of taxation tend to favour alternatives with low initial
costs and high running costs. The total costs of buildings can
thus be influenced considerably by the form of taxation.

   Thus there is a wide variation of fiscal relief against building
expenditure, ranging from the total absence of relief against
investment in commercial or residential property, through the
general run of investment in industrial property, to the favoured
case of a building treated as plant for tax purposes and situated
in a development area. The case has often been argued that whilst
maintenance expenditure is wholly allowable against liability to
tax, and capital expenditure, subject to the incidence of grants
and allowances, is not allowable, then a given volume of maintenance
work must be less expensive to an organisation than a corresponding
volume of new construction; hence building expenditure is liable
to bias against new construction in favour of maintenance, even
when maintenance would otherwise be uneconomic. If this is so,
the demand for maintenance is increased at the expense of demand
for new construction, which would put the same volume of physical
resources to more productive and less labour-intensive use.
Maintenance saving investment is also stifled; the use of buildings
is prolonged beyond their natural life, existing buildings are
put to uneconomic uses, and the quality of the environment
deteriorates.

7. Inflation

Inflation can be identified in two forms, general inflation and specific inflation. General inflation refers to the increase in the price of a whole range of goods and/or services; thus the retail price index is an index of general inflation. Specific inflation refers to the increase in price of   particular goods or services. Inflation can have a profound effect on the financial performance of alternative design solutions. This is especially true when the rate of increase is high, as has been seen with fossil fuel prices over the 1973-1980 period. It is necessary to estimate the differential rate of inflation when costs increase more quickly than the general rate of inflation, i.e. specific inflation of a design alternative.

Given that the object of financial appraisal is to minimise future expenditure, whilst taking into account the idea of the time value of money, it is obvious that inflation must be taken into account, as in such conditions 'cash' and 'purchasing power' do not have a stable relationship.

8. The place of inflation in the time value of money

Suppose that neither inflation nor capital markets exist. In such circumstances an investor might specify that his time value of money is (say) 5%. That is, he is willing to forego £100 today as long as he receives at least £105 in return in 12 months' time. In fact, cash itself is of little use to anyone as it is just pieces of paper; but the important thing about cash is that it provides purchasing power. Thus when an investor states that his time value of money is 5%, the 'true' interpretation of the situation is that he is willing to give up £100 worth of consumption (or purchasing power) today, as long as he can have at least an extra 5% of consumption in 12 months' time.

If we now have the same situation except that there is a general rate of inflation of 10%, then if the investor gives up £100 today, he would have to receive £110 (i.e. 10% extra cash) in 12 months' time just to be able to consume the identical amount given up 12 months earlier. As the investor requires 5% extra consumption, he will require in cash 5% more than £110: £110 (1 + 0.05) = £115.50.

Thus in conditions where the general rate of inflation is 10% per annum, an individual who has a time value of money of 5% in terms of consumption will have a time value of money in terms of cash of 15½%. Because of the presence of 10% general inflation the investor will need an extra 15½% of cash to enable him to consume 5% more goods/or services.

If the time value of money in consumption or purchasing power terms is referred to as the 'real rate of interest', the time value of money in cash terms is referred to as the market (or money) interest rate, and the general rate of inflation is referred to as the retail price index (RPI), then the following relationship holds:

a) (1 + real interest rate) (1 + RPI)=(1 + market interest rate)

b) Market interest rate = $\left[(1 + \text{real interest rate}) (1 + \text{RPI})\right] - 1$

c) Real interest rate = ($\dfrac{(1 + \text{market interest rate})}{(1 + \text{RPI})}$) - 1

Thus if the real interest rate is 5% and the RPI is 10% then the market or money interest rate is:

$$\left[(1 + 0.05) (1 + 0.10)\right] - 1 = 0.155$$

9.   'Money' and 'real' cashflows

In the face of inflation perhaps appraisal techniques should take the following approach.
Either the 'money' cashflows of the project should be discounted to present value using the market or money discount rate, or the 'money' cashflows of the project should first be discounted to 'general purchasing power' cashflows using the RPI as the discount rate and then these should be further discounted to present value using the real rate of discount.

10.   Conclusion

If the present value is calculated for each building to be compared, then supposedly the most economic solution in terms of total costs can be identified.
Although the arithmetic is easy the main variables which affect the answer are briefly re-stated namely;

   The inflation element in discount rates and the resultant high rates used in calculations tend to favour projects with low initial costs and high running costs.

   The project risk affects the organisation's rate of return and the discount rate.

   Existing taxation regulations can have a prominent effect on the cashflows of a project.

   Differential inflation rates affect the cost of design alternatives.

   All these variables are closely interrelated and more attention must be paid to the 'details' of cost-in-use calculations before the correct balance between initial and running costs can be established.

References

1. Wilson J.P.   Inflation, Deflation, Reflation (1980)
                    Business books.

2. Seeley I.H.   Building Economics (1979), Second edition.
                    The Macmillan Press Ltd.

3. Goodacre P.   Re-appraisal of Cost-in-use Calculations. (1978)
                    Building Technology and Management 16 May, p.p. 6-7.

4. Urien R.   Some thoughts about the economic justification of
                    life cycle costing formulae (1975) Industrial Forum
                    (3-4) p.p. 53 - 62.

5. Drake E.B.   Economics of Maintenance (1969), Quantity Surveyor
                    July/August p.p. 3 - 6.

COSTS IN USE OF SCHOOL BUILDINGS,
WEST MIDLANDS STUDY

Dr A C Sidwell BSc(Hons) ARICS MCIOB MBIM
Mr Mohammed Da'abis MSc

Dr Sidwell is a Lecturer in Construction Management and Building
Economics and Mr Da'abis a research student in the Department of
Construction and Environmental Health, University of Aston

Introduction

Few members of the building team would deny that effective feedback
of maintenance and operating costs of buildings would assist in
improving projects in the future. The performance of buildings is,
to a large extent, an unknown quantity and in extreme cases buildings
may be unusable and require demolition.

At an early stage the design team must consider its strategy -
whether to build traditional or non-traditional, high or low, wide
or narrow etc. Each variable will influence capital costs, maintenance and operating costs.

This study had two objectives: firstly, to establish a data base
of historic costs in use for school buildings and, secondly, to
analyse variations in the data and to identify reasons for high or
low costs. The seven metropolitan boroughs in the West Midlands were
invited to co-operate in the study. Three did so, three were unable
to provide some assistance but did not have detailed maintenance
records, and one declined.

Research Method and Sample

Data on maintenance, running and capital costs for 25 school buildings were collected from three authorities, designated A, B and C*.
Each school was visited. The data on maintenance, general repairs,
vandalism, fire damage and minor works covered the period 1967-1980
with some exceptions where authorities only retained records for six
years. Costs were abstracted under seven operation headings, except
for authority C which was under four headings.

Running costs for cleaning, caretaking, fuel and water costs
covered the period 1975-1980 only.

---

*Footnote: the views, analysis and conclusions in this paper are
 entirely the responsibility of the authors and do not represent
 the views of the Metropolitan Authorities in any way.

Data were also collected on school age, floor area, number of pupils, capital costs and construction. Eighteen of the twenty-five schools were built after 1965, others were built after 1930 and one built in 1889. In some cases, capital costs were not available and were estimated by comparison with costs of other school buildings in the same period. The capital cost of the school built in 1889 was estimated by viewing the change in construction costs every twenty-five years, the result is obviously approximate. All costs were brought to 1980 prices.

The data were divided into three construction types:

   (i)   traditional
  (ii) non-traditional (ie system built)
 (iii) mixed (ie traditional and non-traditional)
and two education levels
   (i) primary schools
  (ii) secondary schools

Details of the sample are given in Table 1

TABLE 1: The sample

Type of School	Traditional	Non-Traditional	Mixed	TOTAL
Primary	11	3	0	14
Secondary	5	3	3	11
TOTAL	16	6	3	25

The following hypothesis was investigated:
- that there are differences between different types of schools (traditional, non-traditional, mixed, primary and secondary and with regard to the following eight categories:-

    (i)   capital costs
   (ii) maintenance costs
  (iii) re-decoration costs
   (iv) minor new works
    (v) energy costs
   (vi) cleaning and caretaking costs
  (vii) vandalism
 (viii) fire damage

The 't' test was applied to the tables of results to indicate their significance.

Results

(i) Capital Costs

Table 2 shows the average capital costs per m^2 of the schools studied. Secondary schools had higher capital costs than primary schools and non-traditional schools were more expensive than traditionally constructed schools.

TABLE 2: Average capital costs in £/m^2 (1980 prices)

Type of School	Traditional	Non-Traditional	Mixed
Primary	230	242	–
Secondary	303	488	355

(ii) Maintenance Costs

The records of the metropolitan authorities were studied and maintenance costs (excluding redecorations) were abstracted for each year under six headings, main structure, internal construction, finishes and fittings, plumbing and drainage, mechanical services, and electrical services. Table 3 shows the average total maintenance costs per year for each type of school and Table 4 shows the average annual costs for each of the six maintenance headings.

Traditional Versus Non-Traditional Construction - Primary Schools

The results indicate a cyclic pattern of expenditure peaking every three years. Non-traditional schools had higher maintenance costs than traditional schools during the period 1968-1975 while traditional schools had higher maintenance costs in the period 1976-1980. The overall decline in expenditure towards 1980 may be associated with expenditure cuts. The two most expensive items were the main structure and mechanical services. The high cost of maintenance to the main structure was caused, particularly for non-traditional buildings, by replacement of defective concrete panels, problems with flat bituminous felt roofs and large areas of glazing. Mechanical services were costly because of the need for repairs, major new installations and modifications to faulty and inadequate systems.

Comparison between the data for all schools and those built during the 1930's confirms that the older schools incur higher maintenance costs.

Secondary Schools

The results indicate a cyclic pattern of expenditure peaking every four years. Non-traditional schools had the highest maintenance costs during the period 1967-1976 and mixed schools (a combination of traditional and non-traditional construction) the highest maintenance costs from 1977-1980. Maintenance costs were caused by replacement

TABLE 3: The average maintenance costs (excluding redecoration) in £/m² per year per type of school (1980 prices)

School	Primary		Secondary		
Type	Tradi-tional	Non-Tradi-tional	Tradi-tional	Non-Tradi-tional	Mixed
Year	Average Maintenance Costs £	Average Maintenance Costs £	Average Maintenance Costs £	Average Maintenance Costs £	Average Maintenance Costs £
1967	2.8100	0.5022	0.6577	1.8700	1.6559
1968	1.5262	2.1084	0.6156	0.9951	0.8023
1969	1.0974	3.9310	0.5677	0.9074	0.6179
1970	0.7602	2.6821	0.5383	1.3777	0.4924
1971	1.1798	1.4724	1.2875	1.3397	0.7003
1972	1.2422	4.7318	0.9310	2.8275	2.4202
1973	2.8203	4.5535	0.8641	3.1354	0.8749
1974	2.3178	1.4610	1.0879	1.5988	1.6880
1975	3.2204	3.3283	1.3438	1.7644	1.6680
1976	3.1616	1.6016	1.3084	1.9293	1.5735
1977	1.7493	0.9062	0.3985	1.4574	2.2318
1978	3.1568	2.3076	2.4159	2.1477	3.0416
1979	2.6639	1.0908	1.9801	2.8695	3.5253
1980	2.7550	1.7935	2.1486	3.0312	1.8162
TOTAL	30.4605	32.3704	16.1451	27.2511	23.1083
AVERAGE	2.1757	2.3121	1.1532	1.9465	1.6505

TABLE 4: The average costs in £/m^2 per year per type of school of various items of maintenance (1980 prices)

School	Primary		Secondary		
Type	Traditional	Non-Traditional	Traditional	Non-Traditional	Mixed
Item	Cost £	Cost £	Cost £	Cost £	Cost £
1 Main Structure	0.8532	1.1906	0.2566	0.7429	0.6137
2 Internal Construction	0.2350	0.3065	0.1712	0.1936	0.2930
3 Finishes and fittings	0.1434	0.1277	0.1316	0.1719	0.1467
4 Plumbing	0.2465	0.2273	0.1504	0.2587	0.1442
5 Mechanical Services	0.5554	0.3035	0.3536	0.4003	0.2450
6 Electrical Services	0.1422	0.1565	0.0898	0.1791	0.2079
TOTAL	2.1757	2.3121	1.1532	1.9465	1.6505

of glazing, ceiling panels, ironmongery, plumbing fittings and modification of partitions. Maintenance to mechanical services included repairs and modifications to heating systems, lifts and swimming pool plant.

Primary Versus Secondary Schools

In the cases studied primary schools had higher maintenance costs than secondary schools, for both traditional and non-traditional construction. This may be explained by the fact that the primary schools were older than the secondary schools in the sample. The density of pupils in primary schools was higher than secondary schools, the average densities being 0.2 and 0.14 pupils per m^2. Finally, the secondary schools had higher initial capital costs than the primary schools, which may have influenced the maintenance costs.

(iii) Redecoration Costs

Redecoration differs from general maintenance in that it is usually done every six or seven years and is a postponable item when funds are limited. Table 5 gives the mean redecoration costs for each type of school and illustrates that primary schools had higher costs than secondary schools, possibly because of the age distribution in the sample. The traditionally constructed primary schools incurred higher redecoration costs than non-traditionally built schools. On the other hand, traditionally constructed secondary schools incurred lower redecoration costs than non-traditionally built schools.

The results for redecoration costs were, to an extent, inconclusive. It is thought that this is because redecoration work may be delayed, and perhaps the often smaller primary schools may be redecorated rather than secondary schools when funds are limited.

(iv) Minor New Works

Many of the school buildings studied incurred expenditure which could not properly be considered as maintenance or repair. These costs were abstracted separately and are given in Table 6. They included modification and alterations to heating systems for the provision of ventilation, internal partitions, services, and safety considerations.

(v) Energy Costs

The term 'energy costs' includes expenditure on water, electricity, gas, oil and solid fuel as used for heating, cooking, lighting and general power requirements of school buildings. They are continuously incurred and vary in relation to price levels and weather conditions. For example, costs peaked during the bad winter of 1978. The results in Table 7 indicate that the non-traditional schools had higher energy costs than traditional schools. Study of the construction detailing of the schools in the sample lead us to the opinion that the higher heating costs in non-traditional schools are influenced by losses through joints between components which, in many cases, were not airtight. One practical difficulty is that pupils often pick out and damage the jointing materials.

TABLE 5: The mean redecoration costs in $£/m^2$ per year per type of school

School	Primary		Secondary		
Type	Traditional	Non-traditional	Traditional	Non-Traditional	Mixed
1967	1.3052	0.0890	0.0209	0.5808	0.1427
1968	1.6604	6.9335	0.0155	0.3378	0.0048
1969	0	0.0013	0.0003	0.0038	0
1970	1.1520	0.0451	0.3966	1.4057	0
1971	2.6800	0.0131	0.0145	0.4282	0.0022
1972	0.7076	0.0861	1.6503	1.6420	1.7012
1973	0.6168	1.3419	0.0215	1.5896	0.5995
1974	0.1486	0.7991	0.0015	0.9631	2.4069
1975	0.2115	0.6978	0.0873	0.0451	1.1063
1976	0.6830	0	0.2167	1.4173	0.0176
1977	1.0337	0.1183	1.6192	0.2674	0.7904
1978	0.7710	1.0670	0.0413	0.2998	0.3633
1979	0.0326	1.3594	0.1037	0.3641	1.1904
1980	0.2181	0.9022	0.0806	0.2230	0.6250
TOTAL	11.2205	13.4538	4.2699	9.5977	8.9503
MEAN	0.8014	0.9609	0.3049	0.6834	0.6393

TABLE 6: The Costs of Minor New Works in £/m^2 per Type of School (1980 prices)

School Type	Primary Traditional	Primary Non-Traditional	Secondary Traditional	Secondary Non-Traditional	Mixed
1967					
1968					
1969					
1970					
1971					
1972					
1973					
1974	.0133			.0003	
1975				.0611	
1976				.0006	
1977	.1963		.1391		.7271
1978	.0014		.0590	.0211	.0600
1979	.0981		.0003	.0319	.1447
1980	.3127		.6169		
TOTAL	.6218		.8153	.1150	.9318

TABLE 7: The Average Energy Costs in £/m^2
Per Year Per Type of School (1980 prices)

School	Primary		Secondary		
Type	Tradi-tional	Non-Tradi-tional	Tradi-tional	Non-Tradi-tional	Mixed
1975	2.0555	2.6498	1.9900	2.7850	2.3017
1976	2.1982	2.8130	1.9876	2.9185	2.3867
1977	3.0827	3.3382	2.2142	3.6094	2.5912
1978	2.9820	3.5365	2.5291	4.2531	3.0472
1979	2.6985	3.5232	2.6588	3.8737	3.0203
1980	3.3265	3.9992	2.9285	4.7571	4.1488
TOTAL	16.3434	19.8594	14.2982	22.1916	17.4959
AVERAGE	2.7239	3.3099	2.3830	3.6986	2.9152

(vi) Cleaning and Caretaking

Cleaning and caretaking costs include wages, consumables, insurances, and supervision. Like energy costs, they are continuous but less variable except by wage and material costs. Costs declined in 1980 and this may have been due to expenditure cuts. The results in Table 8 show a difference between all types of schools. The statistically signfiicant results are:- primary and secondary non-traditional schools were more costly than primary and secondary traditional schools. Primary traditional schools were more costly than secondary traditional schools.

These variations are probably associated with differences in the agre, design and layout of schools and pupil density.

(vii) Vandalism

The incidence of vandalism is generally thought to be influenced by the social and environmental aspects of the school. Clearly, some construction detailing and materials are less susceptible to damage than others. Some schools in the case studies were better than others, and this was not related to whether the construction was traditional or non-traditional. Rather, it was a question of individual detailing; examples were large glazed areas at ground floor, accessible plumbing and drainage fittings, low false ceilings in corridors, flimsy doors and ironmongery.

TABLE 8: The Average Cleaning and Caretaking Costs in £/m^2 Per Year Per Type of School (1980 prices)

School	Primary		Secondary		
Type	Traditional	Non-Traditional	Traditional	Non-Traditional	Mixed
1975	5.6358	6.0485	4.1151	5.1498	6.1571
1976	7.3716	5.9557	4.8274	5.5982	6.6362
1977	4.1693	6.5943	5.0332	6.4126	7.3466
1978	7.4994	6.4302	5.1658	6.4292	7.3481
1979	4.4427	6.3226	4.9357	6.2279	7.1228
1980	5.5970	5.5170	4.3638	5.4671	6.2424
TOTAL	34.7154	36.8682	28.4410	35.2848	40.8528
AVERAGE	5.7859	6.1447	4.7401	5.8808	6.8088

Table 9 gives the average costs of vandalism; clearly, secondary schools are more costly than primary schools.

(viii) Fire Damage

Many school fires are started deliberately and are therefore vandalism. However, the data are recorded separately here in Table 10. The costs of minor fire damage may be met out of general maintenance funds but major fires require separate authorisation.

Only a few fires occurred in the schools studied, though records were available since 1967. The incidence of fire seems to be a relatively recent problem.

Conclusions

This paper reports research in the West Midlands to compile data on the maintenance costs of school buildings. The data were examined in detail to test the hypothesis that there are differences in costs between different types of schools.

The results clearly show that, for the sample, schools of non-traditional construction have higher capital costs, general maintenance costs, energy costs, cleaning and caretaking costs than schools of traditional construction.

Further, that the secondary schools had higher capital costs and energy costs than primary schools, whereas primary schools had higher general maintenance, redecoration and cleaning and caretaking costs than secondary schools.

TABLE 9: The Average Costs of Vandalism in £/m^2
Per Year Per Type of School (1980 prices)

School	Primary		Secondary		
Type	Traditional	Non-Traditional	Traditional	Non-traditional	Mixed
1967					
1968					
1969					
1970					
1971	.0064	.0123	.0316	.0144	.0114
1972		.0172	.0019	.0169	.0487
1973	.0007		.0175	.0127	.0083
1974	.0015		.0065	.0004	.0132
1975	.0233		.0527	.0021	.0065
1976	.0111		.0053	.0491	.0298
1977	.0187	.0004	.0238	.0271	.0217
1978	.0102		.0353	.0191	.0478
1979	.0101	.0039	.0369	.0222	.0051
1980	.0100		.0487	.1387	.0030
TOTAL	.0920	.0338	.2602	.3063	.1955

TABLE 10: The Average Costs of Fire Damage in £/m^2 Per
Year (since 1967) Per Type of School (1980 prices)

School	Primary		Secondary		
Type	Traditional	Non-Traditional	Traditional	Non-Traditional	Mixed
1977	.1231			.0016	1.7996
1978		.2963			
1979				.0377	
1980			.1910		
TOTAL	.1231	.2963	.1910	.0393	1.7996

Keyword Index